Table of Contents

Introduction

Genetics professors all agree: the way to learn genetics is by solving many problems. Do as many problems as you can each day, keeping up with the problems relevant for that day's lecture or reading material. Don't wait until the night before an exam to try all the problems or you will be overwhelmed and will become very frustrated. You can only develop your problem-solving skills by practice. Genetics requires a set of analytical skills that may not have been called upon in your other biology courses. You will have to solve word problems, think in terms of experiments, and make decisions about what of the information you have learned is useful in answering a particular question. Many of the analytical and logical skills that you develop will be broadly applicable in other disciplines. Your professors can ask so many different types of questions that you must develop your ability to reason through problems rather than trying to learn a few rote approaches to problem solving.

Types of Problems Encountered in the Book:

Classical Genetics Problems: these problems, seen most often in Chapters 2-5 and 12-14, are like puzzles. Try to avoid getting frustrated by any one problem; if you do, take a breather and come back to the problem later. Always begin working on these problems by reviewing the basic principles needed for their solution.

Experimental Questions are found more often in Chapters 6-11 and 16-18, the chapters on molecular analyses. Such questions usually require a good understanding of the theory and practice underlying a variety of techniques. Make sure you have a good handle on the fundamental methodologies such as gel electrophoresis and nucleic acid hybridization. When would each be used? How is each done? What materials do you need (cells, DNA, RNA or protein)?

Social and Ethical Issues: solutions have not been provided for these case histories because there are no obvious, easy, "right" answers to many of the ethical issues. Many people are involved or impacted by each issue (we call these stakeholders), and each has valid concerns and beliefs that cause them to hold particular viewpoints. Your appreciation for these complex ethical issues will benefit from discussion that clarifies the viewpoints of all stakeholders and attempts to find ground for compromises where possible.

In discussing what can be done in the situations described in the Social and Ethical Issues, consider the following:

♦   Who are the players?

Study Guide/Solutions Manual

for use with

Genetics
From Genes to Genomes

Second Edition

Leland H. Hartwell
Fred Hutchinson Cancer Research Center

Leroy Hood
The Institute for Systems Biology

Michael L. Goldberg
Cornell University

Ann E. Reynolds
University of Washington

Lee M. Silver
Princeton University

Ruth C. Veres

Prepared by

Debra Nero
Cornell University

Boston   Burr Ridge, IL   Dubuque, IA   Madison, WI   New York   San Francisco   St. Louis
Bangkok   Bogotá   Caracas   Kuala Lumpur   Lisbon   London   Madrid   Mexico City
Milan   Montreal   New Delhi   Santiago   Seoul   Singapore   Sydney   Taipei   Toronto

The McGraw·Hill Companies

Study Guide/Solutions Manual for use with
GENETICS: FROM GENES TO GENOMES, SECOND EDITION
HARTWELL/HOOD/GOLDBERG/REYNOLDS/SILVER/VERES

Published by McGraw-Hill Higher Education, an imprint of The McGraw-Hill Companies, Inc.,
1221 Avenue of the Americas, New York, NY 10020. Copyright © The McGraw-Hill Companies,
Inc., 2004. All rights reserved.

This book is printed on acid-free paper.

1 2 3 4 5 6 7 8 9 0 VNH VNH 0 9 8 7 6 5 4 3

ISBN 0-07-246258-2

www.mhhe.com

- Who is involved in making the decisions?

- Who will be impacted by the decision made or action taken?

- What values are important for each of these players?

- What are the options?

A course of action may have to be a compromise for individuals involved.

In **Chapter 19** and the **Genetic Portraits** (References A-E at the back of the book) the problems allow you to integrate your accumulated knowledge and think like a scientist pursuing a specific line of research using genetic analysis. You will need to draw on all your resources to decide if you need to do crosses or analyze molecular events or a combination of both.

How this Student Guide will help you:

- **Synopsis:** The synopsis is a distillation of each chapter's content. It is good practice to try first to define terms and summarize information for yourself. Try to determine the main focus of the chapter and think about this information in the context of everything else you have learned up to that point.

- **Significant Elements:** This section reminds you of the types of problems and applications that you should be able to answer given the material in the chapter.

- **Problem Solving: Tips and How to Begin:** These sections include items to remember and useful ways to approach the problems. Don't let any of the problems overwhelm you. Rewrite the information given in a format that will be useful for you. In other words, learn to DIAGRAM THE CROSS OR EXPERIMENT. Read the problem carefully and closely - **wording is important.** You must understand the terms used in order to interpret the information that has been given and to understand correctly what is being asked.

- **Solutions to Problems:** The Student Guide contains the detailed solutions to all the problems found in the Genetics: From Genes to Genomes textbook, and outlines the reasoning leading to these solutions.

What additional things can you do to help yourself learn genetics?

- Become an active learner as you read new material. Stop every few pages to ask yourself: How would I summarize what I just read? How does it apply to what I have already learned? What do I not understand of this material?

- There is a large amount of new vocabulary, so it is useful to make a list of the genetics terms that are new to you along with the definitions. You need not only to learn the vocabulary, but you also need to be able to think in these terms – that is, you must become fluent in the language of genetics. Try to weave together the new vocabulary in a narrative describing the main points of the chapter. If you encounter difficulty when trying to explain a particular concept, you probably don't have a sufficiently solid understanding of that concept. The first problem of each chapter will help you evaluate your comfort with key vocabulary terms.

♦ Read the problem carefully and closely - **wording is important.** You must understand the terms used in order to interpret the information that has been given and to correctly understand what is being asked.

♦ Answer the question asked! Always go back to the original problem and make sure you have done this.

♦ Don't let the problem overwhelm you.

♦ Go to your professor with questions about a difficult concept. Put some effort into describing your difficulties, as this sometimes leads you to discover the answer. Being able to identify exactly what confuses you is a very valuable general skill that is particularly important for genetics.

♦ Approach problem solving as a challenging game or puzzle, not drudgery. When you are in the right frame of mind you will see that solving genetic puzzles can be fun!

♦ Always remember always that the problems presented in this book are not a punishment, but instead are designed to help you learn and understand genetics, one of the essential basics of all biological and medical science.

Good luck in the course!

Chapter 2   Mendel's Breakthrough: Patterns, Particles and Principles of Heredity

Synopsis:

Chapter 2 covers the basic principles of inheritance, first described by Mendel, that form the foundation of the Laws of Segregation and Independent Assortment. You will see in chapter 4 how these laws relate to chromosome segregation during meiosis. Chapter 2 contains most of the essential terminology used to describe inheritance. You should become very familiar with and fluent in the use of these terms because they will be used in increasingly sophisticated ways in subsequent chapters. A good way to assure that you have a solid grasp of the meanings of the new terms is to pretend you are describing each word or phenomenon to a friend or relative who is not a science major. Often giving an example of each term is useful. The first problem at the end of the chapter is also a useful gauge of how well you know these terms.

A few of the terms defined in this chapter are very critical yet often misunderstood. Learn to be precise about the way in which you use these terms:

genes and *alleles* of genes - a gene determines a trait; and there are different alleles or forms of a gene. The <u>pea color gene</u> has two alleles: the <u>yellow allele</u> and the <u>green allele</u>;

genotype and *phenotype* - genotype is the genetic makeup of an organism (written as alleles) and phenotype is what the organism looks like;

homozygous and *heterozygous* - when both alleles of a gene are the same, we say the organism is homozygous for that gene; if the two alleles are different, the organism is heterozygous;

dominant and *recessive* - the dominant allele is the one that controls the phenotype in the heterozygous genotype.

Significant Elements:

After reading the chapter and thinking about the concepts you should be able to:

♦ When describing genotypes of individuals, remember there are <u>two alleles of each gene</u>. If you are describing <u>gametes</u> in a problem, remember there is only <u>one allele</u> of each gene per gamete.

♦ Recognize if a trait is dominant or recessive by considering the phenotype of the F_1 generation.

♦ Recognize ratios:

in the Study Guide monohybrid ratio refers to any ratio involving one gene, e.g. the phenotypic monohybrid ratio from a cross between two individuals heterozygous for one gene (3 dominant :1 recessive);

in the Study Guide dihybrid ratio refers to any ratio involving 2 genes, for e.g. a phenotypic dihybrid ratio from a cross between two individuals heterozygous for two genes (9:3:3:1).

♦ Recognize the need for and be able to set up a test cross.

♦ Determine probabilities using the basic rules of probability:

Product rule: If two outcomes must occur together, the probability of one outcome AND the other occurring is the product of the two individual probabilities. (The final outcome is the result of two independent events.) So, the probability of getting a 4 on one die AND a 4 on the second die is the product of the two individual probabilities.

Sum rule: If there is more than one way in which an outcome can be produced, the probability of either one OR the other occurring is the sum of the individual probabilities. In this case, the outcomes are mutually exclusive.

♦ Draw and interpret pedigrees as in **Figures 2.20 and 2.21**.

♦ Set up Punnett squares by determining the gametes produced by the parents and the probabilities of the potential offspring as in **Figures 2.11 and 2.15**.

♦ Use and interpret branched line diagrams as in **Figure 2.17** and Problems 2-5b and 2-16a.

Problem Solving - How to Begin:

♦ As you work on problems in the first few chapters, you will begin to recognize some similarities in the type of problem. The majority of problems in the first few chapters involve genetic crosses. It is best to BE CONSISTENT in your approach and formatting to solve such problems. To begin, rewrite the information given in a format that will be useful in solving the problem: DIAGRAM THE CROSS.

phenotype of one parent x phenotype of other parent → phenotype(s) of progeny

♦ Use the THREE ESSENTIAL QUESTIONS to determine basic information about the genotypes and/or phenotypes of the parents and/or offspring. These questions are distilled from many years of helping students figure out how to solve problems. They are designed to force you to focus on the underlying genetic basis of the information in a problem. Each of the questions has an identifying characteristic that helps you answer the question - see the Hints for an idea of what sort of information allows you to formulate specific answers. As you go through further chapters the Hints will be further refined.

THREE ESSENTIAL QUESTIONS (3EQ):

1. How many genes are involved in the cross?

2. For **each gene** involved in the cross: what are the phenotypes associated with the gene? Which phenotype is the dominant one and why? Which phenotype is the recessive one and why?

3. For **each** gene involved in the cross: is it X-linked or autosomal?

At this point, only questions 3EQ #1 and 3EQ #2 may be applied. The material that is the basis of question 3EQ #3 will be covered in Chapter 4.

<u>Hints</u>:

For 3EQ #1 look for the number of different phenotypes or phenotypic classes in the progeny. In Chapter 2 each gene has only 2 phenotypes.

For 3EQ #2 if the parents of a cross are true-breeding, look at the phenotype of the F_1 individuals. Also, look at the F_2 progeny – the 3/4 portion of the 3:1 phenotypic monohybrid ratio is the dominant one.

Solutions to Problems:

2-1. a. **4**; b. **3**; c. **6**; d. **7**; e. **11**; f. **13**; g. **10**; h. **2**; i. **14**; j. **9**; k. **12**; l. **8**; m. **5**; n. **1**.

2-2. There are **several advantages to using peas** for the study of inheritance. **(1) Peas have a fairly rapid generation time** (at least two generations per year if grown in the field, three or four generations per year if grown in greenhouses). **(2) Peas can either self-fertilize or be artificially crossed** by an experimenter. **(3) Peas produce large numbers of offspring** (hundreds per parent). **(4) Peas can be maintained as pure-breeding lines**, simplifying the ability to perform subsequent crosses. **(5) Because peas have been maintained as inbred stocks, two easily distinguished, discrete forms of many phenotypic traits are known. (6) Peas are easy and inexpensive to grow**.

In contrast, studying **genetics in humans has several disadvantages. (1) The generation time of humans is long** (roughly 20 years). **(2) There is no self-fertilization in humans, and it is not ethical to manipulate crosses. (3) Humans produce only a small number of offspring per mating** (usually one) or per individual (almost always less than 20). **(4) Although people that are homozygous for a trait (analogous to pure-breeding) exist, homozygosity cannot be maintained** because mating with another individual is needed to produce the next generation. **(5) Because human populations are not inbred, most human traits show a continuum of phenotypes**; only a few traits have two very distinct forms. **(6) People require a lot of expensive care to "grow".**

There is nonetheless one advantage to the study of genetics in humans. Because many inherited traits result in disease syndromes, and because the world's population now exceeds 6 billion, **a very large number of individuals with variant phenotypes can be recognized. Thus, the number of genes identified in this way is rapidly increasing**.

2-3. The **dominant trait (short tail) is easier to eliminate** from the population by selective breeding. **You can recognize every animal that has inherited the short tail allele**, because only one such dominant allele is needed to see the phenotype. **Short-tailed mice can be prevented from mating**. The recessive dilute coat color allele, on the other hand, can be passed unrecognized from generation to generation in heterozygous mice (carriers). The heterozygous mice do not express the phenotype, so they cannot be distinguished from homozygous dominant mice with normal coat color. Only the homozygous recessive mice express the dilute phenotype and could be prevented from mating; the heterozygotes could not.

2-4.

a. The *Aa bb CC DD* woman can produce 2 genetically different eggs that vary in their allele of the first gene (*A* or *a*). She is homozygous for the other 3 genes and can only make eggs with the *b C D* alleles for these genes. Thus, using the product rule (because the inheritance of each gene is independent), she can make $2 \times 1 \times 1 \times 1 = $ **2 different types of gametes**: (*A b C D* and *a b C D*).

b. Using the same logic, an *AA Bb Cc dd* woman can produce $1 \times 2 \times 2 \times 1 = $ **4 different types of gametes**: *A* (*B* or *b*) (*C* or *c*) *d*.

c. A woman of genotype *Aa Bb cc Dd* can make $2 \times 2 \times 1 \times 2 = $ **8 different types of gametes**: (*A* or *a*) (*B* or *b*) *c* (*D* or *d*).

d. A woman who is a quadruple heterozygote can make $2 \times 2 \times 2 \times 2 = $ **16 different types of gametes**: (*A* or *a*) (*B* or *b*) (C or *c*) (*D* or *d*).

2-5.

a. The combination of alleles in the egg and sperm allows only one genotype: ***aa Bb Cc DD Ee***.

b. Because the inheritance of each gene is independent, you can use the product rule to determine the number of different types of gametes that are possible: $1 \times 2 \times 2 \times 1 \times 2 = 8$ (as in problem 2-4). To figure out the types of gametes, consider the possibilities for each gene separately and then the possible combinations of genes in a consistent order. For each gene the possibilities are: *a*, (*B* : *b*), (*C* : *c*), *D*, and (*E* : *e*). The possibilities can be determined using the product rule. Thus for the first 2 genes [*a*] × [*B* : *b*]gives [*a B* : *a b*] × [*C* : *c*] gives [*a B C* : *a B c* : *a b C* : *a b c*] × [*D*] gives [*a B C D* : *a B c D* : *a b C D* : *a b c D*] × [*E* : *e*] gives [***a B C D E* : *a B c D E* : *a B C D***

$e : aBcDe : abCDE : abCDe : abcDE : abcDe$]. This problem can also be visualized with a branch diagram:

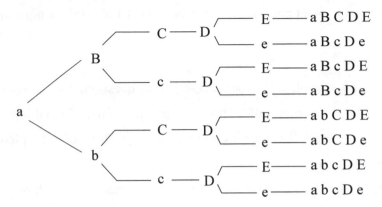

2-6.

a.   In this problem, the first of the **3 Essential Questions** (3EQ #1) is answered for you (there is only 1 gene), as is 3EQ # 2 (the dominant allele is dimple D and the recessive allele is nondimple d; this dominance relationship will be symbolized in the form dimple D > nondimple d from now on.) Next, diagram the cross. In all cases, the male parent is written first for consistency.

   nondimpled $\times$ dimpled $\rightarrow$ proportion F_1 with dimple?

   Therefore: $dd \times Dd^* \rightarrow$ **1/2 dimple** : 1/2 nondimple

   (*Note that the dimpled woman in this cross had a dd [nondimpled] mother, so the woman's genotype MUST be heterozygous.)

b.   dimpled $\times$ nondimple $\rightarrow$ nondimple F_1

       $D? \times dd \rightarrow dd$

   Because they have a nondimpled child (dd), the husband must also have a d allele to contribute to the offspring. **The husband has genotype Dd.**

c.   dimpled $\times$ nondimpled $\rightarrow$ eight F_1, all dimpled

       $D? \times dd \rightarrow 8\ D\text{-}$

   The D allele in the children must come from their father. The father could be either DD or Dd but **it is more probable that his genotype is DD**. We cannot rule out the heterozygous genotype (Dd). However, the probability that all 8 children would inherit the D allele from a Dd parent is only $(1/2)^8$ or 1/256.

2-7. Because two phenotypes result from the mating of two cats of the same phenotype, the short-haired parent cats must have been heterozygous. The phenotype expressed in the heterozygotes (the parent cats) is the dominant phenotype. Therefore, **short-hair is dominant to long-hair**.

2-8.

a.   3EQ #1 is answered in the problem (1 gene); 3EQ #2 <u>is</u> the question! Two affected individuals have an affected child and a normal child. This is not possible if the affected individuals were homozygous for a recessive allele conferring piebald spotting. Therefore, **the piebald trait must be the dominant phenotype.**

b.   If the trait is dominant, the piebald parents could be either homozygous (*PP*) or heterozygous (*Pp*). However, because the two affected individuals have an unaffected child *(pp)*, **they both must be heterozygous (*Pp*).**

Diagram the cross:

Spotted $\times$ spotted $\rightarrow$ 1 spotted : 1 normal

Pp $\times$ Pp $\rightarrow$ 1 Pp : 1 pp

2-9.

a.   The only unambiguous cross is:

homozygous recessive $\times$ homozygous recessive $\rightarrow$ all homozygous recessive

The only cross that fits this criteria is: dry $\times$ dry $\rightarrow$ all dry. Therefore, **dry is the recessive phenotype (*ss*) and sticky is the dominant phenotype (*S-*).**

b.   A 1:1 ratio comes from a testcross of heterozygous sticky (*Ss*) $\times$ dry (*ss*). However, **the sticky x dry matings here include both the *Ss* $\times$ *ss* AND the homozygous sticky (*SS*) $\times$ dry (*ss*).** A 3:1 ratio comes from crosses between two heterozygotes, *Ss* $\times$ *Ss*. However, the *SS* individuals are also sticky. Thus the sticky $\times$ sticky matings in this human population are a mix of matings between two heterozygotes (*Ss* $\times$ *Ss*), between two homozygotes (*SS* $\times$ *SS*) and between a homozygote and heterozygote (*SS* $\times$ *Ss*). **The 3:1 ratio of the heterozygote cross is therefore obscured by being combined with results of the two other crosses.**

2-10.

a.   **1/6** because a die has 6 different sides.

b.   There are three possible even numbers (2, 4, and 6). The probability of obtaining any one of these is 1/6. Because the 3 events are mutually exclusive, use the sum rule: 1/6 + 1/6 + 1/6 = 3/6 = **1/2**.

c.   You must roll either a 3 or a 6, so 1/6 + 1/6 = 2/6 = **1/3**.

d.   Each die is independent of the other, thus the product rule is used: 1/6 × 1/6 = **1/36**.

e.   The probability of getting an even number on one die is 3/6 = 1/2 (see part b). This is also the probability of getting an odd number on the second die. This result could happen either of 2 ways – you could get the odd number first and the even number second, or vice-versa. Thus the probability of both occurring is 1/2 × 1/2 × 2 = **1/2.**

f.   The probability of any specific number on a die = 1/6. The probability of the same number on the other die =1/6. The probability of both occurring at same time is 1/6 × 1/6 = 1/36. The same probability is true for the other 5 possible numbers on the dice. Thus the probability of any of these mutually exclusive situations occurring is 1/36 + 1/36 + 1/36 + 1/36 + 1/36 + 1/36 = 6/36 = **1/6**.

g.   The probability of getting two numbers both over four is the probability of getting a 5 or 6 on one die (1/6 + 1/6 = 1/3) and 5 or 6 on the other die (1/3). The results for the two dice are independent events, so 1/3 × 1/3 = **1/9**.

2-11. The first two parts of this problem involve the probability of occurrence of two independent traits: the sex of a child and galactosemia. The parents are heterozygous for galactosemia, so there is a 1/4 chance that a child will be affected (that is, homozygous recessive). The probability that a child is a girl is 1/2. The probability of an affected girl is therefore 1/2 × 1/4 = 1/8.

a.   Fraternal (non-identical) twins result from two independent fertilization events and therefore the probability that both will be girls with galactosemia is the product of their individual probabilities (see above); 1/8 × 1/8 = **1/64**.

b.   For identical twins, one fertilization event gave rise to two individuals. The probability that both are girls with galactosemia is **1/8**.

For parts c-g, remember that each child is an independent genetic event. The sex of the children is not at issue in these parts of the problem.

c.   Both parents are carriers (heterozygous), so the probability of having an unaffected child is 3/4. The probability of 4 unaffected children is 3/4 × 3/4 × 3/4 × 3/4 = **81/256**.

d.   The probability that at least one child is affected is all outcomes except the one mentioned in part c. Thus, the probability is 1 - 81/256 = **175/256**.

e. The probability of an affected child is 1/4 while the probability of an unaffected child is 3/4. Therefore $1/4 \times 1/4 \times 3/4 \times 3/4 = \mathbf{9/256}$.

f. The probability of 2 affected and 1 unaffected in any one particular birth order is $1/4 \times 1/4 \times 3/4 = 3/64$. There are 3 mutually exclusive birth orders that could produce 2 affecteds and 1 unaffected – unaffected child first born, unaffected child second born, and unaffected child third born. Thus, there is a $3/64 + 3/64 + 3/64 = \mathbf{9/64}$ chance that 2 out of 3 children will be affected.

g. The phenotype of the last child is independent of all others, so the probability of an affected child is **1/4**.

2-12.

a. The probability of any phenotype in this cross depends only on the gamete from the heterozygous parent alone. The probability that a child will resemble the quadruply heterozygous parent is thus $1/2A \times 1/2B \times 1/2C \times 1/2D = 1/16$. The probability that a child will resemble the quadruply homozygous recessive parent is $1/2a \times 1/2b \times 1/2c \times 1/2d = 1/16$. **The probability that a child will resemble either parent is then 1/16 + 1/16 = 1/8**. This cross will produce 2 different phenotypes for each gene or $2 \times 2 \times 2 \times 2 = \mathbf{16\ potential\ phenotypes.}$

b. The probability of a child resembling the recessive parent is 0; the probability of a child resembling the dominant parent is $1 \times 1 \times 1 \times 1 = 1$. The probability that a child will resemble one of the two parents is 0+1=**1**. **Only 1 phenotype is possible** in the progeny (dominant for all 4 genes), as $(1)^4 = 1$.

c. The probability that a child would show the dominant phenotype for any one gene is 3/4 in this sort of cross (remember the 3/4 : 1/4 monohybrid ratio of phenotypes), so the probability of resembling the parent for all four genes is $(3/4)^4 = \mathbf{81/256}$. There are 2 phenotypes possible for each gene, so $(2)^4 = \mathbf{16\ different\ kinds\ of\ progeny.}$

d. All progeny will resemble their parents because all of the alleles from both parents are identical, so the **probability = 1**. There is only 1 phenotype possible for each gene in this cross; because $(1)^4=1$, the child can have only **one possible phenotype** when considering all four genes.

2-13. Do a **test cross** between your normal winged fly (*W-*) and a short winged fly that must be homozygous recessive (*ww*). The possible results are diagrammed here; the first genotype in each cross is that of the normal winged fly whose genotype was originally unknown:

> *WW* x *ww* → **all *W-* (normal wings)**

or *Ww* x *ww* → **1/2 *W-* (normal wings) : 1/2 *ww* (short wings)**.

2-14. Diagram the crosses:

closed $\times$ open $\rightarrow$ F_1 open $\rightarrow$ F_2 145 open : 59 closed

F_1 open $\times$ closed $\rightarrow$ 81 open : 77 closed

The results of the crosses all fit the pattern of inheritance of a single gene, with the closed trait being recessive. The first cross is a cross like Mendel did with his pure-breeding plants, although we don't know from the information provided if the starting plants were pure-breeding or not. The F_1 result of this cross shows that open is dominant and that the parental cucumber plants were indeed true-breeding homozygotes. The self-fertilization of the F_1 resulting in a ratio of 3 open : 1 closed ratio supports this hypothesis. The second cross is a test cross. It confirms that the F_1 plants from the first cross are heterozygous hybrids, because there is a 1:1 ratio of open and closed progeny. Thus, all the data is consistent with open being the dominant trait.

2-15. Diagram the cross:

black $\times$ red $\rightarrow$ 1 black : 1 red

No, you cannot tell how coat color is inherited from the results of this one mating. In effect, this was a test cross – a cross between animals of different phenotypes resulting in offspring of two phenotypes. This does not indicate whether red or black is the dominant phenotype. To determine which phenotype is dominant, remember that an animal with a recessive phenotype must be homozygous. Thus, **if you mate several red horses to each other and also mate several black horses to each other, the crosses that always yield only offspring with the parental phenotype must have been between homozygous recessives.** For example, if all the black $\times$ black matings result in only black offspring, black is recessive. Some of the red $\times$ red crosses (that is, crosses between heterozygotes) would then result in both red and black offspring in a ratio of 3:1. To establish this point, you might have to do several red $\times$ red crosses, because some of these crosses could be between red horses homozygous for the dominant allele. You could of course ensure that you were sampling heterozygotes by using the progeny of black $\times$ red crosses (such as that described in the problem) for subsequent black $\times$ black or red $\times$ red crosses.

2-16. The F_1 must be heterozygous for all the genes because the parents were pure-breeding (homozygous). The appearance of the F_1 establishes that the dominant phenotypes for the four traits are tall, purple flowers, axial flowers and green pods.

a. From a heterozygous $F_1 \times F_1$ cross, both dominant and recessive phenotypes can be seen. Thus, you expect $2 \times 2 \times 2 \times 2 = 16$ different phenotypes when considering the four traits together. The possibilities can be determined using the product rule with the pairs of phenotypes for each gene, because the traits are inherited independently. Thus: [tall : dwarf] × [green : yellow] gives [tall green : tall yellow : dwarf green : dwarf yellow] × [purple : white] gives [tall green purple : tall yellow purple : dwarf green purple : dwarf yellow purple : tall green white : tall yellow white : dwarf green white : dwarf yellow white] × [terminal : axial] which gives **tall green purple terminal : tall yellow purple terminal : dwarf green purple terminal : dwarf yellow purple terminal : tall green white terminal : tall yellow white terminal : dwarf green white terminal : dwarf yellow white terminal : tall green purple axial : tall yellow purple axial : dwarf green purple axial : dwarf yellow purple axial : tall green white axial : tall yellow white axial : dwarf green white axial : dwarf yellow white axial**. The possibilities can also be determined using the branch method shown below, which might in this complicated problem be easier to track.

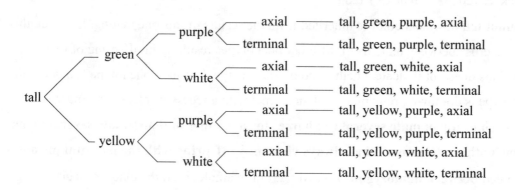

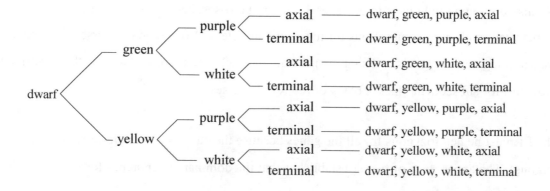

b. Designate the alleles: T = tall, t = dwarf; G = green; g = yellow; P = purple, p = white; A = axial, a = terminal. The cross $Tt\ Gg\ Pp\ Aa$ (an F$_1$ plant) $\times$ $tt\ gg\ pp\ AA$ (the dwarf parent) will produce 2 phenotypes for the tall, green and purple genes, but only 1 phenotype (axial) for the fourth gene or $2 \times 2 \times 2 \times 1 = $ **8 different phenotypes**. The first 3 genes will give a 1/2 dominant : 1/2 recessive ratio of the phenotypes (for example 1/2 T : 1/2 t) as this is in effect a test cross for each gene. Thus, the proportion of each phenotype in the progeny will be $1/2 \times 1/2 \times 1/2 \times 1 = $ 1/8.

Using either of the methods described in part a, the progeny will be **1/8 tall green purple axial : 1/8 tall yellow purple axial : 1/8 dwarf green purple axial : 1/8 dwarf yellow purple axial : 1/8 tall green white axial : 1/8 tall yellow white axial : 1/8 dwarf green white axial : 1/8 dwarf yellow white axial**.

2-17. Apply the 3EQ to each cross separately. Remember that 4 phenotypic classes in the progeny means there are 2 genes controlling the phenotypes. Determine the phenotypic ratio for each gene separately. A 3:1 monohybrid ratio tells you which phenotype is dominant and that both parents were heterozygous for the trait; in contrast, a 1:1 ratio results from a test cross where the dominant parent was heterozygous.

a. 3EQ#1 - there are 2 genes in this cross (4 phenotypes). 3EQ#2 - one gene controls purple : white with a monohybrid ratio of $94 + 28 = 122$ purple : $32 + 11 = 43$ white or ~3 purple : 1 white. The second gene controls spiny : smooth with a monohybrid ratio of $94 + 32 = 1\ 26$ spiny : $28 + 11 = 39$ smooth or ~3 spiny : 1 smooth. Thus, designate the alleles P = **purple**, p = **white**; S = **spiny**, s = **smooth**. Therefore, this is a straightforward dihybrid cross: $Pp\ Ss \times Pp\ Ss\ \rightarrow$ **9 $P\text{-}\ S\text{-}$: 3 $P\text{-}\ ss$: 3 $pp\ S\text{-}$: 1 $pp\ ss$**.

b. The 1 spiny : 1 smooth ratio indicates a test cross for the pod shape gene. Because all progeny were purple, at least one parent plant must have been homozygous for the P allele of the flower color gene. The cross was **either $PP\ Ss \times P\text{-}\ ss$ or $P\text{-}\ Ss \times PP\ ss$**.

c. This is similar to part b, except that here all the progeny were spiny so at least one parent must have been homozygous for the S allele. The 1 purple : 1 white test cross ratio indicates that the parents were **either $Pp\ S\text{-} \times pp\ SS$ or $Pp\ SS \times pp\ S\text{-}$**.

d. Looking at each trait individually, there are $89 + 31 = 120$ purple : $92 + 27 = 119$ white. A 1 purple:1 white monohybrid ratio denotes a test cross. For the other gene, there are $89 + 92 = 181$ spiny : $31 + 27 = 5\ 8$ smooth, or a 3 spiny : 1 smooth ratio indicating that the parents were both heterozygous for the S gene. The genotypes of the parents were $pp\ Ss \times Pp\ Ss$.

e. There is a 3 purple : 1 white ratio among the progeny, so the parents were both heterozygous for the *P* gene. All progeny have smooth pods so the parents were both homozygous recessive *ss*. The genotypes of the parents are ***Pp ss*** × ***Pp ss***.

f. There is a 3 spiny : 1 smooth ratio, indicative of a cross between heterozygotes (*Ss* × *Ss*). All progeny were white so the parents must have been homozygous recessive *pp*. The genotypes of the parents are ***pp Ss*** x ***pp Ss***.

2-18. Diagram the cross:

yellow round × yellow round → 156 yellow round : 54 yellow wrinkled

The monohybrid ratio for seed shape is 156 round : 54 wrinkled = 3 round : 1 wrinkled. The parents must therefore have been heterozygous (*Rr*) for the pea shape gene. All the offspring are yellow and therefore have the *Yy* or *YY* genotype. The parent plants were **Y- Rr x YY Rr** (that is, you know at least one of the parents must have been *YY*).

2-19. Diagram the cross:

smooth black ♂ × rough white ♀ → F$_1$ rough black

→ F$_2$ 8 smooth white : 25 smooth black : 23 rough white : 69 rough black

a. Since only one phenotype was seen in the first cross, we can assume that the parents were true breeding, that the F$_1$ generation consists of heterozygous animals, and that the phenotype of the F$_1$s indicates the dominant allele. Therefore, ***R* = rough, *r* = smooth; *B* = black, *b* = white**. In the F$_2$ generation, consider each gene separately. For the coat texture, there were 8 + 25 = 33 smooth : 23 + 69 = 92 round, or a ratio of ~1 smooth : 3 round. For the coat color, there were 8 + 23 = 31 white : 25 + 69 = 94 black, or about ~1 white : 3 black, so the F$_2$ progeny support the conclusion that the F$_1$ animals were heterozygous for both genes.

b. An F$_1$ male is heterozygous for both genes, or *Rr Bb*. The smooth white female must be homozygous recessive; that is, *rr bb*. Thus, *Rr Bb* x *rr bb* → 1/2 *Rr* (rough) : 1/2 *rr* (smooth) and 1/2 *Bb* (black) : 1/2 *bb* (white). The inheritance of these genes is independent, so apply the product rule to find the expected phenotypic ratios among the progeny, or **1/4 rough black : 1/4 rough white : 1/4 smooth black : 1/4 smooth white**.

2-20. Three characters (genes) are being analyzed in this cross. While we can usually tell which alleles are dominant from the phenotype of the heterozygote, we are not told the phenotype of the heterozygote (that is, the original pea plant that was selfed). Instead, use the monohybrid phenotypic ratios to determine which allele is dominant and which is recessive for each gene. Consider height first. There are $272 + 92 + 88 + 35 = 487$ tall plants and $93 + 31 + 29 + 11 = 164$ dwarf plants. This is a ratio of ~3 tall : 1 dwarf, indicating that **tall is dominant**. Next consider pod shape, where there are $272 + 92 + 93 + 31 = 488$ inflated pods and $88 + 35 + 29 + 11 = 163$ flat pods, or approximately 3 inflated : 1 flat, so **inflated is dominant**. Finally, consider flower color. There were $272 + 88 + 93 + 29 + 11 = 493$ purple flowers and $92 + 35 + 31 + 11 = 169$ white flowers, or ~3 purple : 1 white. Thus, **purple is dominant**.

2-21.

a. First diagram the cross, and then figure out the monohybrid ratios for each gene:

   $Aa\ Tt$ x $Aa\ Tt$ $\rightarrow$ 3/4 A- (achoo) : 1/4 aa (non-achoo) and 3/4 T- (trembling) : 1/4 tt (non-trembling).

   The probability that a child will be A- (and have achoo syndrome) is independent of the probability that it will lack a trembling chin, so the probability of a child with with achoo syndrome but without trembling chin is 3/4 A- $\times$ 1/4 tt = **3/16**.

b. The probability that a child would have neither dominant trait is 1/4 $aa \times$ 1/4 tt = **1/16**.

2-22. Diagram the cross:

   $YY\ rr \times\ yy\ RR$ $\rightarrow$ all $Yy\ Rr$ $\rightarrow$ 9/16 Y- R- (yellow round) : 3/16 Y- rr (yellow wrinkled) : 3/16 $yy\ R$- (green round) :1/16 $yy\ rr$ (green wrinkled).

Each F_2 pea results from a separate fertilization event. The probability of 7 yellow round F_2 peas is $(9/16)^7 = 4{,}782{,}969/268{,}435{,}456 = $ **0.018**.

2-23.

a. **Recessive** - two unaffected individuals have an affected child (**aa**). Therefore the parents involved in the consanguineous marriage must both be carriers (Aa).

b. **Dominant** - the trait is seen in each generation and every affected person (A-) has an affected parent. Note that III-3 is unaffected (aa) even though both his parents are - this would not be possible for a recessive trait. The term carrier is not applicable, because everyone with a single A allele shows the trait.

c. **Recessive** - two unaffected, carrier parents (Aa) have an affected child (**aa**), as in part a.

2-24. Diagram the cross, where P is the normal pigmentation allele and p is the albino allele:

   normal $\times$ normal $\rightarrow$ albino

$$P? \quad \times \quad P? \quad \rightarrow \quad pp$$

An albino must be homozygous recessive *pp*. The parents are normal in pigmentation and therefore could be *PP* or *Pp*. Because they have an albino child, **they must both be carriers (*Pp*). The probability that their next child will have the *pp* genotype is 1/4.**

2-25. Diagram the cross! In humans, this is usually done as a pedigree. Remember that the affected siblings must be *cf cf*.

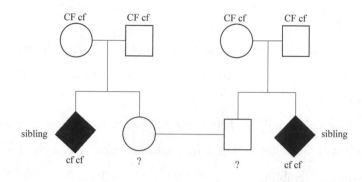

a.   The probability that II-2 is a carrier is **2/3**. Both families have an affected sibling, so both sets of parents (that is, all the people in generation I) must have been carriers. Thus, the expected genotypic ratio in the children is 1/4 affected : 1/2 carrier : 1/4 homozygous normal. II-2 is NOT affected, so she cannot be normal. Of the 3/4 remaining possibilities, 2/4 are that she is heterozygous. There is therefore a 2/3 chance that she is a carrier.

b.   The probability that II-2 × II-3 will have an affected child is 2/3 (the probability that the mother is a carrier as seen in part a) × 2/3 (the probability the father is a carrier using the same reasoning) × 1/4 (the probability that two carriers can produce an affected child) = **1/9**.

c.   The probability that both parents are carriers and that their child will be a carrier is 2/3 × 2/3 × 1/2 = 2/9 (using the same reasoning as in part b, except asking that the child be a carrier instead of affected). However, it is also possible for *CfCf* × *Cfcf* parents to have children that are carriers. Remember that there are 2 possible ways for this particular mating to occur – homozygous father × heterozygous mother or vice versa. Thus the probability of this sort of mating is 2 × 1/3 (the probability that a particular parent is *CfCf*) × 2/3 (the probability that the other parent is *Cfcf*) × 1/2 (the probability such a mating could produce a carrier child = 2/9. The probability that a child could be carrier from either of these two scenarios (where both parents are carriers or where only one parent is a carrier) is the sum of these mutually exclusive events, or 2/9 + 2/9 = **4/9**.

2-26.

a.   Because the disease is rare the affected father is most likely to be heterozygous (*Hh*). There is a **1/2** chance that the son inherited the *H* allele from his father and will thus develop the disease.

b.   The probability of an affected child is: 1/2 (the probability that Joe is *Hh*) × 1/2 (the probability that the child inherits the *H* allele if Joe is *Hh*) = **1/4**.

2-27. The trait is **recessive** because pairs of unaffected individuals (I-1 × I-2 as well as II-3 × II-4) had affected children (II-1, III-1, and III-2). There are also two cases in which an unrelated individual must have been a carrier (II-4 and either I-1 or I-2), so the disease allele appears to be common in the population.

2-28.

a.   The inheritance pattern seen in Figure 2.20 could be caused by a **rare dominant mutation**. In this case, the affected individuals would be heterozygous (*Hh*) and the normal individuals would be *hh*. Any mating between an affected individual x an unaffected individual would give 1/2 normal (*hh*) : 1/2 affected (*Hh*) children. However, the same pattern of inheritance could be seen if the disease were caused by a **common recessive mutation**. In the case of a common recessive mutation, all the affected individuals would be *hh*. Because the mutant allele is common in the population, most or even all of the unrelated individuals could be assumed to be carriers (*Hh*). Matings between affected and unaffected individuals would then also yield phenotypic ratios of progeny of 1/2 normal (*Hh*) : 1/2 affected (*hh*).

b.   **Determine the phenotype of the 14 children of III-6 and IV-8.** If the disease is due to a **recessive allele**, then III-6 and IV-8 must be homozygotes for this recessive allele, and **all their children must have the disease.** If the disease is due to a **dominant mutation**, then III-6 and IV-8 must be heterozygotes (because they are affected but they each had one unaffected parent), and **1/4 of their 14 children would be expected to be unaffected.**

Alternatively, you could **look at the progeny of matings between unaffected individuals** in the pedigree such as III-1 and an unaffected spouse. If the disease were due to a **dominant mutation,** these matings would all be homozygous recessive × homozygous recessive and would **never give affected children.** If the disease is due to a **recessive mutation**, then many of these individuals would be carriers, and if the trait is common then at least some of the spouses would also be carriers, so such matings **could give affected children.**

2-29. Diagram the cross by drawing a pedigree.

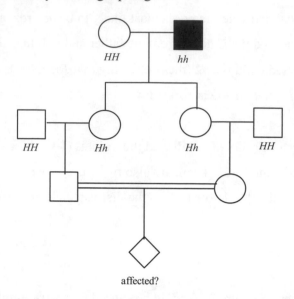

affected?

a. Assuming the disease is very rare, the first generation is *HH* unaffected (I-1) × *hh* affected (I-2). Thus, both of the children (II-2 and II-3) must be carriers (*Hh*). Again assuming this trait is rare in the population, those people marrying into the family (II-1 and II-4) are homozygous normal (*HH*). Therefore, the probability that III-1 is a carrier is 1/2; III-2 has the same chance of being a carrier. Thus the probability that a child produced by these two first cousins would be affected is 1/2 (the probability that III-1 is a carrier) × 1/2 (the probability that III-2 is a carrier) × 1/4 (the probability the child of two carriers would have an *hh* genotype) = **1/16** = 0.0625.

b. If 1/10 people in the population are carriers, then the probability that II-1 and II-4 are *Hh* is 0.1 for each. In this case an affected child in generation IV can only occur if III-1 and III-2 are both carriers. III-1 can be a carrier as the result of 2 different matings: (i) II-1 homozygous normal × II-2 carrier or (ii) II-1 carrier × II-2 carrier. (Note that II-2 must be a carrier because of the normal phenotype and the fact that one parent was affected.) The probability of III-1 being a carrier is thus the probability of mating (i) × the probability of generating a *Hh* child from mating (i) + the probability of mating (ii) × the probability of generating an *Hh* child from mating (ii) = 0.9 (the probability II-1 is *HH*, which is the probability for mating [i]) × 1/2 (the probability that III-1 will inherit *h* in mating [i]) + 0.1 (the probability II-1 is *H*, which is the probability for mating [ii]) × 2/3 (the probability that III-1 will inherit *h* in mating [ii]; remember that III-1 is known not to be *hh*) = 0.45 + 0.067 = 0.517. The chance that III-2 will inherit *h* is exactly the same. Thus, the probability that IV-1 is *hh* = 0.517 (the probability III-1 is *Hh*) × 0.517 (the probability that III-2 is *Hh*) × 1/4 (the probability the child of two carriers will be *hh*) = **0.067**. This number is slightly higher than the answer to part a, which was 0.0625, so the increased likelihood that II-1 or II-4 is a carrier makes it only slightly more likely that IV-1 will be affected.

2-30. If polycistic kidney disease is dominant, then the child is *Pp* and inherited the *P* disease allele from one parent or the other, yet phenotypically the parents are *pp*. Perhaps one of the parents is indeed *Pp,* but this parent does not show the disease phenotype for some reason. As we will see in the next chapter, such situations are not uncommon: the unexpressed dominant allele is said to have **incomplete penetrance** in these cases. Alternatively, it could be that both parents are indeed *pp* and the *P* allele inherited by the child was due to a **spontaneous mutation** during the formation of the gamete in one of the parents; again, we will discuss this topic in the next chapter. It is also possible that the **father of the child is not the male parent of the couple**. In this case the biological father must have the disease.

2-31. Diagram the cross(es):

midphalangeal × midphalangeal → 1853 midphalangeal : 209 normal

$$M? \quad\quad x \quad\quad M? \quad\quad \to M? : mm$$

The following genotypes are possible:

MM	×	*MM*	→	all *MM*
Mm	×	*MM*	→	all *M-*
MM	×	*Mm*	→	all *Mm*
Mm	×	*Mm*	→	3/4 *M-* : 1/4 *mm*

The 209 normal children must have arisen from the last cross, so approximately 3 × 209 = 630 children should be their *M-* siblings. Thus, about 840 of the children or **~40% came from the last mating** and the other 60% of the children were the result of one or more of the other matings. This problem illustrates that much care in interpretation is required when the results of many matings in mixed populations are reported.

2-32.

a. Both diseases are known to be rare, so normal people marrying into the pedigree are assumed to be homozygous normal. **Nail-patella (*N*) syndrome is dominant** because all affected children have an affected parent. **Alkaptonuria (*a*) is recessive** because the affected children are the result of a consanguineous mating between 2 unaffected individuals (III-3 × III-4). Genotypes: **I-1 *nn Aa*; I-2 *Nn AA* (or I-1 *nn AA* and I-2 *Nn Aa*); II-1 *nn AA*; II-2 *nn Aa*; II-3 *nn A-*; II-4 *nn A-*; II-5 *Nn Aa*; II-6 *nn AA*; III-1 *nn AA*; III-2 *nn A-*; III-3 *nn Aa*; III-4 *Nn Aa*; III-5 *nn***

A-; **III-6** *nn A-*; **IV-1** *nn A-*; **IV-2** *nn A-*; **IV-3** *Nn A-*; **IV-4** *nn A-*; **IV-5** *Nn aa*; **IV-6** *nn aa*;
IV-7 *nn A-*.

b. The cross is *nn A-* (IV-2) × *Nn aa* (IV-5). The ambiguity in the genotype of IV-2 is due to the
   uncertainty of her father's genotype (III-2). His parents' genotypes are *nn AA* (II-1) x *nn Aa* (II-2)
   so there is a 1/2 chance III-2 is *nn AA* and a 1/2 chance he is *nn Aa*. Thus, for each of the
   phenotypes below you must consider both possible genotypes for IV-2. For each part below,
   calculate the probability of the child inheriting the correct gametes from IV-2 × the probability of
   obtaining the correct gametes from IV-3 to give the desired phenotype. If both the possible IV-2
   genotypes can produce the needed gametes, you will need to sum the two probabilities.

   ❶ for the child to have both syndromes (*N- aa*), IV-2 would have to contribute an *n a*
   gamete. This could only occur if IV-2 were *nn Aa*. The probability IV-2 is *nn Aa* is 1/2,
   and the probability of receiving an *n a* gamete from IV-2 if he is *nn Aa* is also 1/2. The
   probability that IV-5 would supply an *N a* gamete is also 1/2. Thus, 1/2 × 1/2 × 1/2 = 1/8.
   There is no need to sum probabilities in this case because IV-2 cannot produce an *n a*
   gamete if his genotype is *nn AA*.

   ❷ for the child to have only nail-patella syndrome (*N- A-*), IV-2 would have to provide an *n*
   A gamete and IV-5 an *N a* gamete. This could occur if IV-2 were *nn Aa*; the probability is
   1/2 (the probability IV-2 is *Aa*) x 1/2 (the probability of an *A* gamete if IV-2 is *Aa*) × 1/2
   (the probability of an *N a* gamete from IV-5] = 1/8. This could also occur if IV-2 were *nn*
   AA. Here, the probability is 1/2 (the probability IV-2 is *nn AA*) x 1 (the probability of an
   n A gamete if IV-2 is *nn AA*) x 1/2 (the probability of an *N a* gamete from IV-5] = 1/4.
   Summing the probabilities for the two mutually exclusive IV-2 genotypes, 1/8 + 1/4 = **3/8**.

   ❸ for the child to have just alkaptonuria (*nn aa*), IV-2 would have to contribute an *n a*
   gamete. This could only occur if IV-2 were *nn Aa*. The probability IV-2 is *nn Aa* is 1/2,
   and the probability of receiving an *n a* gamete from IV-2 if he is *nn Aa* is also 1/2. The
   probability that IV-5 would supply an *n a* gamete is also 1/2. Thus, 1/2 × 1/2 × 1/2 = **1/8**.
   There is no need to sum probabilities in this case because IV-2 cannot produce an *n a*
   gamete if his genotype is *nn AA*.

   ❹ the probability of neither defect is 1 – (sum of the first 3) = 1 - (1/8 + 3/8 + 1/8) = 1 - 5/8
   = **3/8**. You can make this calculation because there are only the four possible outcomes
   and you have already calculated the probabilities of three of them.

Chapter 3   Extensions to Mendel: Complexities in Relating Genotype to Phenotype

Synopsis:

This chapter builds on the principles of segregation and independent assortment that you learned in Chapter 2. An understanding of those basic principles will help you understand the more complex inheritance patterns in Chapter 3. While the basic principles for the inheritance of alleles of one or more genes hold true, the expression of the corresponding phenotypes is more complicated.

Chapter 3 describes several examples of **single gene** inheritance in which phenotypic ratios are different from the complete dominance examples in Chapter 2 (see <u>Significant Elements</u> below and **<u>Table 3.1</u>**):

- ◆ incomplete dominance;
- ◆ codominance;
- ◆ dominance series of multiple alleles
- ◆ lethal alleles;
- ◆ pleiotropy.

Also introduced in this chapter are examples in which **two or more interacting genes** determine the phenotype. Remember from Chapter 2 that crosses involving two genes have a 9 *A-B-* : 3 *A- bb* : 3 *aa B-* : 1 *aa bb* dihybrid F_2 phenotypic ratio. If two genes interact then the dihybrid ratio seen is a modification of the 9:3:3:1. This modification involves the same genotypes for the progeny. Because of the affect of an allele of one of the genes on the other gene (**epistasis**), the four classes can get added together in different combinations. Multigene inheritance (>2 genes involved) leads to even more phenotypic classes. With more genes controlling a trait you see a continuous range of phenotypes instead of discrete traits, as in **<u>Figure 3.20</u>**.

Penetrance and **expressivity** are terms used to describe some cases of variable phenotypic expression in different individuals. Penetrance describes the fraction of individuals with a mutant genotype who are affected while expressivity describes the extent to which individuals with a mutant genotype are affected.

Significant Elements:

After reading the chapter and thinking about the concepts you should be able to:

♦ Novel phenotypes arise when there is codominance or incomplete dominance. The novel phenotype will appear in the F_1 generation. In the F_2 generation, this same phenotype <u>must</u> be the largest component of the 1:2:1 monohybrid ratio.

♦ Realize that if you see a series of crosses involving different phenotypes for a certain trait, for example coat color, and each individual cross gives a monohybrid ratio, then all the phenotypes are controlled by one gene with many alleles, that is, the problem involves an **allelic series** as in <u>**Figure 3.6**</u>. It is important to write a dominance hierarchy for the alleles of the gene, e.g. $a = b > c$. Thus, a is codominant or incompletely dominant to b and both a and b are completely dominant to c.

♦ Understand that **lethal mutations** are almost always recessive alleles, see <u>**Figure 3.9**</u>. If there is a recessive lethal allele present in a cross, you can never make that allele homozygous. Therefore the cross must have involved parents heterozygous for the lethal allele. Instead of the expected 1:2:1 ratio in the progeny, one of the 1/4 classes is lethal, so the monohybrid phenotypic ration will be 2/3 heterozygous phenotype : 1/3 other (viable) homozygous phenotype caused by homozygosity for the other allele. The recessive lethal allele may be pleiotropic and show a different, dominant phenotype, as in <u>**Figure 3.10**</u>.

♦ Remember that **epistasis involves two genes**. In epistasis, <u>none of the progeny die</u>. All are present, but instead of four phenotypic classes in a 9:3:3:1 phenotypic dihybrid ratio, you will see an epistatic variation where 2 or 3 of the phenotypes have been summed together, for example 9:3:4 or 9:7.

Problem Solving Tips:

♦ Do enough problems so you can distinguish single and two gene traits on the basis of inheritance patterns. Look for the number of classes in the F_2 generation to identify single gene inheritance (3:1, 1:2:1, 1:1 or 2:1) versus 2 gene inheritance (9:3:3:1 or an epistatic variation). In the Study Guide 'monohybrid ratio' is used in a more general sense than in the text to refer to any ratio that is based on the segregation of the alleles of a single gene (3:1, 1:2:1, 1:1 or 2:1). Likewise, in the Study Guide dihybrid ratio refers to any ratio based on the segregation of the alleles of 2 genes (9:3:3:1, 1:3:4, 15:1, etc).

♦ Be able to derive the monohybrid phenotypic ratios for incomplete dominance/codominance and lethal alleles involving inheritance of a single gene.

♦ It is critical that you understand the 9:3:3:1 phenotypic dihybrid ratio involves the 4 classes 9/16 *A- B-* : 3/16 *A- bb* : 3/16 *aa B-* : 1/16 *aa bb*, where a dashed line (-) indicates either a dominant or recessive allele.

♦ If the phenotype involves 2 genes, be able to propose ways in which two genes interact based on offspring ratios. Do not merely memorize the altered ratios (see **Table 3.2**); instead think through what the combinations of alleles mean.

♦ Remember the product rule of probability and use it to determine proportions of genotypes or phenotypes for independently assorting genes.

Problem Solving - How to Begin:

THREE ESSENTIAL QUESTIONS (3EQ):

#1. How many genes are involved in the cross?

#2. For **each gene** involved in the cross: what are the phenotypes associated with the gene? Which phenotype is the dominant one and why? Which phenotype is the recessive one and why?

#3. For **each** gene involved in the cross: is it X-linked or autosomal?

At this point, only questions #1 and #2 may be applied. The material that is the basis of question #3 will be covered in Chapter 4.

Hints:

BE CONSISTENT. Always diagram the crosses and write out the genotypes. Set the problems up the same way. Note the repetitive approach to many of the problems in this chapter. Make sure you always distinguish between genotypes and phenotypes when working the problems.

To answer $3EQ$#1, look for the number of phenotypic classes in the F_2 progeny. Two phenotypes usually means 1 gene, 4 phenotypes MUST be due to 2 genes. Three phenotypes is ambiguous - this could result from 1 gene with codominance or incomplete dominance, or from 2 genes with epistasis. Use the ratio of phenotypes to distinguish between these possibilities. One gene with codominance/incomplete dominance MUST give 1:2:1 while 2 genes with 3 phenotypes will be an epistatic variation of a 9:3:3:1.

For $3EQ$#2 when one gene is involved, look at the phenotype of the F_1 individuals. If the phenotype of the F_1 progeny is like one of the parents, then that phenotype is the dominant one. Also, examine the F_2 progeny – the 3/4 portion of the 3:1 phenotypic monohybrid ratio is the dominant one. If the phenotype of the F1 progeny is unlike either parent, then it may be the alleles of the gene are codominant or incompletely dominant. In this case, the novel F_1 phenotype will be seen again in the largest class of F_2 progeny.

After you answer 3EQ#1 and #2 to the best of your ability, use the answers to assign genotypes to the parents of the cross. Then follow the cross through, figuring out the expected phenotypes and genotypes in the F_1 and F_2 generations. Remember to assign the expected phenotypes in a manner consistent with those initially assigned to the parents. Next, compare your predicted results to the observed data you were given. If the 2 sets of information match, then your initial genotypes were correct! In many cases there may be two possible set of genotypes for the parents. If your predicted results do not match the data given, try the other set of genotypes for the parents. See problems 3-22 and 3-39a for examples illustrating this issue.

Solutions to Problems:

3-1. a. **2**; b. **6**; c. **11**; d. **8**; e. **7**; f. **9**; g. **120**; h. **3**; i. **5**; j. **4**; k. **1**; l. **10**.

3-2. The problem states that the intermediate pink phenotype is caused by incomplete dominance for the alleles of a single gene. We suggest that you employ genotype symbols that can show the lack of complete dominance; the obvious R for red and r for white does not reflect the complexity of this situation. In such cases we recommend using a base letter as the gene symbol and then employing superscripts to show the different alleles. To avoid any possible misinterpretations, it is always advantageous to include a separate statement making the complexities of the dominant/recessive complications clear. Designate the two alleles f^r = red and f^w = white, so the possible genotypes are $f^r f^r$ = red; $f^r f^w$ = pink; and $f^w f^w$ = white. Note that the phenotypic ratio is the same as the genotypic ratio in incomplete dominance.

a.   Diagram the cross: $f^r f^w$ x $f^r f^w$ → **1/4 $f^r f^r$ (red) : 1/2 $f^r f^w$ (pink) : 1/4 $f^w f^w$ (white)**.

b. $f^w f^w$ x $f^r f^w$ → **1/2 $f^r f^w$ (pink) : 1/2 $f^w f^w$ (white)**.

c. $f^r f^r$ x $f^r f^r$ → **1 $f^r f^r$ (red)**.

d. $f^r f^r$ x $f^r f^w$ → **1/2 $f^r f^r$ (red) : 1/2 $f^r f^w$ (pink)**.

e. $f^w f^w$ x $f^w f^w$ → **1 $f^w f^w$ (white)**.

f. $f^r f^r$ x $f^w f^w$ → **1 $f^r f^w$ (pink)**.

The cross shown in **part f** is the most efficient way to produce pink flowers, because all the progeny will be pink.

3-3. yellow $\times$ yellow $\rightarrow$ 38 yellow : 22 red : 20 white

Three phenotypes in the progeny show that the yellow parents are not true breeding. The ratio of the progeny is close to 1/2 : 1/4 : 1/4. This is the result expected for crosses between individuals heterozygous for incompletely dominant genes. Thus:

$$c^r c^w \text{ x } c^r c^w \rightarrow \text{ 1/2 } c^r c^w \text{ (yellow) : 1/4 } c^r c^r \text{ (red) : 1/4 } c^w c^w \text{ (white).}$$

3-4. A cross between individuals heterozygous for an incompletely dominant gene give a ratio of 1/4 (one homozygote) : 1/2 (heterozygote with the same phenotype as the parents) : 1/4 (other homozygote). Because the problem already states which genotypes correspond to which phenotypes, you know that the color gene will give a monohybrid phenotypic ratio of 1/4 red : 1/2 purple : 1/4 white, while the shape gene will give a monohybrid phenotypic ratio of 1/4 long : 1/2 oval : 1/4 round. Because the inheritance of these two genes is independent, use the product rule to generate all the possible phenotype combinations (note that there will be $3 \times 3 = 9$ classes) and their probabilities, thus generating the dihybrid phenotypic ratio for two incompletely dominant genes: **1/16 red long : 1/8 red oval : 1/16 red round : 1/8 purple long : 1/4 purple oval : 1/8 purple round : 1/16 white long : 1/8 white oval : 1/16 white round**. As an example, to determine the probability of red long progeny, multiply 1/4 (probability of red) $\times$ 1/4 (probability of long) = 1/16. If you have trouble keeping track of the 9 possible classes, it may be helpful to list the classes in the form of a branch diagram.

Phenotype	*Probability of phenotype*
red, long	$1/4 \times 1/4 = 1/16$
red, oval	$1/4 \times 1/2 = 1/8$
red, round	$1/4 \times 1/4 = 1/16$
purple, long	$1/2 \times 1/4 = 1/8$
purple, oval	$1/2 \times 1/2 = 1/4$
purple, round	$1/2 \times 1/4 = 1/8$
white, long	$1/4 \times 1/4 = 1/16$
white, oval	$1/4 \times 1/2 = 1/8$
white, round	$1/4 \times 1/4 = 1/16$

3-5. white long $\times$ purple short $\rightarrow$ 301 long purple : 99 short purple : 612 long pink : 195 short pink : 295 long white : 98 short white

Deconstruct this dihybrid phenotypic ratio for two genes into separate constituent monohybrid ratios for each of the 2 traits, flower color and pod length. For flower color note that there are 3 phenotypes: 301 + 99 purple : 612 +195 pink : 295 + 98 white = 400 purple : 807 pink : 393 white =

1/4 purple : 1/2 pink : 1/4 white. This is a typical monohybrid ratio for an incompletely dominant gene, so **flower color is caused by an incompletely dominant gene with c^p giving purple when homozygous, c^w giving white when homozygous, and the $c^p c^w$ heterozygotes being pink.** For pod length, the phenotypic ratio is 301 + 612 + 295 long : 99 + 195 + 98 short = 1208 long : 392 short = 3/4 long : 1/4 short. This 3:1 ratio is that expected for a cross between individuals heterozygous for a gene in which one allele is completely dominant to the other, so **pod shape is controlled by 1 gene with long (*L*) completely dominant to short (*l*).**

3-6.

a. A person with sickle-cell anemia is a homozygote for the sickle-cell allele: $Hb^S Hb^S$.

b. The child must be homozygous $Hb^S Hb^S$ and therefore must have inherited a mutant allele from each parent. Because the parent is phenotypically normal, he/she must be a **carrier with genotype $Hb^S Hb^A$.**

c. Each individual has two alleles of every gene, including the β-globin gene. If an individual is heterozygous, he/she has two different alleles. Thus, if each parent is heterozygous for different alleles, there are **four possible alleles** that could be found in the five children. This is the maximum number of different alleles possible (barring the very rare occurrence of a new, novel mutation in a gamete that gave rise to one of the children). If one or both of the parents were homozygous for any one allele, the number of alleles distributed to the children would of course be less than four.

3-7. Remember that the gene determining ABO blood groups has 3 alleles and that $I^A = I^B > i$.

a. The O phenotype means the girl's genotype is *ii*. Each parent contributed an *i* allele, so her parents could be ***ii* (O) or $I^A i$ (A) or $I^B i$ (B).**

b. A person with the B phenotype could have either genotype $I^B I^B$ or genotype $I^B i$. The mother is A and thus could not have contributed an I^B allele to this daughter. Instead, because the daughter clearly does not have an I^A allele, the mother must have contributed the *i* allele to this daughter. The mother must have been an $I^A i$ heterozygote. The father must have contributed the I^B allele to his daughter, so he could be **either $I^B I^B$, $I^B i$, or $I^B I^A$.**

c. The genotypes of the girl and her mother must both be $I^A I^B$. The father must contribute either the I^A or the I^B allele, so there is only one phenotype and genotype that would exclude a man as her father - **the O phenotype (genotype *ii*).**

3-8. To approach this problem, look at the mother/child combinations to determine what alleles the father must have contributed to each child's genotype.

a. The father had to contribute I^B, N, and Rh^- alleles to the child. The only male fitting these requirements is **male c** whose phenotype is B, MN, and Rh^+ (note that the father must be Rh^+Rh^- because the daughter is Rh^-).

b. The father had to contribute i, N, and Rh^- alleles. The father could be either male c (O MN Rh^+) or male d (B MN Rh^+). As we saw previously, male c is the only male fitting the requirements for the father in part a. Assuming one child per male as instructed by the problem, the father in part b must be **male d**.

c. The father had to contribute I^A, M, and Rh^- alleles. Only **male b** (A M Rh^+) fits these criteria.

d. The father had to contribute either I^B or i, M, and Rh^-. Three males have the alleles required: these are male a, male c, and male d. However, of these three possibilities, only **male a** remains unassigned to a mother/child pair.

3-9. Designate the alleles: p^m (marbled) $> p^s$ (spotted) $= p^d$ (dotted) $> p^c$ (clear).

a. Diagram the crosses:

1. $p^m p^m$ (homozygous marbled) $\times$ $p^s p^s$ (spotted) $\rightarrow$ $p^m p^s$ (marbled F_1)

2. $p^d p^d$ x $p^c p^c$ $\rightarrow$ $p^d p^c$ (dotted F_1)

3. $p^m p^s$ x $p^d p^c$ $\rightarrow$ 1/4 $p^m p^d$ (marbled) : 1/4 $p^m p^c$ (marbled) : 1/4 $p^s p^d$ (spotted dotted) : 1/4 $p^s p^c$ (spotted) = **1/4 spotted dotted : 1/2 marbled : 1/4 spotted**.

b. The F_1 from cross 1 are **marbled ($p^m p^s$)** from the first cross **and dotted ($p^d p^c$)** from the second cross as shown in part a.

3-10. Designate the gene p (for pattern; this is a different p gene than that in the previous problem 3-9 because the two problems involve different plant species). There are 7 alleles, p^1-p^7, with p^7 being the allele that codes for absence of pattern and $p^1 > p^2 > p^3 > p^4 > p^5 > p^6 > p^7$.

a. There are **7 different patterns** possible. These are associated with the following genotypes: p^1-, $p^2 p^a$ (where $p^a = p^2, p^3, p^4 ... p^7$), $p^3 p^b$ (where $p^b = p^3, p^4, p^5 ... p^7$), $p^4 p^c$ (where $p^c = p^4, p^5, p^6$, and p^7), $p^5 p^d$ (where $p^d = p^5, p^6$, and p^7), $p^6 p^e$ (where $p^e = p^6$ and p^7), and $p^7 p^7$.

b. The phenotype dictated by the allele p^1 has the greatest number of genotypes associated with it = **7** ($p^1 p^1$ $p^1 p^2$ $p^1 p^3$, etc.). **The absence of pattern** is caused by just one genotype, $p^7 p^7$.

c. This finding suggests that **the allele determining absence of pattern (p^7) is very common** in these clover plants with the $p^7 p^7$ genotype is the most frequent in the population. The other alleles are present, but are much less common in this population.

3.11.

a.   First analyze these crosses for the answers to the 3EQ (<u>see Problem Solving Hints above</u>). All of the crosses have results that can be explained by one gene - either a 3:1 phenotypic monohybrid ratio showing that one allele is completely dominant to the other, or a 1:1 ratio showing that a test cross was done for a single gene, or all progeny with the same phenotype as the parents. You can thus conclude that all of the coat colors are controlled by the alleles of **one gene, with chinchilla (C) > himalaya (c^h) > albino (c^a).**

b.   1. $c^h c^a \times c^h c^a$

2. $c^h c^a \times c^a c^a$

3. $Cc^h \times C(c^h$ or $c^a)$

4. $CC \times c^h c^h$

5. $Cc^a \times Cc^a$

6. $c^h c^h \times c^a c^a$

7. $Cc^a \times c^a c^a$

8. $c^a c^a \times c^a c^a$

9. $Cc^h \times c^h(c^h$ or $c^a)$

10. $Cc^a \times c^h c^a$.

c. Cc^h (from cross 9) $\times Cc^a$ (from cross 10) $\rightarrow$ 1/4 CC (chinchilla) : 1/4 Cc^a (chinchilla) : 1/4 Cc^h (chinchilla) : 1/4 $c^h c^a$ (himalaya) = **3/4 chinchilla : 1/4 himalaya**.

3.12. You know from the incompatibility system that each plant must be heterozygous for two different alleles of the incompatibility gene. A '−' in the chart indicates no seeds were produced in a particular mating, which means the parents have an incompatibility allele in common. A '+' in the chart means the parents do not have <u>any</u> alleles in common so that they can produce seeds. Thus, looking at the results in the table, plants 2, 3, and 5 must have one (or both) of the alleles from plant 1 since they are incompatible with plant 1. Arbitrarily designate two different alleles for plant 1 as i^1 and i^2. Plants 2, 3 and 5 must have either allele i^1 or i^2 or both. Plant 4 has neither allele present in plant 1. Let's give plant 4 alleles i^3 and i^4. Furthermore, plant 2 does not share alleles with plants 3, 4 or 5 so it must have the allele found in plant 1 that plant 5 does not contain. Designate i^1 as the allele shared between plants 1 and 2 and designate i^2 as the allele shared between plants 1 and 3 and 5. Plant 3 shares an allele with plant 5, i^2, but it does not share an allele with plant 4. Call the second allele in plant 3 i^5. Plants 4 and 5 share an allele, say i^3. We still don't know a second allele for plant 2, but it is not any of those carried by plants 3, 4, or 5, so it must be another allele i^6. Thus, **there are**

six alleles total.

plant	genotype
1	i^1i^2
2	i^1i^6
3	i^2i^5
4	i^3i^4
5	i^2i^3

3-13. Diagram the cross. Figure out an expected monohybrid ratio for each gene separately, then apply the product rule to generate the expected dihybrid ratio. Also recall that the albino phenotype is epistatic to all other coat colors.

$A^yA\ Cc \times A^yA\ cc \rightarrow$ monohybrid ratio for the A gene alone: 1/4 A^yA^y (dead) : 1/2 A^yA (yellow) : 1/4 AA (agouti) = 2/3 A^yA (yellow) : 1/3 AA (agouti); monohybrid ratio for the C gene: 1/2 Cc (non-albino) : 1/2 cc (albino).

Overall there will be 2/6 $A^yA\ Cc$ (yellow) : 2/6 $A^yA\ cc$ (albino) : 1/6 $AA\ Cc$ (agouti) : 1/6 $AA\ cc$ (albino) = **2/6 $A^yA\ Cc$ (yellow) : 3/6 -- cc (albino) : 1/6 $AA\ Cc$ (agouti)**.

3-14.

a. The 2/3 montezuma : 1/3 wild type phenotypic ratio, and the statement that montezumas are never true-breeding, together suggest that there is a **recessive lethal allele** of this gene. When there is a recessive lethal, crossing two heterozygotes results in a 1:2:1 genotypic ratio, but one of the 1/4 classes of homozygotes do not survive. **The result is the 2:1 phenotypic ratio as seen in this cross. Both the montezuma parents were therefore heterozygous, *Mm*.** The M allele must confer the montezuma coloring in a dominant fashion, but homozygosity for M is lethal.

b. Designate the alleles: M = montezuma, m = greenish; F = normal fin, f = ruffled. Diagram the cross: ***MmFF* x *mmff*** $\rightarrow$ expected monohybrid ratio for the M gene alone: 1/2 Mm (montezuma) : 1/2 mm (wild type); expected monohybrid ratio for the F gene alone: all Ff. The expected dihybrid ratio = **1/2 *Mm Ff* (montezuma) : 1/2 *mm Ff* (greenish, normal fin)**.

c. ***MmFf* × *Mm Ff*** $\rightarrow$ expected monohybrid ratio for the M gene alone: 2/3 montezuma (Mm) : 1/3 greenish (mm); expected monohybrid ratio for the F gene alone: 3/4 normal fin (F-) : 1/4 ruffled (ff). The expectations when considering both genes together is: **6/12 montezuma normal fin : 2/12 montezuma ruffled fin : 3/12 green normal fin : 1/12 green ruffled fin.**

3-15.

a. Based upon the Comprehensive Example in the textbook at the end of Chapter 3, we can deduce some information about the genotypes of the parents from their phenotypes. The rest we have to deduce based on the phenotypes of the progeny. The yellow parent must have an A^y allele, but we don't know the second allele of the A gene (A^y-). A^y is epistatic to the B gene so we don't know what alleles this yellow mouse has at the B gene (we'll leave these alleles as *??*). Since this mouse does show color we know it is not *cc* (albino), so it must have at least one C allele (C-). The brown agouti parent has at least one A allele (A −); it must be *bb* at the B gene; and since there is color it must also be C-. The mating between these two can be represented as A^y- *?? C*- × *A*- *bb C*-. Now consider the progeny. Because one pup was albino (*cc*), the parents must both be *Cc*. A brown pup (*bb*) indicates that both parents had to be able to contribute a *b* allele, so we now know the first mouse must have had at least one *b* allele. The fact that this brown pup was non-agouti means both parents carried an *a* allele. The black agouti progeny tells us that the first mouse must have also had a *B* allele (but it was yellow because A^y is epistatic to *B*). The complete genotypes of the mice are therefore: **$A^y a$ Bb Cc x Aa bb Cc.**

b.   Think about each gene individually, then the effect of the other genes in combination with that phenotype. *C*- leads to a phenotype with color; *cc* gives albino (which is epistatic to all colors determined by the other genes). The possible genotypes of the progeny of this cross for the *A* gene are $A^y A$, $A^y a$, *Aa* and *aa*, giving yellow, yellow, agouti and non-agouti phenotypes, respectively. Since yellow is epistatic to *B*, non-albino mice with A^y will be yellow regardless of the genotype of the *B* gene. *Aa* is agouti; with the *aa* genotype there is no yellow on the hair (non-agouti). The type of coloration depends on the *B* gene. For *B* the offspring could be *Bb* (black) or *bb* (brown). In total, **six different coat color phenotypes are possible: albino (-- -- cc), yellow (A^y(A or a) -- C), brown agouti (A- bb C-), black agouti (A- B- C-), brown (aa bb C-), black (A- B- C-).**

3-16. walnut x single → F$_1$ walnut x F$_1$ walnut → 93 walnut : 29 rose : 32 pea : 11 single

a.   3EQ#1 - four F$_2$ phenotypes means there are 2 genes, *A* and *B*. Both genes affect the same structure, the comb. The F$_2$ phenotypic dihybrid ratio among the progeny is close to 9:3:3:1, so there is no epistasis. Because walnut is the most abundant F$_2$ phenotype, it must be the phenotype due to the *A- B-* genotype. Single combs are the least frequent class, and are thus *aa bb*. Now assign genotypes to the cross. If the walnut F$_2$ are *A- B-*, then the original walnut parent must have been *AA BB*:

    AA BB × *aa bb* → *Aa Bb* (walnut) → 9/16 *A- B-* (walnut) : 3/16 *A- bb* (rose) : 3/16 *aa B-* (pea) : 1/6 *aa bb* (single).

b.   Diagram the cross, recalling that the problem states the parents are homozygous:

    AA bb (rose) × *aa BB* (pea) → *Aa Bb* (walnut) → **9/16 *A- B-* (walnut) : 3/16 *A- bb* (rose) : 3/16 *aa B-* (pea) : 1/6 *aa bb* (single).** This F$_2$ is in identical proportions as the F$_2$ generation in part a.

c. Diagram the cross: *A- B-* (walnut) x *aa B-* (pea) → 12 *A- B-* (walnut) : 11 *aa B-* (pea) : 3 *A- bb* (rose) : 4 *aa bb* (single). Because there are pea and single progeny, you know that the walnut parent must be *Aa*. The 1 *A-* : 1 *aa* monohybrid ratio in the progeny also tells you the walnut parent must have been *Aa*. Because some of the progeny are single, you know that both parents must be *Bb*. In this case, the monohybrid ratio for the *B* gene is 3 *B-* : 1 *bb*, so both parents were *Bb*. The original cross must have been **Aa Bb x aa Bb**. You can verify that this cross would yield the observed ratio of progeny by multiplying the probabilities expected for each gene alone. For example, you anticipate that 1/2 the progeny would be *Aa* and 3/4 of the progeny would be *Bb,* so $\frac{1}{2} \times \frac{3}{4} = 3/8$ of the progeny should be walnut; this is close to the 12 walnut chickens seen among 30 total progeny.

d. Diagram the cross: *A- B-* (walnut) x *A- bb* (rose) → all *A- B-* (walnut). The progeny are all walnut, so **the walnut parent must be BB**. No pea progeny are seen, so **both parents cannot be Aa, so one of the two parents must be AA**. This could be either the walnut or the rose parent or both.

3-17. black × chestnut → bay → black : bay : chestnut : liver

Four phenotypes in the F_2 generation means there are two genes determining coat color. The F_1 bay animals produce four phenotypic classes, so they must be doubly heterozygous, *Aa Bb*. Crossing a liver colored horse to either of the original parents resulted in the parent's phenotype. The liver horse's alleles do not affect the phenotype, suggesting the recessive genotype *aa bb*. Though it is probable that the original black mare was *AA bb* and the chestnut stallion was *aa BB,* each of these animals only produced 3 progeny, so it cannot be definitively concluded that these animals were homozygous for the dominant allele they carry. Thus, **the black mare was A- bb, the chestnut stallion was aa B-, and the F_1 bay animals are Aa Bb. The F2 horses were: bay (A- B-), liver (aa bb), chestnut (aa B-), and black (A- bb).**

3-18.

a. Because unaffected individuals had affected children, the **trait is recessive**. From affected individual II-1, you know the mutant allele is present in this generation. The trait was passed on through II-2 who was a carrier. All children of affected individuals III-2 × III-3 are affected, as predicted for a recessive trait. However, generation V seems inconsistent with recessive inheritance of a single gene. This result <u>is</u> consistent with two different genes involved in hearing with a defect in either gene leading to deafness. **The two family lines shown contain mutations in two separate genes, and the mutant alleles of both genes determining deafness are recessive.**

b. Individuals in generation V are doubly heterozygous (*Aa Bb*), having inherited a dominant and recessive allele of each gene from their parents (*aa BB* × *AA bb*). The people in generation V are not affected because **the product of the dominant allele of each gene is sufficient for normal function.** This is an example of the complementation of two genes.

3-19. green x yellow → green → 9 green : 7 yellow

a. Two phenotypes in the F2 generation could be due to one gene or to two genes with epistasis. If this is one gene, then $GG \times gg \rightarrow Gg \rightarrow 3/4$ *G-* (green) : 1/4 *gg* (yellow). The actual result is a 9:7 ratio, not a 3:1 ratio. 9:7 is an epistatic variant of the 9:3:3:1 phenotypic dihybrid ratio, so there are **2 genes** controlling color. The genotypes are:

  AA BB **(green)** × *aa bb* **(yellow)** → *Aa Bb* **(green)** → **9/16 *A- B-* (green) : 3/16 *A- bb* (yellow) : 3/16 *aa B-* (yellow) : 1/16 *aa bb* (yellow).**

b. **F₁** *Aa Bb* × *aa bb* → 1/4 *Aa Bb* (green) : 1/4 *aa Bb* (yellow) : 1/4 *Aa bb* (yellow) : 1/4 *aabb* (yellow) = 1/4 green : 3/4 yellow.

3-20.

a. **No**, a single gene cannot account for this result. While the 1:1 ratio seems like a testcross, the fact that **the phenotype of one class of offspring (linear) is not the same as either of the parents** argues against this being a testcross.

b. The appearance of four phenotypes means **two genes** are controlling the phenotypes.

c. The 3:1 ratio suggests that **two alleles of one gene** determine the difference between the wild-type and scattered patterns.

d. The true-breeding wild-type fish are homozygous by definition, and the scattered fish have to be homozygous recessive according to the ratio seen in part c, so the cross is: *bb* **(scattered)** × *BB* **(wild type)** → **F₁** *Bb* **(wild type)** → **3/4 *B-* (wild type) : 1/4 *bb* (scattered).**

e. The inability to obtain a true-breeding nude stock suggests that the nude fish are heterozygous (*Aa*) and that the *AA* genotype dies. Thus *Aa* **(nude)** × *Aa* **(nude)** → **2/3 *Aa* (nude) : 1/3 *aa* (scattered).**

f. Going back to the linear cross from part b, the fact that there are four phenotypes led us to propose two genes were involved. The 6:3:2:1 ratio looks like an altered 9:3:3:1 ratio in which some genotypes may be missing, as predicted from the result in part e that *AA* animals do not survive. The 9:3:3:1 ratio results from crossing double heterozygotes, so **the linear parents are doubly heterozygous *Aa Bb*. The lethal phenotype associated with the *AA* genotype produces the 6:3:2:1 ratio. The phenotypes and corresponding genotypes of the progeny of the linear × linear cross are: 6 linear, *Aa B-* : 3 wild-type, *aa B-* : 2 nude, *Aa bb* : 1**

scattered, *aa bb*. Note that the *AA BB, AA Bb, AA Bb,* and *AA bb* genotypes are missing due to lethality.

3-21. Dominance relationships are between <u>alleles of the same gene</u>. **Only one gene is involved when considering dominance relationships. Epistasis involves two genes**. The alleles of one gene affect the phenotypic expression of the second gene.

3-22.

a.   white x white $\rightarrow$ F_1 white $\rightarrow$ 126 white:33 purple

3EQ#1 - At first glance this inheritance seems to involve only one gene, as true breeding white parents give white F_1s. However, if this were true, then the F_2 MUST be totally white as well! The result that some F_2 plants are purple shows that this is NOT controlled by 1 gene! It must be **due to 2 genes**. What ratio is 126 : 33? Usually, you would divide through by the smallest number, yielding in this case 3.8 white : 1 purple, which is neither a recognizable monohybrid nor dihybrid ratio. This is NOT the correct way to convert raw numbers to a ratio. In crosses controlled by 2 genes there must be 16 genotypes in the F_2 progeny, even though the phenotypes may not be distributed in the usual 9/16 : 3/16 : 3/16 : 1/16 ratio. If the 159 F_2 progeny are divided equally into 16 genotypes, then there are 159/16 = ~10 F_2 plants/genotype. The 126 white F_2s therefore represent 126/10 = 13 genotypes. Likewise the 33 purple plants represent 33/10 = 3 genotypes. The correct F_2 dihybrid phenotypic ratio is thus 13 white : 3 purple.

You can now assign genotypes to the parents in the cross. Because the parents are homozygous (true-breeding) and there are 2 genes controlling the phenotypes, there are two possible ways to set up the genotypes of the parents: *AA BB* **(white)** $\times$ *aa bb* **(white)** $\rightarrow$ *Aa Bb* **(white, same as *AA BB* parent)** $\rightarrow$ **9** *A- B-* **(white) : 3** *A- bb* **(unknown phenotype) : 3** *aa B-* **(unknown phenotype) : 1** *aa bb* **(white, same as *aa bb* parent)**. If you assume that *A- bb* is white and *aa B-* is purple (or vice versa), then this is a match for the observed data presented in the cross above (9 + 3 + 1 = 13 white : 3 purple).

Alternatively, you could try to diagram the cross as *AA bb* $\times$ *aa BB* $\rightarrow$ *Aa Bb* (whose phenotype is unknown as this is NOT a genotype seen in the parents) $\rightarrow$ 9 *A-B-* (unknown phenotype) : 3 *A- bb* (white like the *AA bb* parent) : 3 *aa B-* (white like the *aa BB* parent) : 1 *aa bb* (unknown phenotype). Such a cross cannot give an F_2 phenotypic ratio of 13 white : 3 purple. The only F_2 classes that could be purple are *A- B-*, but this is impossible because it is too large and because the F_1 flies must then have been purple, or the *aa bb* class which is too small. Therefore, the first set of possible genotypes (written in bold above) is the best fit for the observed data.

b.   white F_2 × self white F_2 → 3/4 white : 1/4 purple. Assume that the *aa B-* class is purple in part a above. A 3:1 monohybrid ratio means the parents are both heterozygous for one gene with purple due to the recessive allele. The second gene is not affecting the ratio, so both parents must be homozygous for the same allele of that gene. Thus the cross must be: ***Aa BB*** (white) × *Aa BB* (white self cross) → 3/4 *A- BB* (white) : 1/4 *aa BB* (purple).

c.   purple F_2 × self → 3 purple : 1 white. Again, the selfed parent must be heterozygous for one gene and homozygous for the other gene. Because purple is *aa B-*, the genotypes of the purple F_2 plant must be ***aa Bb***.

d.   white F_2 × white F_2 (not a self cross) → 1/2 purple : 1/2 white. The 1:1 monohybrid ratio means a test cross was done for one of the genes. The second gene is not altering the ratio in the progeny, so the parents must be homozygous for that gene. If purple is *aa B-*, then the genotypes of the parents must be ***aa bb* (white)** × ***Aa BB* (white)** → 1/2 *Aa Bb* (white) : 1/2 *aa Bb* (purple).

3-23. $I^A I^B\ Ss$ x $I^A I^A\ Ss$ → expected monohybrid ratio for the *I* gene of 1/2 $I^A I^A$: 1/2 $I^A I^B$; expected ratio for the *S* gene considered alone of 3/4 *S-* : 1/4 *ss*. Use the product rule to generate the phenotypic ratio for both genes considered together and then assign phenotypes, remembering that all individuals with the *ss* genotype look like type O. The phenotypic ratio for both genes is: 3/8 $I^A I^A$ *S-* : 3/8 $I^A I^B$ *S-* : 1/8 $I^A I^A$ *ss* : 1/8 $I^A I^B$ *ss* = 3/8 A : 3/8 AB : 1/8 O : 1/8 O = **3/8 Type A : 3/8 Type AB : 2/8 Type O**.

3-24.

a.   *Aa Bb* × *Aa Bb* → 9 *A- B-* : 3 *A- bb* :3 *aa B-* : 1 *aa bb*

Since the defect in enzyme is only seen if <u>both</u> genes are defective, only *aa bb* will result in abnormal progeny, giving a phenotypic dihybrid ratio of **15 normal : 1 abnormal**.

b.   *Aa Bb Cc* × *Aa Bb Cc* → the dihybrid ratio for *A* and *B* is 9 *A- B-* : 3 *A- bb* :3 *aa B-* : 1 *aa bb*; while the monohybrid ratio for *C* is 3 *C-* : 1 *cc*. Use the product rule to generate the expected phenotypic trihybrid ratio. Remember that the <u>only</u> abnormal genotype will be *aa bb cc*, which will occur with a probability of 1/16 × 1/4 = 1/64. The expected phenotypic trihybrid ratio is thus **63/64 normal : 1/64 abnormal**.

3-25.

a.   blood types: **I-1 AB; I-2 A; I-3 B; I-4 AB; II-1 O; II-2 O; II-3 AB; III-1 A; III-2 O**.

b.   genotypes: **I-1 *Hh* $I^A I^B$; I-2 *Hh* $I^A i$ (or $I^A I^A$); I-3 *H-* $I^B I^B$ (or $I^B i$); I-4 *H-* $I^A I^B$; II-1 *H-* *ii*; II-2 *hh* $I^A I^A$ (or $I^A i$ or $I^A I^B$); II-3 *Hh* $I^A I^B$; III-1 *Hh* $I^A i$; III-2 *hh* $I^A I^A$ (or $I^A I^B$ or $I^A i$ or $I^b i$ or $I^B I^B$)**

At first glance, you find inconsistencies between expectations and what could be inherited from a parent. For example, I-1 (AB) × I-2 (A) could not have an O child (II-2). The epistatic *h* allele (which causes the Bombay phenotype) could explain these inconsistencies. If II-2 has an O phenotype because she is *hh*, her parents must both have been *Hh*. The Bombay phenotype would also explain the second seeming inconsistency of two O individuals (II-1 and II-2) having an A child. II-2 could have received an *A* allele from one of her parents and passed this on to III-1 together with one *h* allele. Parent II-1 would have to contribute the *H* allele so that the *A* allele would be expressed; the presence of *H* means that II-1 must also be *ii* in order to be type O. A third inconsistency is that individuals II-2 and II-3 could not have an *ii* child since II-3 has the $I^A I^B$ genotype, but III-2 has the O phenotype. This could also be explained if II-3 were *Hh* and III-2 is *hh*.

3-26. The difference between pleiotropic mutations and traits determined by several genes would be seen if crosses were done using pure-breeding plants (wild type x mutant), then selfing the F_1 progeny. If **several genes** were involved there **would be several different combinations of the petal color, markings and stem position phenotypes** in the F_2 generation. If all 3 traits were due to an allele present at **one gene**, the **three phenotypes would always be inherited together** and the F_2 plants would be either yellow, dark brown, erect OR white, no markings and prostrate.

3-27.

a. *Aa Bb Cc* × *Aa Bb Cc* → 9/16 *A- B-* × 3/4 *C-* : 9/16 *A- B-* × 1/4 *cc* :3/16 *A- bb* × 3/4 *C-* : 3/16 *A- bb* × 1/4 *cc*: 3/16 *aa B-* × 3/4 *C-* :3/16 *aa B-* × 1/4 *cc* : 1/16 *aa bb* × 3/4 *C-* : 1/16 *aa bb* × 1/4 *cc* = 27/64 *A- B- C-* (wild type) :9/64 *A- B- cc* : 9/64 *A- bb C-* : 3/64 *A- bb cc* : 9/64 *aa B-C-* : 3/64 *aa B-cc* : 3/64 *aa bb C-* : 1/64 *aa bb cc* = **27/64 wildtype : 37/64 mutant**.

b. Diagram the crosses:

  1. unknown male × *AA bb cc* → 1/4 wild type (*A- B- C-*) : 3/4 mutant
  2. unknown male × *aa BB cc* → 1/2 wild type (*A- B- C-*) : 1/2 mutant
  3. unknown male × *aa bb CC* → 1/2 wild type (*A- B- C-*) : 1/2 mutant

The 1:1 ratio in test crosses 2 and 3 is expected if the unknown male is heterozygous for one of the genes that are recessive in the test cross parent. The 1 wild type : 3 mutant ratio arises when the male is heterozygous for two of the genes that are homozygous recessive in the test cross parent. (If you apply the product rule to 1/2 *B-* : 1/2 *bb* and 1/2 *C-* : 1/2 *cc* in the first cross, then you find 1/4 *B- C-*, 1/4 *B- cc*, 1/4 *bb C-*, and 1/4 *bb cc*. Only *B- C-* will be wild type, the other 3 classes will be mutant). Thus the unknown male must be *Bb Cc*. In test cross 1 the male could be either *AA* or *aa*. Crosses 2 and 3 show that the male is only heterozygous for one of the recessive genes in each case - gene *C* in test cross 2 and gene *B* in test cross 3. In order to get wild type progeny in both crosses, the male must be *AA*. Therefore the genotype of the unknown male is **AA Bb Cc**.

3-28.

a. Diagram one of the crosses:

   white-1 $\times$ white-2 $\rightarrow$ red F_1 $\rightarrow$ 9 red : 7 white

   Even though there are only 2 phenotypes in the F_2, this is not controlled by one gene - the 9:7 ratio shows that this is an epistatic variation of 9:3:3:1, so there are 2 genes controlling these phenotypes. Individuals must have at least one dominant allele of both genes in order to get the red color. Thus the genotypes of the two white parents in this cross are *aa BB* $\times$ *AA bb*. The same conclusions hold for the other 2 crosses. If white-1 is mutant in gene *A* and white-2 is mutant in gene *B*, then white-3 must be mutant in gene *C*. Therefore, **three genes are involved**.

b. **White-1 is *aa BB CC*; white-2 is *AA bb CC* and white-3 is *AA BB cc*.**

c. *aa BB CC* (white-1) $\times$ *AA bb CC* (white-2) $\rightarrow$ *Aa Bb CC* (red) $\rightarrow$ 9/16 *A- B- CC* (red) : 3/16 *A- bb CC* (white) : 3/16 *aa B- CC* (white) : 1/6 *aa bb CC* (white).

3-29.

a. Analyze each cross by answering 3EQ#1 and #2. In cross 1 – there are 2 genes because there are 3 classes in the F2 showing an epistatic 12:1:3 ratio, and LR is the doubly homozygous recessive class. In cross 2 – only 1 gene is involved because there are 2 phenotypes in a 3:1 ratio; WR>LR. In cross 3 – again, there is only 1 gene (2 phenotypes in a 1:3 ratio); DR>LR. In cross 4 - 1 gene (2 phenotypes, 3:1 ratio); WR>LR. In cross 5 - 2 genes (as in cross 1, there is a 12:1:3 ratio of three classes); LR is the double homozygous recessive. In total, **there are 2 genes controlling these phenotypes in foxgloves.**

b. Remember that all four starting strains are true-breeding. In cross 1 the parents can be assigned the following genotypes: ***AA BB* (WR-1)** $\times$ ***aa bb* (LR)** $\rightarrow$ *Aa Bb* (WR) $\rightarrow$ 9 *A- B-* (WR) : 3 *A- bb* (WR; this class displays the epistatic interaction) : 3 *aa B-* (DR) :1 *aa bb* (LR). The results of cross 2 suggested that DR differs from WR-1 by one gene, so **DR is *aa BB***; cross 3 confirms these genotypes for DR and LR. Cross 4 introduces WR-2, which differs from LR by one gene and differs from DR by 2 genes, so **WR-2 is *AA bb***. Cross 5 would then be *AA bb* (WR-2) $\times$ *aa BB* (DR) $\rightarrow$ *Aa Bb* (WR) $\rightarrow$ 9 *A- B-* (WR) : 3 *A- bb* (WR) : 3 *aa B-* (DR) : 1 *aa bb* (LR) = 12 WR : 3 DR : 1 LR.

c. WR from the F_2 of cross 1 $\times$ LR $\rightarrow$ 253 WR : 124 DR : 123 LR. Remember from part b that LR is *aa bb* and DR is *aa B-* while WR can be either *A- B-* or *A- bb* = *A- ??*. The experiment is essentially a test cross for the WR parent. The observed monohybrid ratio for the *A* gene is 1/2 *Aa* : 1/2 *aa* (253 *Aa* : 124 + 123 *aa*), so the WR parent must be *Aa*. The DR and LR classes of progeny show that the WR parent is also heterozygous for the *B* gene (DR is *Bb* and LR is *bb* in these progeny). Thus, the cross is ***Aa Bb* (WR)** × ***aa bb* (LR)**.

3-30. In **Figure 3.21b** the A^0 and B^0 alleles are non-functional. The A^1 and B^1 alleles each have the same effect on the phenotype (plant height in this example). Thus, the shortest plants are $A^0A^0B^0B^0$, and the tallest plants are $A^1A^1B^1B^1$. **The phenotypes are determined by the total number or A^1 and B^1 alleles in the genotype**. Thus, $A^1A^0B^0B^0$ plants are the same phenotype as $A^0A^0B^0B^1$. In total there will be 5 different phenotypes: 4 '0' alleles (total = 0); 1 '1' allele + 3 '0' alleles (total = 1); 2 '1' alleles + 2 '0' alleles (total = 2); 3 '1' alleles + 1 '0' allele (total = 3); and 4 '1' alleles (total = 4).

In **Figure 3.16** the *a* allele = *b* allele = no function (in this case no color = white). If the *A* allele has the same level of function as a *B* allele then you would see 5 phenotypes as was the case for Figure 3.21b. But since there are a total of 9 phenotypes, this cannot be true so *A≠B*. Notice that *aa Bb* is lighter than *Aa bb* even though both genotypes have the same number of dominant alleles. Thus **an *A* allele has more effect on coloration than a *B* allele**. If you assume, for example, that *B* = 1 unit of color and *A* = 1.5 unit of color, then 16 genotypes lead to 9 phenotypes.

3-31.

a. Answer 3EQ#1 and #2 for all 5 crosses. Cross 1 - 1 gene, red>blue. Cross 2 - 1 gene, lavender>blue. Cross 3 - 1 gene, codominance/incomplete dominance (1:2:1), bronze is the phenotype of the heterozygote. Cross 4 - 2 genes with epistasis (9 red : 4 yellow : 3 blue). Cross 5 - 2 genes with epistasis (9 lavender : 4 yellow : 3 blue). **In total there are 2 genes. One gene controls blue (c^b), red (c^r) and lavender (c^l) where $c^r = c^l > c^b$. The second gene controls the yellow phenotype: *Y* seems to be colorless (or has no effect on color), so the phenotype is determined by the alleles of the *c* gene. The *y* allele makes the flower yellow, and is epistatic to the *c* gene.**

b. **cross 1 – c^rc^r *YY* (red) x c^bc^b *YY* (blue) → c^rc^b *YY* (red) → 3/4 c^r- *YY* (red) : 1/4 c^bc^b *YY* (blue)**

   cross 2 – c^lc^l *YY* (lavender) x c^bc^b *YY* (blue) → c^lc^b *YY* (lavender) → 3/4 c^l- *YY* (lavender) : 1/4 c^bc^b *YY* (blue)

   cross 3 – c^lc^l *YY* (lavender) x c^rc^r *YY* (red) → c^lc^r *YY* (bronze) → 1/4 c^lc^l *YY* (lavender) : 1/2 c^lc^r *YY* (bronze) : 1/4 c^rc^r *YY* (red)

   cross 4 – c^rc^r *YY* x c^bc^b *yy* (yellow) → c^rc^b *Yy* (red) → 9/16 c^r- *Y*- (red) : 3/16 c^r- *yy* (yellow) : 3/16 c^bc^b *Y*- (blue): 1/16 c^bc^b *yy* (yellow)

   cross 5- c^lc^l *yy* (yellow) x c^bc^b *YY* (blue) → c^lc^b *Yy* (lavender) → 9/16 c^l- *Y*- (lavender) : 3/16 c^l- *yy* (yellow) : 3/16 c^bc^b *Y*- (blue) : 1/16 c^bc^b *yy* (yellow)

c. $c^r c^r$ *yy* (yellow) x $c^l c^l$ *YY* (lavender) → $c^r c^l$ *Yy* (bronze) → monohybrid ratio for the *c* gene is 1/4 $c^r c^r$: 1/2 $c^r c^l$: 1/4 $c^l c^l$ and monohybrid ratio for the *Y* gene is 3/4 *Y-* : 1/4*y*. Using the product rule, these generate a dihybrid ratio of **3/16 $c^r c^r$ *Y-* (red) : 3/8 $c^r c^l$ *Y-* (bronze) : 3/16 $c^l c^l$ *Y-* (lavender) : 1/16 $c^r c^r$ *yy* (yellow) : 1/8 $c^r c^l$ *yy* (novel genotype) : 1/16 $c^l c^l$ *yy* (yellow).** You expect the $c^r c^l$ *yy* genotype to be yellow as *y* is normally epistatic to the *c* gene. However, you have no direct evidence from the data in any of these crosses that this will be the case, so it is possible that this genotype could cause a different and perhaps completely new phenotype.

3-32.

a.   The pattern in both families **looks like a recessive trait** since unaffected individuals have affected progeny and the trait skips generations. For example, in the Smiths II-3 must be a carrier, but in order for III-5 to be affected II-4 must also be a carrier. **If the trait is rare (as is this one) you wouldn't expect two heterozygotes to marry by chance as many times as required by these pedigrees. The alternative explanation is that the trait is dominant but not 100% penetrant.**

b.   Assuming this is a dominant but not completely penetrant trait, **individuals II-3 and III-6 in the Smiths' pedigree individual and II-6 in the Jeffersons' pedigree** must carry the dominant allele but not express it in their phenotypes.

c.   If the trait were common, **recessive inheritance** is the more likely mode of inheritance.

d.   **None**; in cases where two unaffected parents have an affected child, both parents would be carriers of the recessive trait.

3-33. Diagram the cross:

   D1d1 D2d2 d3d3 × *d1d1 D2d2 D3d3* → calculate the expected monohybrid ratios for each gene (here we consider only the phenotype for each of the three genes for simplicity rather than the corresponding genotypes): 1/2 *D1* : 1/2 *d1*; 3/4 *D2* : 1/4 *d2*; 1/2 *D3* : 1/2 *d3*. Use the product rule to determine the expected dihybrid ratio for D1 and D2 = 3/8 *D1 D2* : 1/8 *D1 d2* : 3/8 *d1 D2* : 1/8 *d1 d2*. Then multiply in the third gene to obtain the expected trihybrid ratio considering all three genes simultaneously = 3/16 *D1 D2 D3* (normal) : 3/16 *D1 D2 d3* (deaf 1 gene) : 1/16 *D1 d2 D3* (deaf 1 gene) : 1/16 *D1 d2 d3* (deaf 2 genes) : 3/16 *d1 D2 D3* (deaf 1 gene) : 3/16 *d1 D2 d3* (deaf 2 genes) : 1/16 *d1 d2 D3* (deaf 2 genes) : 1/16 *d1 d2 d3* (deaf 3 genes). The totals are 3/16 normal : 7/16 deaf due to 1 gene : 5/16 deaf due to 2 genes : 1/16 deaf due to 3 genes. Now apply the product rule to account for the incompletely penetrant lethality for those mutant at 2 or 3 of the genes. For the double mutant individuals, 1/4 die and 3/4 are alive and deaf. Thus 5/16 double mutant x 3/4 alive and deaf = 15/64 live deaf children with double mutations. For the triple mutant individuals 3/4 die and 1/4 are alive and deaf. Thus 1/16 triple mutant × 1/4 alive and deaf = 1/64 live, deaf triple mutants. In total, there are 3/16 normal + 7/16 single mutant deaf + 15/64 live double mutant deaf + 1/64 live triple mutant deaf children = 12/64 normal + 28/64 single mutant deaf + 15/64 double mutant deaf + 1/64 triple mutant deaf = 56/64 live children and 8/64 dead fetuses. Of the live-born children, **44/56 would be deaf. This means that there is a 78.6% chance that any live-born child of these two parents would be deaf**.

3-34. The hairy × hairy → 2/3 hairy : 1/3 normal cross tells us that the hairy flies are heterozygous, that the hairy phenotype is dominant to normal, and that the homozygous hairy progeny are lethal (that is, hairy is a recessive lethal). Thus, hairy is *Hh*, normal is *hh,* and the lethal genotype is *HH*. Normal flies therefore should be ***hh* (normal-1)** and a cross with hairy (*Hh*) would be expected to always give 1/2 *Hh* (hairy) : 1/2 *hh* (normal) as seen in cross 1. In cross 2, the progeny MUST for the same reasons be 1/2 *Hh* : 1/2 *hh*, yet they ALL appear normal. This suggests the normal-2 stock has another mutation that suppresses the hairy wing phenotype in the *Hh* progeny. The hairy parent must have the recessive alleles of this suppressor gene (*ss*), while the normal-2 stock must be homozygous for the dominant allele (*SS*) that suppresses the hairy phenotype. Thus cross 2 is ***hh SS* (normal-2)** x *Hh ss* (hairy) → 1/2 *Hh Ss* (normal because hairy is suppressed) : 1/2 *hh Ss* (normal). In cross 3, the normal-3 parent is heterozygous for the suppressor gene: ***hh Ss* (normal-3)** × *Hh ss* (hairy) → the expected ratios for each gene alone are 1/2 *Hh* : 1/2 *hh* and 1/2 *Ss* : 1/2 *ss*, so the expected ratio for the two genes together is 1/4 *Hh Ss* (normal) : 1/4 *Hh ss* (hairy) : 1/4 *hh Ss* (normal) : 1/4 *hh ss* (normal) = 3/4 normal : 1/4 hairy. In cross 4 you see a 2/3 : 1/3 ratio again, as if you were crossing hairy x hairy. After a bit of trial-and-error examining the remaining possibilities for these two genes, you will be able to demonstrate that this cross was ***Hh Ss* (normal-4)** × *Hh ss* (hairy) → expected ratio for the individual genes are 2/3 *Hh* : 1/3 *hh* and 1/2 *Ss* : 1/2 *ss*, so the expected ratio for the two genes together from the product rule is 2/6 *Hh Ss* (normal) : 2/6 *Hh ss* (hairy) : 1/6 *hh Ss* (normal) : 1/6 *hh ss* (normal) = 2/3 normal : 1/3 hairy.

3-35.

a. The most likely mode of inheritance is a **single gene with incomplete dominance** such that $f^n f^n$ = normal (<250 mg/dl), $f^n f^a$ = intermediate levels of serum cholesterol (250-500 mg/dl) and $f^a f^a$ homozygotes = elevated levels (>500 mg/dl). Some of the individuals in the pedigrees do not fit this hypothesis. In t 2 of the families, two normal parents have a child with intermediate levels of serum cholesterol: **Family 2 –I-2 $f^n f^n$ × I-3 $f^n f^n$ → 3 $f^a f^n$ children; and Family 4 - I-1 $f^n f^n$ × I-2 $f^n f^n$ → 2 $f^n f^a$ children**.

b. **Factors other than just the genotype are involved in the expression of the phenotype. Such factors could include diet, level of exercise, and other genes**.

Chapter 4   The Chromosome Theory of Inheritance

Synopsis:

Chapter 4 is extremely critical for understanding basic genetics because it connects Mendel's Laws with chromosome behavior during meiosis. While you may have learned mitosis and meiosis in your basic biology class, now is the time to make sure you understand these processes in the context of inheritance. The physical basis for inheritance is chromosome segregation during meiosis. You should have an increased understanding of the importance of meiosis for genetic diversity through both independent assortment and recombination.

Genes are located on chromosomes and travel with them during cell division and gamete formation.

In the first division of meiosis, homologous chromosomes in germ cells segregate from each other, so each gamete receives one member of each matched pair, as predicted by Mendel's first law, as in **Figure 4.13**. Also, during the first meiotic division the independent alignment of each pair of homologous chromosomes results in the independent assortment of genes carried on different chromosomes, as predicted by Mendel's second law. The second meiotic division generates gametes with a haploid number of chromosomes (n). Fertilization of an egg and a sperm restores the diploid number of chromosomes (2n) to the zygote.

The experiments that showed the correlation between chromosome behavior and inheritance using X-linked genes in *Drosophila* are described in this chapter. X-linked traits have characteristic inheritance patterns recognized in pedigrees or in results of reciprocal crosses.

Significant Elements:

After reading the chapter and thinking about the concepts you should be able to:

♦ Understand homologs and alleles in meiosis as seen in **Figure 4.12**. Think of a good analogy. For example, think of the road or street that you live on as a copy of a chromosome. Nearby there is another copy of the same street (homologous chromosome). The 2 copies of the street are very similar, but not identical. For instance, any building (gene) found on one copy will be found in the same position on the other copy (alleles). The 2 copies of your residence are identical on both homologous streets (your gene is homozygous). Your next-door neighbor's residence has minor differences between the 2 copies - the front door is green on one and yellow on the other (heterozygous). What will happen to these 2 streets during meiosis? During mitosis?

♦ draw chromosome alignments during metaphase of mitosis, meiosis I, meiosis II

♦ describe how chromosome behavior explains the laws of segregation and independent assortment

- identify sex-linked inheritance patterns - see 3EQ#3 below. Determine genotypes in sex-linked pedigrees and probabilities of specific genotypes and phenotypes

- If you truly understand meiosis, you can explain how the results seen in the genotype of a child enable you to figure out if non-disjunction occurred in meiosis I or meiosis II and the parent in which it occurred. Your explanation will include sister and non-sister chromatids.

- Understand the differences between sex determination in humans and *Drosophila*, see **Table 4.1**.

Problem Solving Tips:

- Keep clear the distinction between sister chromatids (identical, replicated copies of a chromosome) and homologs (chromosomes carrying the same genes but different alleles).

- Compare and contrast mitosis and meiosis as in **Table 4.2**.

- Two features that tip you off about X-linked inheritance (especially in pedigrees) are criss-cross inheritance (inheritance of a characteristic from mother to son and father to daughter) and when you see different numbers of male and female progeny for a particular phenotype, you should suspect that the gene is located on a sex chromosome. See Problem 4-19.

- Remember that sons receive their X chromosome from their mother and have to pass on their X chromosome to their daughter.

Problem Solving - How to Begin:

THREE ESSENTIAL QUESTIONS (3EQ):

1. How many genes are involved in the cross?

2. For **each gene** involved in the cross: what are the phenotypes associated with the gene? Which phenotype is the dominant one and why? Which phenotype is the recessive one and why?

3. For **each** gene involved in the cross: is it X-linked or autosomal?

   From this point on, all 3 questions are valid.

   Hints:

   For 3EQ#1. look for the number of phenotypes in the progeny.

   For 3EQ#2. if the parents of a cross are true-breeding, look at the phenotype of the F_1

individuals. Also, look at the monohybrid ratios in the F_2 progeny – the 3/4 portion of the 2

phenotypes of progeny is the dominant one.

   For 3EQ#3. Determining whether or not a gene is X-linked is more subtle. In general X-linkage is seen as a clear phenotypic difference between the sexes of a cross. This is NOT a difference in the absolute numbers of males and females of a certain phenotype, but instead a phenotype that is present in one sex and totally absent in the other sex. This difference between the sexes will be seen in either the F_1 generation <u>or</u> the F_2 generation but **not** in both generations in the same cross. It is not possible to make a definitive conclusion about X-linkage based on just one generation of a cross - you **must** see the data from both the F_1 and F_2 progeny. If the sex difference is seen in the F_1 generation then the female parent had the X-linked phenotype. If the sex difference is seen in the F_2 generation then the male parent had the X-linked trait. X-linked genes usually show a 1:1 monohybrid ratio.

After you answer questions 3EQ#1, #2 and #3 to the best of your ability, use the answers to assign genotypes to the parents of the cross. Then follow the cross through, figuring out the expected phenotypes and genotypes in the F_1 and F_2 generations. Remember to assign the expected phenotypes based on those initially assigned to the parents. Next, compare your predicted results to the observed data you were given. If the 2 sets of information match then your initial genotypes were correct! In many cases there may be two possible set of genotypes for the parents. If your predicted results do not match the data given, try the other set of genotypes for the parents.

Solutions to Problems:

4-1. a. **13**; b. **7**; c. **11**; d. **10**; e. **12**; f. **8**; g. **9**; h. **1**; i. **6**; j. **15**; k. **3**; l. **2**; m. **16**; n. **4**; o. **14**; p. **5**.

4-2.

a. **Mitosis produces 2 daughter cells each with 14 chromosomes ($2n$, diploid).** Mitosis maintains the chromosome number.

b. **Meiosis produces 4 cells (n, haploid), each with 7 chromosomes** (one half the number of chromosomes of the starting cell).

4-3. A diploid number of 46 means there are 23 homologous pairs of chromosomes.

a. A child receives **23 chromosomes from the father**.

b. Each somatic cell has **44 autosomes** (22 pairs) and **2 sex chromosomes** (1 pair).

c. **A human ovum (female gamete) contains 23 chromosomes** - one of each homologous pair (22 autosomes and one X chromosome).

d. **One sex chromosome (an X chromosome) is present in a human ovum.**

4-4. The diploid sporophyte contains 7 homologous pairs of chromosomes. One chromosome in each pair came from the male gamete and the other from the female gamete. When the diploid cell undergoes meiosis one homolog of each pair ends up in the gamete. For each homologous pair, the probability that the gamete contains the homolog inherited from the father = 1/2. **The probability that a gamete contains only the homologs inherited from the father is $(1/2)^7 = 0.78\%$.**

4-5.

a. The trait of ivory eyes is recessive and brown is the dominant phenotype; females have 2 alleles of every gene and males have only one. Diagram the cross: ivory ♀ (*bb*) x brown ♂ (*B*) → fertilized eggs are **Bb** ♀ **(brown)**; unfertilized eggs are **b** ♂ **(ivory)**.

b. The cross is *Bb* F1 ♀ x *B* ♂ → fertilized eggs are **1/2 Bb** ♀ **(brown) : 1/2 BB** ♀ **(brown) = all brown** ♀ **progeny**; unfertilized eggs are **1/2 B** ♂ **(brown) : 1/2 b** ♂ **(ivory)**.

4-6.

a. **G_1, S, G_2 and M (<u>see Figure 4.7</u>).**

b. **G_1, S and G_2 are all part of interphase.**

c. **G_1 is the time of major cell growth that precedes DNA synthesis and chromosome replication. Chromosome replication occurs during S phase. G2 is another phase of cell growth after chromosome replication during which the cell synthesizes many proteins needed for mitosis.**

4-7. a. **iii**; b. **i**; c. **iv**; d. **ii**; e. **v**.

4-8. Remember that this textbook uses the convention that the number of chromosomes = number of separate centromeres (<u>see Figures 4.13 and 4.14</u>). Thus, after DNA synthesis each chromosome has replicated and so has 2 chromatids held together at the replicated by attached centromere. This structure is one chromosome with two chromatids. The centromeres separate at anaphase in mitosis or at anaphase II in meiosis; at this point, the two chromatids now become two chromosomes. For an overview of egg formation in humans <u>see Figure 4.18</u>; for an overview of sperm formation in humans <u>see Figure 4.19</u>.

a. **96 chromosomes** with 1 chromatid each = **96 chromatids**;

b. **48 chromosomes** with 2 chromatids each = **96 chromatids**;

c. **24 chromosomes** with 1 chromatid each = **24 chromatids**;

d. **48 chromosome**s unreplicated in G_1 = **48 chromatids**;

e. **48 chromosomes** with 2 chromatids each (replicated in G2) = **96 chromatids**;

f. **48 chromosomes** unreplicated in G_1 preceding meiosis = **48 chromatids**;

g. **48 chromosomes** with 2 chromatids each = **96 chromatids**;

h.   **48 chromosomes** unreplicated before S = **48 chromatids**;

i.   **24 chromosomes** with 1 chromatid each = **24 chromatids**;

j.   **48 chromosomes** with 2 chromatids each = **96 chromatids**;

k.   **24 chromosomes** with 2 chromatids each = **48 chromatids**;

l.   **24 chromosomes** with 1 chromatid each = **24 chromatids**;

m.   **24 chromosomes** with 1 chromatid each = **24 chromatids**.

4-9.

a.   **mitosis, meiosis I, meiosis II**

b.   **mitosis, meiosis I**

c.   **mitosis**

d.   Mitosis is obviously excluded, but the rest of the answer depends on whether your definition of ploidy counts chromosomes or chromatids (review the answer to problem 4-8). Meiosis I in a diploid organism produces daughter cells with n chromosomes but $2n$ chromatids; meiosis II produces haploid daughter cells with n chromosomes or n chromatids. To avoid potential confusion, geneticists usually use the terms "n", "$2n$", and "ploidy" only to describe cells with unreplicated chromosomes. Thus there are **2 possible answers: meiosis II (and meiosis I, see explanation)**.

e.   **meiosis I**

f.   **none**

g.   **meiosis I**

h.   **meiosis II, mitosis**

i.   **mitosis, meiosis I**

4-10. Remember that the problem states that all cells are from the same organism. This determines the designation of mitosis, meiosis I and II. The stage of the cell cycle can be inferred from the morphology of the spindle, the presence or absence of the nuclear membrane, and whether homologous chromosomes (or sister chromatids) are paired (or connected through a centromere) or separated. **The n number is 3 chromosomes.**

a.   **anaphase of meiosis I**

b.   **metaphase of mitosis** (not meiosis II! because there are 6 chromosomes or 2n in this cell)

c.   **telophase of meiosis II**

d.   **anaphase of mitosis**

e.   **metaphase of meiosis II**

4-11. When trisomy 21 cells undergo meiosis there is an extra unpaired chromosome 21. The paired chromosome 21s are directed to opposite spindle poles, but the extra chromosome 21 segregates randomly into one of the products of meiosis I. When the cell with the extra chromosome 21 undergoes meiosis II, both products will have an extra chromosome 21. If one of these products of meiosis II becomes the ovum, a trisomic child would be produced after fertilization. There is **a 2/4 chance** that an ovum will contain the extra chromosome 21.

4-12.

a. metaphase of mitosis:

b. metaphase of meiosis I: (Note the pairing of homologous chromosomes and the two possible alignments of the 2 non-homologous chromosome pairs.)

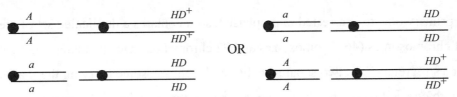

OR

c. metaphase of meiosis II: Shown below is only one of the two products of meiosis I from the cell diagrammed at the left in part b; the other product of this meiosis I would have chromosomes bearing a and HD^+. The other alignment in meiosis I (shown at the right in part b) will give A HD^+ and a HD daughter cells.

4-13.

a. The cell is in **metaphase/early anaphase of meiosis I in a male**, assuming the heterogametic sex (that is, the sex with two different sex chromosomes) in *Tenebrio* is male. Note the heteromorphic chromosome pair in the center of the cell.

b. It is not possible to distinguish **centromeres, telomeres or sister chromatids**, among other structures.

c. $n = 5$.

4-14. The genetic reshuffling that occurs during meiosis due to the independent alignment of maternal and paternal chromosomes and recombination between homologs could lead to **novel collections of alleles of their genes that could help an individual survive.**

4-15.

a. **400 sperm** are produced from 100 primary spermatocytes;

b. **200 sperm** are produced from 100 secondary spermatocytes;

c. **100 sperm** are produced from 100 spermatids;

d. **100 ova** are formed from 100 primary oocytes. Remember that although each primary oocyte will produce three or four meiotic products (depending on whether the first polar body undergoes meiosis II), only one will become an egg (ovum);

e. **100 ova** develop from 100 secondary oocytes - the other 100 haploid products are polar bodies;

f. **No ova** are produced from polar bodies.

4-16. The primary oocyte is arrested in prophase I, so it contains a duplicated set of the diploid number of chromosomes (46 chromosomes and 92 chromatids). During meiosis I, the homologous chromosomes segregate into two separate cells, so the chromosome carrying the *A* alleles will segregate into one cell while the chromosome carrying the *a* alleles will segregate into the other cell. One of these cells becomes the secondary oocyte (containing 23 chromosomes each with 2 chromatids and more of the cytoplasm) and the other becomes the polar body. **The genotype of the dermoid cyst that develops from a secondary oocyte could be either *AA* or *aa*.** Note that the attached sister chromatids in the secondary oocyte must separate from each other before the first mitosis leading to cyst formation.

4-17. Remember that ZW is ♀, ZZ is ♂ and WW is lethal.

a. The ZW eggs would give rise to **only ZW females**.

b. Cells resulting from meiosis will be 1/2 Z : 1/2 W. Upon chromosomal duplication they would become ZZ or WW. ZZ cells develop into males and WW is lethal, so **only males** are produced by this mechanism.

c. After eggs have gone through meiosis I, they will contain either a replicated Z or a replicated W chromosome. If the sister chromatids separate to become the chromosomes you will have 1/2 ZZ:1/2 WW. Again **only ZZ males** are produced since WW cells are not viable.

d. Meiosis of a ZW cell produces 4 haploid products, 2 Z : 2 W, one of which is the egg and the other 3 are the polar bodies. If the egg is Z then it has 1/3 chance of fusing with a Z polar body and 2/3 chance of fusing with a W polar body = 1/3 ZZ ($\male$) : 2/3 ZW ($\female$). If the egg is W then the fusion products will be 1/3 WW (lethal) : 2/3 ZW ($\female$). Because these 2 types of eggs are mutually exclusive, you add these 2 probabilities: 1/2 (probability of Z egg) × (1/3 ZZ : 2/3 ZW) + 1/2 (probability of W egg) × (1/3 WW : 2/3 WZ) = 1/6 ZZ : 2/6 ZW + 1/6 WW : 2/6 WZ = 1/6 ZZ ($\male$) : 4/6 ZW ($\female$) : 1/6 WW (lethal) = **1/5 ZZ ($\male$) : 4/5 ZW ($\female$).**

4-18. In birds males are ZZ, females are ZW. Diagram the crosses between true-breeding birds:

yellow $\female$ × brown $\male$ → brown $\female$ and $\male$; brown $\female$ × yellow $\male$ → yellow $\female$ and brown $\male$

Answer the 3EQ#1 - there are 2 phenotypes in the second cross, so there is 1 gene controlling color in canaries. Question #2 - because the parents are true-breeding, the first cross shows that brown > yellow. Question #3 - these two crosses are reciprocal crosses and the progeny show a sex-associated difference in phenotypes - both sexes are the same phenotype (brown) in the first cross but different phenotypes in the second cross (brown $\female$ and yellow $\male$). This indicates the **trait is sex-linked**. Also, the second cross shows crisscross inheritance - brown females × yellow males → brown sons and yellow daughters. This is also characteristic of a sex-linked trait. **Therefore, the alleles of the gene are Z^B (brown allele on Z chromosome) and Z^b (yellow allele on Z chromosome)**; the W chromosome does not carry the feather color gene. The first cross was Z^bW x Z^BZ^B → Z^BW (brown females) and Z^BZ^b (brown males). The second cross was Z^BW x Z^bZ^b → Z^BZ^b (brown males) and Z^bW (yellow females).

4-19. wild type $\female$ x yellow vestigial $\male$ → F_1 wild type $\female$ and $\male$ → F_2 16 yellow vestigial $\male$: 48

yellow $\male$: 15 vestigial $\male$: 49 wild type $\male$: 31 vestigial $\female$: 97 wild type $\female$

3EQ#1 - there are 4 different phenotypes in the F_2, so 2 genes involved. 3EQ#2 - one gene determines body color and the F_1 shows that wild type is dominant to yellow; the other gene determines wing length and the F_1 shows that wild type is dominant to vestigial. This conclusion is reinforced by the F_2 progeny where there are 48 + 49 + 97 normal wing individuals : 16 + 15 + 31 vestigial winged flies = 194 normal : 61 vestigial = 3 : 1. 3EQ#3 - in the F_2 there are wild type and vestigial males and females in a 3:1 ratio, so wing shape is an autosomal trait; in the F_2 there are yellow males but no yellow females, so body color is an X-linked trait. Thus **vestigial is the**

recessive allele of an autosomal gene and yellow is the recessive allele of an X-linked gene. The original cross was thus $y^+/y^+ ; vg^+/vg^+$ ♀ x $y/Y ; vg/vg$ ♂.

4-20. The answer to this question depends upon when the nondisjunction occurs. Nondisjunction in meiosis I in the male results in one secondary spermatocyte that contains both sex chromosomes (the X and the Y) and the other secondary spermatocyte that lacks sex chromosomes. If the red-eyed males are $X^{w+}Y$, then the sperm that are formed after nondisjunction in meiosis I are $X^{w+}Y$ and nullo (O) sex chromosome. The X^wX^w female makes X^w eggs, so fertilization will produce **X^wO (white-eyed, sterile male) and $X^{w+}X^wY$ (red-eyed female)** progeny. If nondisjunction occurred in meiosis II in the male, the sperm would be $X^{w+}X^{w+}$ or YY and nullo (O) sex chromosome. After fertilization of the X^w eggs, the zygotes would be $X^{w+}X^{w+}X^w$ (lethal) or **X^wYY (fertile white-eyed males) and X^wO (sterile white-eyed males)**. Notice that the normal progeny of this cross will be $X^{w+}X^w$ (red-eyed females) and X^wY (white-eyed males), which are indistinguishable from the nondisjunction progeny in terms of their eye colors.

4-21. In birds, females are the heterogametic sex, (ZW); Z^B represents the Z chromosome with the barred allele, Z^b is non-barred.

a. Z^BW (barred hen) x Z^bZ^b(non-barred rooster) → **Z^bW (non-barred females) and Z^BZ^b (barred males)**.

b. Z^BZ^b $F_1 \times Z^bW$ → **Z^BW (barred) and Z^bW (non-barred) females and Z^BZ^b and Z^bZ^b (barred and non-barred) males**.

4-22. Pedigrees 1-4 show examples of each of the four modes of inheritance.

a.   **Pedigree 1** represents a recessive trait because two unaffected individuals have affected children. If the trait were X-linked, then I-1 would have to be affected in order to have an affected daughter. The trait is **autosomal recessive**. **Pedigree 2** represents recessive inheritance (see part a). This is **X-linked recessive** inheritance as autosomal recessive inheritance was already accounted for in part a. This conclusion is supported by the data showing that the father (I-1) is unaffected and only sons show the trait in generation II, implying that the mother must be a carrier. **Pedigree 3** shows the inheritance of a dominant trait because affected children always have an affected parent (remember that all four diseases are rare). The trait must be **autosomal dominant** because the affected father transmits it to a son. **Pedigree 4** represents an **X-linked**

dominant trait as characterized by the transmission from affected father to all of his daughters but none of his sons.

b. **Pedigree 1** - both parents are carriers for this autosomal recessive trait so there is a **1/4 chance** that the child will be affected (*aa*). **Pedigree 2** – individual I-2 is a carrier for this X-linked trait. The probability is 1/2 that she will pass on X^a to her daughter II-5. The unaffected father (I-1) contributes a normal X^A chromosome, so the probability that is II-5 is a carrier = 1/2. The **probability of an affected son** = 1/2 (probability II-5 is a carrier) × 1/2 (probability II-5 contributes X^a) × 1/2 (probability of Y from father II-6) = **1/8**. The **probability of an affected daughter = 0** because II-6 must contribute a normal X^A. **Pedigree 3** - for an autosomal dominant trait there is a **1/2 chance** that the heterozygous mother (II-5) will pass on the mutant allele to a child of either sex. **Pedigree 4** - the father (I-1) passes on the mutant X chromosome to all his daughters and none of his sons. Therefore II-6 does not carry the mutation, as shown by his normal phenotype. The **probability of an affected child = 0**.

4-23.

a. Unaffected individuals have affected children, so albinism is **recessive**.

b. If albinism were X-linked, then I-9 would have to be an affected hemizygote in order to have an affected daughter. As this is not the case, albinism is **autosomal**.

c. *aa*; d. *Aa;* e. *Aa;* f. *Aa*; g. *Aa*; h. *Aa.*

4-24.

a. If the trait is an autosomal dominant, then the fathers must be heterozygous *Rr* because they have unaffected daughters. The mothers would then be homozygous normal, *rr*. The probability of an affected male child from such matings = 1/2 (probability of inheriting *R*) × 1/2 (probability the child will be male) = 1/4. The probability of an unaffected daughter = 1/4. The probability of having 6 affected sons and 5 unaffected daughters = $(1/4)^6 \times (1/4)^5 = (1/4)^{11} = 2 \times 10^{-7}$ or extremely unlikely!

b. This trait could be an example of **Y-linked inheritance or it could represent sex-limited expression of the mutant allele** (that is, the allele is dominant in males and recessive or otherwise unexpressed in females, like male-pattern baldness). There is no direct evidence to support one hypothesis over the other. However there are no known examples of Y-linked inheritance other than maleness, while there are several examples of sex-influenced traits that are male dominant and female recessive.

4-25.

a. Hypertrichosis is almost certainly an **X-linked dominant trait**.

b. The trait is **dominant because every affected child has an affected parent**. If the trait were recessive then several of the people marrying into the family would have to be carriers, yet this trait is rare. Hypertrichosis is highly likely to be **X-linked because two affected males (II-4 and IV-3) pass the trait on to all of their daughters but none of their sons**. Using the same logic as in problem 4-24a, there is only a $(1/4)^{13}$ chance that the trait is autosomal dominant.

c. **III-2 had 4 husbands and III-9 had 6 husbands!**

4-26. When nondisjunction occurs in meiosis I (MI), there are two types of affected gametes. Half of the gametes receive no sex chromosomes (nullo) while the rest of the gametes receive one copy of each of the sex chromosomes in the parent. Thus MI nondisjunction in a male will give XY sperm and nullo sperm. If the nondisjunction occurs in MII, then half of the affected gametes are nullo while the rest receive 2 copies of one of the original sex chromosomes. MII nondisjunction in a male will therefore give either XX sperm or YY sperm and nullo sperm with no sex chromosomes.

a. The boy received both an X^B and a Y from his father. Because both the X and Y segregated into the same gamete, the nondisjunction occurred in **meiosis I in the father**.

b. The son could have received both the X^A and X^B from the mother and the Y from the father, or alternatively the X^B from the mother and the X^A and Y from the father. You can determine that the nondisjunction occurred in **meiosis I, but you cannot determine in which parent**.

c. The son has two X^A chromosomes and a Y. Therefore the nondisjunction occurred in the **mother, but because she is homozygous $X^A X^A$ you cannot determine during which meiotic division**.

4-27. Color-blindness in this family is an X-linked recessive condition, because two unaffected parents have affected children, and because two unrelated individuals would have to carry rare alleles if the trait were autosomal. Since the males are hemizygous with only one allele for this trait, their phenotype directly represents their genotype. **II-2 and III-3 are affected, so they must be $X^{cb}Y$.** Now consider the parents of II-2. **I-2 is normal and therefore $X^{CB}Y$. I-1 must be a carrier so her genotype is $X^{CB}X^{cb}$. II-1 can be either $X^{CB}X^{CB}$ or $X^{CB}X^{cb}$** (but she is more likely to be a normal homozygote if the trait is rare). **II-3 must be a carrier ($X^{CB}X^{cb}$) since she had an affected son. II-4 is $X^{CB}Y$; III-1 must be $X^{CB}X^{cb}$** (because she has normal color vision yet she must have received X^{cb} from her father); **III-2 is either $X^{CB}X^{CB}$ or $X^{CB}X^{cb}$; III-4 is an unaffected male and therefore must be $X^{CB}Y$.**

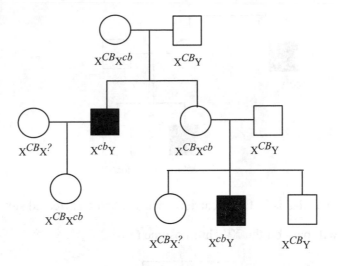

4-28.

a.   Draw the pedigree:

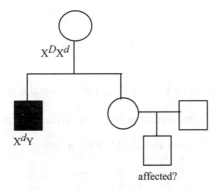

I-1 must have been heterozygous for the *d* allele. The probability that II-2 will have an affected son = 1/2 (the probability that II-1 inherited the X^d chromosome from I-1) × 1/2 (the probability that II-1's son receives the X^d chromosome from her if she is X^DX^d) = **1/4**.

b.

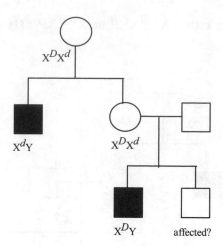

We now know that II-2 is in fact a carrier since she had an affected son. Therefore the probability is **1/2** that she will pass on the X^d chromosome to III-2.

c.

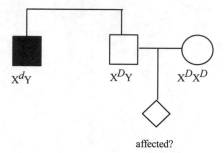

affected?

The mother of these two men was a carrier. She passed on the X^d chromosome to the affected son and she passed on the X^D chromosome to the unaffected son. There is **no chance** that the unaffected man will pass the disease allele to his children.

d.

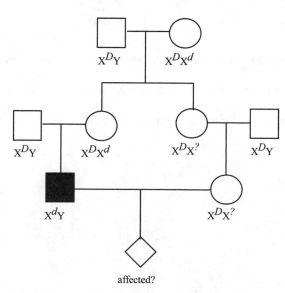

The probability of IV-1 being an affected son = 1/2 (probability that II-3 is a carrier) x 1/2 (probability that III-2 inherits X^d) × 1/2 (probability that IV-1 inherits X^d) × 1/2 (probability IV-1 inherits the Y from III-1) = **1/16**. **The probability that IV-1 is an affected girl** = 1/2 (probability that II-3 is a carrier) x 1/2 (probability that III-2 inherits X^d) × 1/2 (probability that IV-1 inherits X^d) × 1/2 (probability IV-1 inherits X^d from III-1) = **1/16**. **The chance that IV-1 is unaffected** = 1 - (1/16 probability of an affected male + 1/16 probability of an affected female) = 1 - 2/16 = **7/8**.

e.

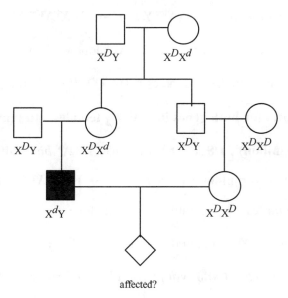

affected?

II-2 must be a carrier in order for her son to inherit the disease. II-3 is the brother and he is normal, so his genotype must be $X^D Y$. Therefore III-2, his daughter, must be homozygous normal. The probability that the fetus (diamond) will be an **affected boy = 0**; the probability of an **affected daughter = 0**; the probability of an **unaffected child = 100%.**

4-29.

a. The father contributed the *i* or I^A, Rh^+, *M* or *N*, and $Xg^{(a-)}$ alleles. Only **alleged father #3** fits these criteria.

b. Alleged father #1 was ruled out based on his X chromosome genotype. He is $Xg^{(a+)}$ and can not have an XX daughter who is $Xg^{(a-)}$. If the daughter was XO (Turner), she could have inherited the one X from either mother or father. If the X came from her mother (who must have been a heterozygote for the two alleles of *Xg*), then her father did not contribute an X chromosome. Therefore **alleged fathers #1 and #3** both fit the criteria for paternity.

4-30. Brown eye color (b) and scarlet (s) are autosomal recessive mutations while vermilion (v) is an X-linked recessive trait. The genes interact such that both brown, vermilion double mutants and brown, scarlet double mutants are white-eyed. When diagramming a cross involving more than one gene you <u>must</u> start with a genotype for each parent that includes information on <u>both genes</u>. Then figure out the genotype of the F_1 progeny. In order to predict the F_2 results, find the expected ratio for each gene separately, and then cross multiply the ratios to generate the F_2. Diagram the following crosses:

a.   vermilion ♀ ($X^v X^v b^+ b^+$) × brown ♂ ($X^{v+}Y\ bb$) → **F₁ $X^v X^{v+}\ b^+ b$ (wild type females) x**

   $X^v Y\ b^+ b$ (vermilion males) → F_2 ratio for brown alone = 3/4 b^+- : 1/4 bb; the ratio for

   vermilion alone in both F_2 females and males = 1/2 X^{v+} : 1/2 X^v (the other chromosome will be

   either X^v or Y); **the ratio for both genes in both F_2 females and males = 3/8 $X^{v+}\ b^+$- (wild**

   type) : 3/8 $X^v\ b^+$ (vermilion) : 1/8 $X^{v+}\ bb$ (brown) : 1/8 $X^v\ bb$ (white).

b.   brown ♀ ($X^{v+} X^{v+}\ bb$) × vermilion ♂ ($X^v Y\ b^+ b^+$) → **F1 $X^v X^{v+}\ b^+ b$ (wild type females)** ×

   $X^{v+}Y\ b^+ b$ (wild type males) → F_2 ratio for brown alone = 3/4 b^+- : 1/4 bb; ratio for vermilion

   alone in the F_2 females = 1 X^{v+}- and in the males = 1/2 X^{v+}: 1/2 X^v. The ratio for both genes in

   the **F_2 females is 3/4 X^{v+}- b^+- (wild type) : 1/4 X^{v+}- bb (brown)** and the dihybrid ratio in the

   F2 males is 3/8 $X^{v+}Y\ b^+$- (wild type) : 3/8 $X^v Y\ b^+$- (vermilion) : 1/8 $X^{v+}Y\ bb$ (brown) : 1/8

   $X^v Y\ bb$ (white).

c.   scarlet ♀ ($b^+ b^+\ ss$) × brown ♂ ($bb\ s^+ s^+$) → **F₁ $b^+ b\ s^+ s$ (wild type,** males and females are

   the same because both genes are autosomal) → F_2 monohybrid ratio for scarlet = 3/4 s^+- : 1/4

   ss and for brown = 3/4 b^+-:1/4 bb. **The F_2 dihybrid ratio** (which hold for both sexes) **= 9/16 s^+-**

   b^+- (wild type) : 3/16 s^+- bb (brown) : 3/16 $ss\ b^+$- (scarlet) : 1/6 $ss\ bb$ (white).

d.   brown ♀ ($bb\ s^+ s^+$) × scarlet ♂ ($b^+ b^+\ ss$) → **F₁ $b^+ b\ s^+ s$ (wild type)** → **F₂ as in part c**

   above.

4-31. The ultraviolet component of sunlight can damage the DNA in skin cells. In normal people (and mice) this damage is almost always repaired by a series of DNA repair enzymes acting in a complicated set of steps. Thus, normal people and mice are fairly resistant to the deleterious effects of ultraviolet light. *Xeroderma pigmentosum* is a phrase used to describe the inability to repair the ultraviolet-generated damage to DNA. In both mice and humans mutations are known in several different genes that give rise to the phenotype known as x*eroderma pigmentosum* (XP). Note that the true-breeding parental strains of mice used in this problem are independently derived. Diagram the cross:

XP short tail ♀ × XP long tail ♂ → F_1 XP^+ short tail ♀ and ♂ → F_2 22 XP short tail ♀ : 28

XP^+ short tail ♀ : 11 XP short tail ♂ : 10 XP long tail ♂ : 14 XP^+ short tail ♂ : 15 XP^+ long tail ♂.

3EQ#1 - there are only two phenotypes for tail length (long and short), so there is one gene controlling this trait. For XP notice that both parents were mutant, but the F_1 progeny are normal! This complementation test (**see Figure 3.15**) indicates that the XP phenotype in each parent is due to a mutation in a different gene. If both parents were mutant in the same gene then the F_1 progeny must show the mutant XP phenotype! Thus **there are two genes controlling XP**. 3EQ#2 - the F_1 progeny show that **short tail (*Sh*) is dominant to long tail (*sh*)** and that **the normal alleles (*XP1⁺***

and *XP2⁺*) are dominant to the mutant alleles (*xp1* and *xp2*) for both of the genes. 3EQ#3 - there are no differences in the phenotypes seen in the sexes in the F1 generation. In the F_2 progeny there are no differences between the sexes for the XP phenotype, but there are NO long tail females. Thus, **XP1 and XP2 are autosomal genes while *Sh* is an X-linked gene**. There are 28 + 14 + 15 XP^+ F_2 progeny : 22 + 11 + 10 XP mutant F_2 progeny = 9 XP^+ : 7 XP. This epistatic phenotypic ratio indicates that the two genes controlling XP assort independently and that a mouse will have the disease if it is homozygous mutant for either or both of these genes.

ShSh xp1xp1 XP2⁺XP2⁺ ♀ × *shY XP1⁺XP1⁺ xp2xp2* ♂ → F_1 *Shsh XP1⁺xp1 XP2⁺xp2* ♀ ×

ShY XP1⁺xp1 XP2⁺xp2 ♂ → F_2 ratio for the *sh* gene alone = 1/4 *Sh*Y males : 1/4 *sh*Y males : 1/2

Sh- females; the F_2 ratio for the two XP genes = 9/16 *XP1⁺- XP2⁺-* (XP^+) : 3/16 *XP1⁺- xp2xp2*

(XP) : 3/16 *xp1xp1 XP2⁺-* (XP) : 1/16 *xp1xp1 xp2xp2* (XP). When these ratios are cross multiplied the final trihybrid phenotypic ratio is: 9/32 *Sh*- (short tail) XP^+ females : 7/32 *Sh*- (short tail) XP females : 9/64 *Sh*Y (short tail) XP^+ males : 7/64 *Sh*Y (short tail) XP males : 9/64 *sh*Y (long tail) XP^+ males : 7/64 *sh*Y (long tail) XP males.

4-32.

a.   Eosin is an X-linked recessive mutation. The cream colored variants were found occasionally in the eosin true-breeding stock. This suggests that the cream flies have the eosin genotype and also have an second mutation that further lightens the eosin phenotype. Diagram the cross:

wild type ♀ × cream ♂ → F$_1$ wild type → F2 104 wild type ♀ : 52 wild type ♂ : 44 eosin

♂ : 14 cream ♂ (8:4:3:1 ratio)

3EQ#1 - there are 3 phenotypes seen in the F$_2$ males. This could be due to one gene with codominance/incomplete dominance, but the phenotypes are not present in a 1:2:1 ratio. Instead the F$_2$ phenotypic ratio seems to be an epistatic modification of 9:3:3:1. Therefore, the eye color must be controlled by 2 genes whose mutant alleles are recessive. The fact that the cream eye color initially arose in the eosin mutant stock suggests that cream is a modifier of eosin. This sort of modifier only alters the mutant allele of the gene it is modifying and has no effect on the wild type allele. **3EQ#2** - the F$_1$ shows you that the wild type phenotype is dominant. Question #3 - as expected, the difference in the phenotypes between the F$_2$ males and females shows that eosin (*e*) is X-linked, but the cream modifier (*cr*) is probably not on the X chromosome (that is, it is autosomal) because some F$_2$ animals are eosin but not cream.

e$^+$*e*$^+$ *cr*$^+$*cr*$^+$ ♀ x *e*Y *crcr* ♂ → **F$_1$** *e*$^+$*e* *cr*$^+$*c* ♀ x *e*$^+$Y *cr*$^+$*cr* ♂ → F$_2$ ratios for each gene alone = 1/2 *e*$^+$- ♀ : 1/4 *e*$^+$Y : 1/4 *e*Y and 3/4 *cr*$^+$- : 1/4 *crcr*. When these are cross multiplied, the **F$_2$ ratio for both genes = 3/8 *e*$^+$- *cr*$^+$- (wild type) ♀ : 1/8 *e*$^+$- *crcr* (wild type) ♀ : 3/16 *e*$^+$Y *cr*$^+$- (wild type) ♂ : 1/16 *e*$^+$Y *crcr* (wild type) ♂ : 3/16 *e*Y *cr*$^+$- (eosin) ♂ : 1/16 *e*Y *crcr* (cream) ♂ = 8 wild type ♀ : 4 wild type ♂ : 3 eosin ♂ : 1 cream ♂.**

b.   *e*Y *cr*$^+$*cr*$^+$ (eosin ♂) x *ee crcr* (cream ♀) → **F$_1$** *ee cr*$^+$*cr* ♀ x *e*Y *cr*$^+$*cr* ♂ → F$_2$ ratios for each gene alone = 1/2 *ee* ♀ : 1/2 *e*Y ♂ and 3/4 *cr*$^+$- : 1/4 *crcr*. **F$_2$ ratios for both genes = 3/8 *ee* *cr*$^+$- (eosin ♀) : 3/8 *e*Y *cr*$^+$- (eosin ♂) : 1/8 *ee crcr* (cream ♀) : 1/8 *e*Y *crcr* (cream ♂).**

c.   *ee cr*$^+$*cr*$^+$ (eosin ♀) x *e*Y *crcr* (cream ♂) → **F$_1$** *ee cr*$^+$*cr* ♀ x *e*Y *cr*$^+$*cr* ♂ → F$_2$ ratios for each gene alone = 1/2 *ee* ♀ : 1/2 *e*Y ♂ and 3/4 *cr*$^+$- : 1/4 *crcr*. **F$_2$ ratios for both genes = 3/8 *ee* *cr*$^+$- (eosin ♀) : 3/8 *e*Y *cr*$^+$- (eosin ♂) : 1/8 *ee crcr* (cream ♀) : 1/8 *e*Y *crcr* (cream ♂).** This is the same result as in part b.

4-33.

a.   The white eye mutation is an X-linked, recessive mutation. Diagram the cross:

white ♂ x purple ♀ → F$_1$ wild type eye color → F$_2$ 3/8 wild type ♀ : 1/8 purple ♀ : 3/16 wild type ♂ : 1/4 white ♂ : 1/16 purple ♂.

33EQ#1 – white and purple eye colors cannot be caused by two different alleles of the same gene because then the F_1 males would be purple due to criss-cross inheritance. This is a also a complementation test and wild type phenotype of the F_1 heterozygotes means there are 2 different genes controlling eye color (**see Figure 3.15**). 3EQ#2 – the mutant allele of one of the genes causes white eye color, the mutant allele of the other gene causes purple eye color. The wild type alleles of both genes contribute to wild type eye color. **In both cases the mutant allele is recessive ($w < w^+$ and $p < p^+$) as demonstrated by the wild-type phenotype of the F_1 animals. You also expect that white (the absence of pigment) should be epistatic to eyes with red or purple pigmentation.** Question #3 – **white is X-linked; purple is an autosomal trait** because it shows no difference between the sexes in either the F_1 or the F_2 generation - there is a 3 p^+- : 1 pp ratio in both males and females in the F2. The original cross was thus: $wY\ p^+p^+$ (white ♂) x $w^+w^+\ pp$ (purple ♀).

b. $ww\ p^+p^+$ (white ♀) x $w^+Y\ pp$ (purple ♂) → **F_1 $w^+w\ p^+p$ (wild type eye color ♀) x $wY\ p^+p$ (white ♂)** → F_2 ratio for white alone = 1/2 w^+w : 1/2 ww females and 1/2 w^+Y : 1/2 wY males; for the purple gene alone the ratio is 3/4 p^+- : 1/4 pp. **The F_2 ratio for both genes in females = 3/8 w^+- p^+- (wild type) : 1/8 w^+- pp (purple) : 3/8 $ww\ p^+$- (white) : 1/8 $ww\ pp$ (white) = 3/8 wild type : 1/2 white : 1/8 purple. The ratio for the males is the same.** Note that in this reciprocal cross there is a phenotypic difference between the sexes in the F1 generation and no difference in the F2 generation.

4-34. Remember that white tigers are shown by <u>unshaded</u> symbols. Test each possibility by assigning genotypes based on the mode of inheritance.

a. **No**. Y linked genes are only expressed in males. Since there are white females, the trait could not be Y-linked.

b. **Yes**. White males would have to have white daughters but not white sons. The outcomes of all matings are consistent with a dominant X-linked allele.

c. **Yes**. The information in the pedigree is consistent with dominant autosomal inheritance.

d. **No**. If this were X-linked recessive inheritance, Kesari would have to be X^WX^W and Tony would be $X^{W^+}Y$. Their son would have to be X^WY and would be white colored, but Bim is not white.

e. **Yes**. There is no data in the cross scheme to allow you to rule out recessive autosomal inheritance in this highly inbred family.

4-35. This is a challenging experimental question. To search for mutations causing nondisjunction in females, you could take potential mutant females that are homozygous for X^y (X^y / X^y) and mate them with X^y / y^+•Y males. If there is no nondisjunction in the females, you would find yellow-bodied female (X^y / X^y) and brown-bodied male (X^y / y^+•Y) progeny. Nondisjunction during meiosis in the female would yield X^y / X^y and nullo-X eggs. X^y / X^y eggs fertilized with a y^+•Y-bearing sperm would be brown bodied females (X^y / X^y / y^+•Y). Nullo-X eggs fertilized by X-bearing sperm would be X^y / O males (yellow-bodied and sterile). The other possible zygotes are X^y / X^y / X^y (triplo-X) and nullo-XY, and both of these combinations are lethal. Thus, **X^y / X^y females yielding elevated frequencies of brown-bodied female and yellow-bodied male progeny from this cross are good candidates to harbor mutations affecting meiotic chromosome segregation.**

Chapter 5   Linkage, Recombination, and the Mapping of Genes on Chromosomes

Synopsis:

This chapter is devoted to a very important topic: linkage of genes. The concept of linkage is the basis for genetic mapping. Genes on the same chromosome are physically connected or linked. Gene pairs that are close together on the same chromosome are genetically linked because the alleles on the same homolog are transmitted together (parental types. into gametes more often than not during meiosis (**see Figures 5.2 and 5.3**).

Gene pairs that assort independently exhibit a recombination frequency of 50% because the number of parental types = the number of recombinants (**see Figure 5.10**). Genes may assort independently either because they are on different chromosomes or because they are far apart on the same chromosome. The recombination frequencies of pairs of genes indicate how often 2 genes are transmitted together. For linked genes, the frequency is less than 50%. The greater the distance between linked genes, the higher the recombination frequency (rf). Recombination frequencies become more inaccurate as the distance between genes increases.

The relationship between relative recombination frequency and distance is used to create genetic maps. The greater the density of genes on the map, and the smaller the distances between the genes, the more accurate the map.

Statistical analysis can help determine whether or not 2 genes assort independently. The Chi square test can also be used to determine how well the outcomes of crosses fit other genetic hypotheses (**see Figure 5.6**).

Tetrad analysis is done in certain yeast in which the meiotic products are kept together in a sac (ascus) so the results of a single meiosis are displayed in each tetrad. This array of 4 spores reveals the relation between genetic recombination and the segregation of chromosomes duringthe 2 meiotic divisions. Solving tetrad analysis problems will increase your understanding of chromosome segregation during meiosis.

In diploid organisms heterozygous for 2 alleles of a gene, rare mitotic recombination between the gene and its centromere can produce genetic mosaics in which some cells are homozygous for one allele of the other (**see Figure 5.26** and Problem 5-31).

Significant Elements:

After reading the chapter and thinking about the concepts you should be able to:

- Determine if genes are linked or not based on the frequency of different types of gametes or progeny.

- Determine whether observed results are statistically consistent with expected results using the chi square test.

- When given the genotypes of parents and information on linkage of genes, list the gametes that can be produced (with and without recombination.

- Beginning with 3 genes, map the genes. Determine the order of the genes and the distance between linked genes based on the % recombination that occurred during meiosis.

- Determine if there is crossover interference in a three point cross.

- Identify different types of tetrads (parental ditype - PD, non-parental ditype - NPD, tetratype-T) and understand what events during meiosis gives rise to these different types of asci.

- Determine linkage between genes given either random spore results or numbers of different types of asci.

- Identify centromere linked genes and determine distance from gene to centromere in *Neurospora* tetrads.

Problem Solving Tips:

This chapter discusses the analysis of linked genes. One way to analyze recombination frequency (rf) between linked genes is using 2 point crosses, as is done in Tetrad Analysis. A more accurate map of linked genes can be derived from analyzing the data from 3 linked genes (3 point crosses). Both of these types of analysis of linked genes lead to maps, but each type of analysis has it's own peculiarities.

- the minimum requirement for detecting recombination is heterozygosity for 2 genes. The recombination events that can be detected are the ones that occur between the 2 genes, giving recombinant gametes instead of parental gametes.

- It is easiest to detect the parental vs recombinant gametes if you do a test cross.

- In a test cross of $aa^+ bb^+ \times aa\ bb$ the expected phenotypic frequencies and classes of progeny are 1 $a^+\text{-} b^+\text{-}$: 1 $aa\ bb$: 1 $a^+\text{-} bb$: 1$aa\ b^+$ if the genes are assorting independently. The genes are genetically linked if you see more parental than recombinant progeny.

- rf = # recombinant progeny / total # progeny = rf $\times$ 100 = mu or cM.

- Recombination frequency is ALWAYS the number of recombinant gametes out of the total number of gametes. Depending on the vagaries of the particular system or problem, we may derive some basic variants of this formula.

♦ When discussing recombination, you must write the genotypes is a way that represents linkage. Remember that there is one allele per homolog, so *aa* becomes *a / a*. When discussing more than one gene on a chromosome you must initially <u>assume</u> an order. While solving a problem, always write the genes in the same arbitrary, assumed order. Thus, a+ b+ c / a b c+ is more accutately written as:

$$\frac{a^+ \qquad b^+ \qquad c}{a \qquad b \qquad c^+}$$

Mapping genes on the X chromosome, using hemizygous male progeny, see Problem 5-5.

♦ The Chi Square calculation is $(o - e)^2/e$

Problem Solving - How to Begin:

THREE ESSENTIAL QUESTIONS (3EQ):

1. How many genes are involved in the cross?

2. For **each gene** involved in the cross: what are the phenotypes associated with the gene? Which phenotype is the dominant one and why? Which phenotype is the recessive one and why?

3. For **each** gene involved in the cross: is it X-linked or autosomal?

The 3EQ are <u>still useful</u>, as you often have to diagram the cross and assign genotypes! For a more thorough list of Hints, review **Solving Problems - How to Begin** in Chapter 4 of this Study Guide.

THREE POINT CROSSES:

♦ In a three point cross, rewrite the classes of progeny (data) assigning genotypes to each trait. These genotypes are based on your answers to the 3EQ!

♦ designate the different gametes or offspring as noncrossover (parental), single crossover or double crossover. The noncrossover classes are those classes of progeny who have one of the intact, non-recombinant homologs from the parent. The noncrossover classes will be represented by the greatest numbers of offspring. The double crossover classes will be represented by the smallest numbers of offspring (<u>**see Figure 5.12**</u>). Sometimes one or both double crossover classes are missing.

♦ The offspring are generated by a heterozygous individual. Therefore, all classes (parental, etc.) will occur as reciprocal pairs of progeny. These reciprocal pairs will be both genetic reciprocals and numerically equivalent.

♦ By examining the pattern of data seen in a problem, you can often start solving the problem with a basic understanding of the linkage relationships of the genes. Some of the more common patterns of data are:

3 linked genes give 8 classes of data that occur as 4 reciprocal pairs;

3 unlinked genes gives 8 classes of data that occur as 4 reciprocal pairs, but all classes are seen in a 1:1:1:1:1:1:1:1 ratio;

4 linked genes gives 16 classes of progeny that occur in 8 reciprocal pairs;

4 unlinked genes gives 16 classes of progeny in a 1:1:1:1:1:1:1:1:1:1:1:1:1:1:1:1 ratio

3 linked genes and 1 gene assorting independently gives 16 classes of data occurring as 8 reciprocal pairs and 4 groups of 4 numerically.

♦ Begin the process of mapping the genes by ordering the genes. To figure out which gene is in the middle of a group of three genes, choose one of the double crossover classes. Compare it to the most similar parental class of progeny where two of the three genes will have the same combination of alleles. The gene that differs is the gene in the middle. See Problem 5-18c for further explanation.

♦ The last step is to determine the distance between the genes on each end and the gene in the middle. Use the formula rf = # recombinants between the 2 genes / total # of progeny.

♦ See Problem 5-18 for an example of using the data to make a map.

♦ See Problem 5-17 for an example of how to generate the data when you start with the map.

♦ See Problem 5-21 for some hints on how to solve more complicated arrangements of genes.

TETRAD ANALYSIS:

♦ Some of the things that can be done here when analyzing rf are based on the biology of the organisms. Thus you must understand some of the basic biology of *Neurospora* and yeast (**see Figure 5.16**) in order to understand the derivation of the Three Easy Rules for Tetrad Analysis.

♦ Rule #1 - If the number of PD tetrads is about equal to the number of NPD tetrads, the genes are unlinked. If the # PD tetrads >>> # NPD tetrads, the genes are linked.

♦ Independent assortment dictates that these two classes of asci will be about equal when genes are unlinked, see **Figure 5.28**.

♦ Rule #2 - Distance between linked genes is determined by calculating recombination frequency using the equation: RF= (NPD + 1)/2T/ total # tetrads × 100.

♦ Rule #3 - Crossovers between a gene and the centromere mean that the two alleles are separated at the second meiotic division (second division segregation) instead of during the first meotic division. Centromere distance can be measured in yeasts with ordered asci, such as *Neurospora* and *Ascobolus*.

♦ These rules are based on the effects of meiosis on two unlinked genes (**see Figure 5.17**) versus the effects on two linked genes (**see Figure 5.19**). Further intricacies of single and double crossovers are discussed in <u>Problem 5-28</u>.

Solutions to Problems:

5-1. a. **8**; b. **4**; c. **1**; d. **11**; e. **2**; f. **5**; g. **6**; h. **3**; i. **10**; j. **12**; k. **9**; l. **7**.

5-2. The null hypothesis is that there is independent assortment of 2 genes yielding a dihybrid phenotypic ratio of 9/16 *R-Y-* : 3/16 *R- yy* : 3/16 *rr Y-* : 1/16 *rr yy*. Use the Chi square (X^2) test to compare Mendel's observed data with the 9:3:3:1 ratio expected for 2 genes that assort independently.

Genotypes	Observed #	Expected #	X^2 square equation	Sum of X^2
R-Y-	315	9/16 (556) = 313	$(315 - 313)^2/313$	0.01
R- yy	108	3/16 (556) = 104	$(108 - 104)^2/104$	0.15
rr Y-	101	3/16 (556) = 104	$(101 - 104)^2/104$	0.09
rr yy	32	1/16 (556) = 35	$(32 - 35)^2/35$	0.26
				0.51

The number of classes is 4, so the degrees of freedom is 4–1 or 3. Using **Table 5-1** in the text, the probability of having obtained this level of deviation by chance alone is between 0.9 and 0.99 (90 - 99%). Thus we can NOT reject the null hypothesis. In other words, **the data are consistent with independent assortment and we therefore conclude that Mendel's data could indeed result from the independent assortment of the 2 genes.**

5-3.

a.   Diagram the cross.

orange (*O- bb*) × black (*oo B-*) → F$_1$ brown (*O- B-*) → 100 brown (*O- B-*) : 25 orange (*O-bb*) : 22 black (*oo B-*) : 13 albino (*oo bb*)

Because the F$_1$ snakes were all brown, we know that the orange snake could not have contributed an *o* allele, or there would have been some black snakes. The orange snake must be *OO bb*. The black snake could not have contributed a *b* allele or there would have been some orange snakes, so the black parent must be *oo BB*. Therefore **the F$_1$ snakes must be *Oo Bb***.

b.   The F$_1$ snakes are heterozygous for both genes (*Oo Bb*). If the two loci assort independently, we expect the F$_2$ snakes to show a 9 orange : 3 black : 3 brown : 1 albino ratio. The total number of F$_2$ progeny is 160. **We expect 90 (160 × 9/16) of these progeny to be brown, 30 (160 × 3/16) to be orange, 30 to be black and 10 (160 × 1/16) to be albino**.

c.

Genotypes	Observed #	Expected #	X^2 square equation	Sum of X^2
O- B-	90	100	$(100–90)^2/90$	1.11
O- bb	30	25	$(25–30)^2/30$	0.83
oo B-	30	22	$(22–30)^2/30$	2.13
oo bb	10	13	$(13–10)^2/10$	0.9
				4.97

There are three degrees of freedom (4 classes − 1) and the *p* value is between 0.5 and 0.1. **The observed values do not differ significantly from the expected.**

d.   There is a 10% - 50% probability that these results would have been obtained by chance if the null hypothesis were true; this is simply another way of writing the meaning of the *p* value.

5-4.

a.   The cross is:

normal (*DD*) × dancer (*dd*) → F$_1$ normal (*Dd*)   F$_2$ 3/4 *D-* (normal) : 1/4 *dd* (dancer)

1/4 of the F$_2$ mice will be dancers if the trait is determined by a single gene with complete dominance.

b.   Diagram the cross:

normal (*AA BB*) × dancer (*aa bb*) → F$_1$ normal (*Aa Bb*)   F$_2$ 15/16 normal (*A- B-* + *A- bb* + *aa B-*) : 1/16 dancer (*aa bb*)

1/16 of the mice would be expected to be dancers given the second hypothesis that dancing mice must be homozygous for the recessive alleles of two genes.

c.   Calculate the chi square values for each situation. Null hypothesis #1: Dancing is caused by the homozygous recessive allele of one gene, so 1/4 of the F$_2$ mice should be dancers. Calculating the expected numbers, 1/4 × 50 mice or 13 should have been dancers, 37 should have been nondancers.

Genotypes	Observed #	Expected #	X^2 square equation	Sum of X^2
nondancers	42	$(3/4) \times 50 = 37$	$(42-37)^2/37$	0.68
dancers	8	$(1/4) \times 50 = 13$	$(8-13)^2/13$	1.92
				2.60

With one degree of freedom the p value is between 0.5 and 0.1 and we cannot reject the null hypothesis. The hypothesis that dancing is caused by the homozygous recessive allele of one gene is therefore a good fit with the data. Null hypothesis #2: Dancing is caused by being homozygous for the recessive alleles of two genes (*aa bb*), so 1/16 of the F_2 mice should be dancers.

Genotypes	Observed #	Expected #	X^2 square equation	Sum of X^2
nondancers	42	47	$(42-47)^2/47$	0.53
dancers	8	3	$(8-3)^2/3$	8.33
				8.86

With one degree of freedom, the p value is < 0.005, so the null hypothesis that two genes control the dancer phenotype is not a good fit; in fact, the hypothesis can be rejected by the criteria employed by most geneticists. **The one gene hypothesis is a better fit with the data.**

5-5. The parents are from true breeding stocks. Diagram the cross:

raspberry eye color ♂ × sable body color ♀ → F_1 wild type eye and body color ♀ × sable body color ♂ (if no mention is made of the eye color, then it is assumed to be wild type) → F_2 216 wild type ♀ : 223 sable ♀ : 191 sable ♂ : 188 raspberry ♂ : 23 wild type ♂ : 27 raspberry sable ♂ 3EQ #1 - the phenotypes seem to be controlled by 2 genes, one for eye color and the other for body color. 3EQ #2 - the F_1 female progeny show that the wild type allele is dominant for body color (s^+ $> s$) and the wild type allele is also dominant for eye color ($r^+ > r$). Question #3 - sable body color is seen in the F_1 males but not the F_1 females, so the s gene is X-linked. In the F_2 generation the raspberry eye color is seen in the males but not in the females, so r is also an X-linked gene. We can now assign genotypes to the true-breeding parents in this cross:

$r s^+ / Y$ (raspberry ♂) × $r^+ s / r^+ s$ ♀ → F_1 $r s^+ / r^+ s$ ♀ (wild type) × $r^+ s / Y$ (sable ♂)

→ [the heterozygous F_1 female can make the following gametes: (parentals) $r s^+$, $r^+ s$ and (recombinants) $r s$, $r^+ s^+$; the F_1 male can make Y and $r^+ s$ gametes] → F_2 will be $r s^+ / r^+ s$ (wild type females), $r^+ s / r^+ s$ (sable females), $r s / r^+ s$ (sable females), $r^+ s^+ / r^+ s$ (wild type females), $r^+ s$ / Y (sable males), $r s^+$ / Y (raspberry males), $r s$ / Y (raspberry sable males), $r^+ s^+$ / Y (wild type males)

The F_1 female is heterozygous for both genes and will therefore make parental and recombinant gametes. The F_1 male is not a true test cross parent, because he does not carry the recessive alleles for both of the X-linked genes. However, this sort of cross <u>can</u> be used for mapping, because F_2 sons receive only the Y chromosome from the F_1 male and are hemizygous for the X chromosome from the F_1 female. Thus the phenotypes in the F_2 males represent the array of parental and recombinant gametes generated by the F_1 female as well as the frequencies of these gametes. Using the F_2 males, the **rf between sable and raspberry = 23 (wild type males) + 27 (raspberry sable males) / 429 (total males) = $0.117 \times 100 = 11.7$ cM**.

5-6. This question asks you to weigh the advantages and disadvantages of using Chi square analysis to test for linkage between two genes using two different assumptions. In assumption #1, as seen in Figure 5.6, the expectation of independent assortment is that the frequency of parentals is the same as the frequency of recombinants. In assumption #2, the expectation of independent assortment is the parental and recombinant progeny should be present in a ratio of 1:1:1:1. **Notice that the null hypothesis is the same in both cases: that the genes are assorting independently.**

Assumption #1, 50 progeny:

Genotypes	Observed #	Expected #	X^2 square equation	Sum of X^2
parental	31	25	$(31 - 25)^2/25$	1.44
recombinant	19	25	$(9 - 25)^2/25$	<u>1.44</u>
				2.88

1 dof (degree of freedom), $0.05 < p < 0.1$ and the null hypothesis cannot be rejected - the data is compatible with the hypothesis that the genes are assorting independently.

Assumption #2, 50 progeny:

Genotypes	Observed #	Expected #	X^2 square equation	Sum of X^2
AB	17	12.5	$(17 - 12.5)2/12.5$	1.62
ab	14	12.5	$(14 - 12.5)2/12.5$	0.18
Ab	8	12.5	$(8 - 12.5)/12.5$	1.62
aB	11	12.5	$(11 - 12.5)2/12.5$	<u>0.18</u>
				3.6

3 dof, $0.10 < p < 0.5$ and the null hypothesis cannot be rejected - the data is compatible with the hypothesis that the genes are assorting independently.

Assumption #1, 100 progeny:

Genotypes	Observed #	Expected #	X^2 square equation	Sum of X^2
parental	62	50	(62 - 50)2/50	2.88
recombinant	38	50	(38 - 50)2/50	2.88
				5.76

1 dof, $0.01 < p < 0.05$ and the null hypothesis is rejected - the genes are not assorting independently, so they must be linked.

Assumption #2, 100 progeny:

Genotypes	Observed #	Expected #	X^2 square equation	Sum of X^2
AB	34	25	(34 - 25)2/25	3.24
ab	28	25	(28 - 25)2/25	0.36
Ab	16	25	(16 - 25)2/25	3.24
aB	22	25	(22 - 25)2/25	0.36
				7.2

3 dof, $0.05 < p < 0.1$ and the null hypothesis cannot be rejected - the data is compatible with the hypothesis that the genes are assorting independently.

In the first experiment with only 50 progeny, $p < 0.05$ using both the 2 class and 4 class assumptions, so neither assumption supports the idea that the genes are linked (although the test with 2 classes is closer to significance). When more progeny are scored, the null hypothesis (that the genes are assorting independently) is rejected when you look at the data as 2 classes but not as 4 classes. This suggests that **using 2 classes is a more sensitive test for linkage than using 4 classes**. There is actually a subtle difference in the null hypotheses: in the 2 class situation the null hypothesis is linkage; in the 4 class situation it is 1:1:1:1, which not only means linkage but also equal viability of the 4 classes. You could imagine a situation in which certain classes are sub-viable, for example any class with the *a* allele. In such a case, you might see linkage with the 2 class test, but you would miss the even more important point that one allele causes reduced viability. **This ability to see the relative viability of the alleles is an advantage to the 4 class method.**

5-7.

a.   *AA BB* × *aa bb* → F$_1$ *A B / a b* (*A B* on one homolog and *a b* on the other homolog). The F1 progeny will produce gametes of the parental types *A B* and *a b;* and of the recombinant types *A b* and *a B*. Because the genes are 40 cM apart, **the recombinants will make up 40% of the gametes (20% *Ab* and 20% *aB*). The parental gametes make up the remaining 60% of the gametes, 30% *A B* and 30% *a b*.** Set up a Punnett square to calculate the frequency of the 4 phenotypes in the F$_2$ progeny. In the Punnett square the phenotypes are shown in parentheses (*A- bb*). **The F$_2$ phenotypic ratio is: 0.59 *A- B-* : 0.16 *A- bb* : 0.16 *aa B-* : 0.09 *aa bb*.**

	0.3 *A B*	0.3 *a b*	0.2 *A b*	0.2 *a B*
0.3 *A B*	*A B / A B* (0.09 *A- B-*)	*A B / a b* (0.09 *A- B-*)	*A B / A b* (0.06 *A- B-*)	*A B / a B* (0.06 *A- B-*)
0.3 *a b*	*a b / A B* (0.09 *A- B-*)	*a b / a b* (0.09 *aa bb*)	*a b / A b* (0.06 *A- bb*)	*a b / a B* (0.06 *aa B-*)
0.2 *A b*	*A b / A B* (0.06 *A- B-*)	*A b / a b* (0.06 *A- bb*)	*A b / A b* (0.04 *A- bb*)	*A b / a B* (0.04 *A- B-*)
0.2 *a B*	*a B / A B* (0.06 *A- B-*)	*a B / a b* (0.06 *aa B-*)	*a B / A b* (0.04 *A- B-*)	*a B / a B* (0.04 *aa B-*)

b.   If the original cross was *AA bb* × *aa BB*, the allele combinations in the F_1 would be *A b / a B*. Parental gametes in this case are 30% *A b* and 30% *a B* and the recombinant gametes are 20% *A B* and 20% *a b*. Set up a Punnett square, as in part a. **The F_2 phenotypic ratio is: 0.54 *A- B-* : 0.21 *A- bb* : 0.21 *aa B-* : 0.04 *aa bb*.**

5-8.

a.   Diagram the cross:

CC DD × *cc dd* → F_1 *C D / c d* × *cc dd* → 903 *Cc Dd*, 897 *cc dd*, 98 *Cc dd*, 102 *cc Dd*. Because the gamete from the homozygous recessive parent is always *c d*, we can ignore one *c* and one *d* allele (the *c d* homolog) in each class of the F_2 progeny. The remaining homolog in each class of F_2 is the one contributed by the doubly heterozygous F_1, the parent of interest when considering recombination. In the F_2 the two classes of individuals with the greatest numbers represent parental gametes (*C D* or *c d* from the heterozygous F_1 parent combining with the *c d* gamete from the homozygous recessive parent). The other two types of progeny result from a recombinant gametes (*C d* or *c D* combining with the *c d* gamete from the homozygous recessive parent). The number of recombinants divided by the total number of offspring x 100 gives the map distance: $(98 + 102)/(903 + 897 + 98 + 102) = 200/2000 = 0.01 \times 100 = 10\%$ rf or **10 map units (mu) or 10 cM**.

b.   *CC dd* × *cc DD* → F_1 *C d / c D* × *c d / c d* → as determined in part a, *c* and *d* are 10 cM apart. Thus the gametes produced by the heterozygous F_1 will be 45% *C d*, 45% *c D*, 5% *C D*, 5% *c d*. After fertilization with *c d* gametes, there would be **45% *Cc dd*, 45% *cc Dd*, 5% *Cc Dd*, 5% *cc dd***. Because this is a test cross, the gametes from the doubly heterozygous F1 parent determine the phenotypes of the progeny.

5-9. To determine the probability that a child will have a particular genotype, look at the gametes that can be produced by the parents. In this example, *A* and *B* are 20 mu apart, so in a doubly heterozygous individual 20% of the gametes will be recombinant and the remaining 80% will be parental. The *aa bb* homozygous man can only produce *a b* gametes. The doubly heterozygous woman, with a genotype of *A B / a b*, can produce 40% *A B*, 40% *a b*, 10% *A b* and 10% *a B* gametes. (Total of recombinant classes = 20%.) **The probability that a child receives the *A b* gamete from the female (and is therefore *A b / a b*) = 10%.**

5-10.

a. Designate the alleles: *H* = Huntington allele, *h* = normal allele; *B* = brachydactyly, *b* = normal fingers. **John's father is *bb Hh*; his mother is *Bb hh*.**

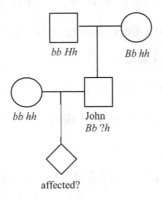

b. We know the John is *Bb* because he has brachydactyly. **His complete genotype could be *Bb Hh* or *Bb hh*.** The probability that John's genotype is *Hh* = 1/2 (chance that he inherited the *H* allele from his father) × 1/3 (probability of >50 years of age for onset of symptoms) = 1/6.

c. The probability that the child will have and express brachydactyly = 1/2 (probability of inheriting *B* from John) × 9/10 (probability of expressing the phenotype) = 0.45. The probability that the child will express Huntington's disease by age 50 = 1/6 (probability that John is *Hh*) × 1/2 (probability that the child will inherit the *H* allele) × 2/3 (probability the child will show symptoms of Huntington's by age 50) = 0.056. **If brachydactyly and Huntington's assort independently the probability that the child will express both phenotypes by age 50 = 0.45 (probability of expressing brachydactyly) × 0.056 (probability of expressing Huntington's by age 50) = 0.025.**

d. If the two loci are linked, the alleles on each of John's homologs will be either *B h / b H* or *B h / b h*. If it is the former (and John has Huntington's), the *B H* gamete could only be produced by recombination. The probability of that specific recombinant gamete = 10% (*b h* constitutes the other 10% of the recombinant gametes). As shown in part a, the probability that John's genotype is $B\ h\ /\ b\ H$ = 1/6. **The probability that John's child will inherit both the Huntington and brachydactyly alleles = 1/6 (probability that John is $B\ h\ /\ b\ H$) × 1/10 (probability of child inheriting $B\ H$ recombinant gamete from John) × 9/10 (probability of expressing brachydactyly) × 2/3 (probability of expressing Huntington's by age 50) = 0.01.**

5-11. Designate the alleles of the genes: A = normal pigmentation and a = albino allele; $Hb\beta^A$ = normal globin and $Hb\beta^S$ = sickle allele.

a. Because both traits are rare in the population, we assume that the parents are homozygous for the wild type allele of the gene dictating their normal traits (that is, they are not carriers). Diagram the cross: $a\ Hb\beta^A\ /\ a\ Hb\beta^A$ (father) $\times$ $A\ Hb\beta^S\ /\ A\ Hb\beta^S$ (mother) $\rightarrow$ $a\ Hb\beta^A\ /\ A\ Hb\beta^S$ (son). Given that the genes are separated by 1 map unit, parental gametes = 99% and recombinant gametes = 1% of the gametes. **The son's gametes will consist of: 49.5% $a\ Hb\beta^A$, 49.5% $A\ Hb\beta^S$, 0.5% $a\ Hb\beta^S$ and 0.5% $A\ Hb\beta^A$.**

b. In this family, the cross is $A\ Hb\beta^A\ /\ A\ Hb\beta^A$ (father) $\times$ $a\ Hb\beta^S\ /\ a\ Hb\beta^S$ (mother) $\rightarrow$ $a\ Hb\beta^S\ /\ A\ Hb\beta^A$ (daughter). **The daughter's gametes will be: 49.5% $a\ Hb\beta^S$, 49.5% $A\ Hb\beta^A$, 0.5% $a\ Hb\beta^A$ and 0.5% $A\ Hb\beta^S$.**

c. The cross is: $a\ Hb\beta^A\ /\ A\ Hb\beta^S$ (son) x $a\ Hb\beta^S\ /\ A\ Hb\beta^A$ (daughter). **The probability of an $a\ Hb\beta^S\ /\ a\ Hb\beta^S$ child (sickle cell and anemic) = 0.005 (probability of $a\ Hb\beta^S$ from son) × 0.495 (probability of $a\ Hb\beta^S$ from daughter) = 0.0025.**

5-12. Diagram the cross:

blue smooth $\times$ yellow wrinkled $\rightarrow$ F$_1$ 1447 blue smooth, 169 blue wrinkled, 186 yellow smooth, 1510 yellow wrinkled.

a. To determine if genes are linked, first predict the results of the cross if the genes are unlinked. In this case, a plant with blue, smooth kernels (A- W-) is crossed to a plant with yellow, wrinkled kernels ($aa\ ww$). Since there are four classes of progeny in the F$_1$, the parent with blue, smooth kernels must be heterozygous for both genes ($Aa\ Ww$). From this cross, **we would predict equal numbers of all four phenotypes in the F$_1$ if the genes were unlinked. Since the numbers are very skewed, with the smaller classes representing recombinant offspring, the genes are linked. rf = (169 + 186) / (1447 + 169 + 186 + 1510) = 355/3312 = 10.7% = 10.7 cM.**

b. The genotype of the blue smooth parent was *Aa Ww*. The arrangement of alleles in the parent is determined by looking at the phenotypes of the largest classes of progeny (the parental reciprocal pair). Since blue, smooth and yellow wrinkled are found in the highest proportion, *A W* must be on one homolog and *a w* on the other = *A W / a w*.

c. The genotype of the blue, wrinkled F$_1$ is *A w /a w*. This genotype can produce *A w* or *a w* gametes in equal proportions. Recombination can NOT be detected here, as this genotype is only heterozygous for one gene. Notice that recombination between these homologs yields the same two combinations of alleles (*A w* and *a w*) as the parental, so each type of gamete is expected 50% of the time. The yellow smooth F$_1$ plant has a genotype of *a W / a w*. Again since recombination cannot be detected, the frequency of each type is 50%. Thus the cross is *A w / a w* (blue wrinkled) × *a W / a w* (yellow smooth). **Four types of offspring are expected in equal proportions: 1/4 *A w / a W* (blue smooth) : 1/4 *A w / a w* (blue wrinkled) : 1/4 *a w / a W* (yellow smooth) : 1/4 *a w / a w* (yellow wrinkled).**

5-13. Diagram the cross:

CC bb (brown rabbits) × *cc BB* (albinos) → F$_1$ *Cc Bb* x *cc bb* → 34 black : 66 brown : 100 albino.

a. If the genes are unlinked, the F$_1$ will produce *C B*, *c b*, *C b* and *c B* gametes in equal proportions. A mating to animals that produce only *c b* gametes will produce four genotypic classes of offspring: 1/4 *Cc Bb* (black) : 1/4 *cc Bb* (albino) : 1/4 *Cc bb* (brown) : 1/4 *cc BB* (albino). The ratio would be **1/4 black : 1/2 albino : 1/4 brown**.

b. If *c* and *b* are linked, then the genotype of the F$_1$ class is *C b / c B* and you would expect the parental type *C b* and *c B* gametes to predominate; *c b* and *C B* are the recombinant gametes and are therefore present at lower levels. The parental gametes are represented in the F$_2$ by the *Cc bb* (brown) and *cc Bb* (albino) classes. Since we cannot distinguish between the albinos resulting from fertilization of recombinant gametes (*c b*), and those resulting from parental gametes (*c B*), we have to use the proportion of the *C B* recombinant class and assume that the other class of recombinants (*c b*) is present in equal frequency. Since the crossing-over is a reciprocal exchange, this assumption is reasonable. There were 34 black progeny; assuming that 34 of the 100 albino progeny were the result of recombinant gametes, **the genes are (34 + 34) recombinant / 200 total progeny = 34% rf = 34 cM apart.**

5-14. Diagram the cross:

wild type ♀ × reduced cinnabar ♂ → F$_1$ ♀ x F$_1$ ♂ → 292 wild type, 9 cinnabar, 7 reduced, 92 reduced cinnabar.

Two genes are involved in this cross, but the frequencies of offspring do not look like frequencies expected for two independently assorting genes (9:3:3:1). The genes must be linked. Designate the alleles: cn^+ = wild type, cn = cinnabar; rd^+ = wild type, rd = reduced. The cross is:

$cn^+ rd^+ / cn^+ rd^+$ ♀ × $cn rd / cn rd$ ♂ → F$_1$ $cn^+ rd^+ / cn rd$.

Recombination occurs in *Drosophila* females but <u>not</u> in males. Thus males can only produce the parental c$n^+ rd^+$ or *cn rd* gametes. The females produce both the parental gametes and the recombinant gametes c$n^+ rd$ and *cn rd$^+$*.

female gametes	male gamete $cn^+ rd^+$	male gamete $cn\ rd$
$cn^+ rd^+$ (parental)	$cn^+ rd^+ / cn^+ rd^+$ (wild type)	$cn^+ rd^+ / cn\ rd$ (wild type)
$cn\ rd$ (parental)	$cn\ rd / cn^+ rd^+$ (wild type)	$cn\ rd / cn\ rd$ (cinnabar reduced)
$cn^+ rd$ (recombinant)	$cn^+ rd / cn^+ rd^+$ (wild type)	$cn^+ rd / cn\ rd$ (reduced)
$cn\ rd^+$ (recombinant)	$cn\ rd^+ / cn^+ rd^+$ (wild type)	$cn\ rd^+ / cn\ rd$ (cinnabar)

The reduced flies and the cinnabar flies are recombinant classes. However, there should be an equal number of recombinant types that have a wild-type phenotype because they got a recombinant $cn^+ rd^+$ gamete from the male parent. If we assume that these recombinants are present in the same proportions, then rf = 2 × (7+9)/400 = 8% recombination. **The genes are separated by 8 cM.**

5-15. v^1, v^2, and v^3 are codominant alleles of the DNA variant marker locus, while *D* and *d* are alleles of the disease gene. The marker and the disease locus are linked. Diagram the cross:

$v^1 D / v^2 d$ (father) × $v^3 d / v^3 d$ (mother) → v^2 ? / $v^3 d$.

The fetus <u>must</u> get a $v^3 d$ homolog from the mother. The fetus also received the v^2 allele of the marker. The father could have given a $v^2 d$ (non-recombinant) or a $v^2 D$ (recombinant) gamete.

a. If the *D* locus and the v^1 allele of the marker are 0 mu apart (there is no recombination between them; they appear to be the same gene), **the probability that the child, who received the v^2 allele of the marker, has the *D* allele = 0.**

b. If the distance between the disease locus and marker is 1 mu, 1% of the father's gametes are recombinant between *D* and the marker locus. Half of the recombinant gametes (0.5%) will be v^2 *D;* the other half will be v^1 *d.* Because we are only considering cases where the child has the v^2 marker, **the probability that the child inherited *D* = 0.05 (probability of inheriting v^2 *D* / 0.5 (probability of inheriting v^2) = 0.01 = 1%.**

c. **5%.**

d. **10%.**

e. **50%.**

5-16.

a. Recombination does NOT occur in male *Drosophila.* Therefore, in the cross *A b / a B* ♀ x *A b / a B* ♂ → F₁ the females will make 4 types of gametes (parental *A b* and *a B;* recombinant *A B* and *a b*). The frequencies of the 2 parental gametes will be equal to each other, and the frequency of the *A B* recombinant will be equal to the frequency of its reciprocal recombinant, *a b.*

female gametes	male gamete *A b*	male gamete *a B*
A b [parental]	*A b / A b* (*A- bb*)	*A b / a B* (*A- B-*)
a B [parental]	*a B / A b* (*A- B-*)	*a B / a B* (*aa B-*)
A B [recombinant]	*A B / A b* (*A- B-*)	*A B / a B* (*A- B-*)
a b [recombinant]	*a b / A b* (*A- bb*)	*a b / a B* (*aa B-*)

There is a ratio of 1/4 *A- bb* : 1/2 *A- B-* : 1/4 *aa B-* for the progeny receiving the parental gametes AND the same ratio among the progeny receiving the recombinant gametes. Thus, **the overall phenotypic dihybrid ratio will <u>always</u> be 1/4 *A- bb* : 1/2 *A- B-* : 1/4 *aa B-*, independent of the recombination frequency between the A and B genes.**

This will <u>not</u> be true of the cross *A B* / *a b* ♀ x *A B* / *a b* ♂. The male will make the parental gametes, *A B* and *a b*, while the female will make 4 types of gametes: parental *A B* and *a b*; recombinant *A b* and *a B*. In this case, as in <u>problem 5-7</u>, half of the progeny will look *A- B-* because they received the *A B* gamete from the male parent irrespective of the gamete from the female parent. When the male parent donates the *a b* gamete, then it is the gamete from the female parent that determines the phenotype of the offspring. The progeny that are *A- bb* and *aa B-* have received a recombinant gamete from the female parent (and the *a b* gamete from the male). These classes of progeny can then be used **to estimate the recombination frequency between the *A* and *B* genes: rf = 2(# of *A- bb* + # of *aa B-*)/total progeny**.

b. In mice, recombination occurs in both females and males. Therefore, in the cross *A b* / *a B* ♀ x *A b* / *a B* ♂ both sexes will make the same array of gametes (parental *A b* and *a B;* recombinant *A B* and *a b*). In this case, the only phenotype of progeny with a singular genotype will be *aa bb*. In the cross under consideration here, the *aa bb* phenotype can only arise if both parents donate the *a b* recombinant gamete. Of course, the other recombinant gamete, *A B*, occurs with equal frequency. The probability of the *aa bb* genotype = (probability of an *a b* recombinant gamete)^2. If you know the frequency of the *aa bb* phenotypic class in the progeny, **recombination frequency (frequency of recombinant products) between the *A* and *B* genes = 2($\sqrt{}$#*aa bb*/# total progeny)**.

In the case of the *A B* / *a b* ♀ × *A B* / *a b* ♂ cross in mice, both sexes are making the same parental gametes (*A B* and *a b*) and recombinant gametes (*A b* and *a B*) gametes. Again, the only phenotype of progeny with a singular genotype will be *aa bb*. In this example, this phenotype is the result of the fusion of the *a b* parental gamete from each parent. The frequency of the *aa bb* phenotype = (the frequency of the *a b* gamete) x (the frequency of the *a b* gamete). If you know the frequency of the *aa bb* phenotypic class in the progeny, **the frequency of non-recombinant products between the *A* and *B* genes = 2($\sqrt{}$#*aa bb*)/# total progeny. Recombination frequency = 1 - frequency of non-recombinant products between the *A* and *B* genes.**

5-17. Diagram the cross:

M C S / *M C S* (Virginia strain) × *m c s* / *m c s* (Carolina strain) → F$_1$ *M C S* / *m c s* × *m c s* / *m c s* → ?

Assume no interference, and remember the map of these 3 genes:

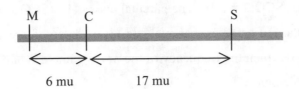

You are being asked to calculate the proportion of the test cross progeny that will have the Virginia parental phenotype. In Chapter 3 we could answer this question by calculating the monohybrid ratios each pair of alleles: $M:m$, $C:c$ and $S:s$ in the heterozygous parent. The test cross parent can only provide the recessive alleles for each gene, so the probabilities of the various phenotypes can be determined by applying the product rule. Unfortunately, this method of arriving at the probability of the progeny phenotypes only works when the genes under discussion are assorting independently! When the 3 genes are linked, as in this problem, to calculate the frequency of a parental class we must calculate first calculate the frequencies of ALL of the phenotypes expected in the test cross progeny!

A parent that is heterozygous for 3 genes will give 8 classes of gametes. In a test cross, the gamete from the heterozygous parent determines the phenotype of the progeny. When the 3 genes are genetically linked, these 8 classes will be found as 4 reciprocal pairs. In other words, each meiotic event in the heterozygous parent must give 2 reciprocal products that occur in equal frequency. For instance, a meiosis with no recombination will produce the parental gametes, $M\,C\,S$ and $m\,c\,s$ in equal frequency. This particular reciprocal pair will also be the most likely event and so the most frequent pair of products. The least probable meiotic event is a double crossover which is a recombination event in the region between M and C (region 1) and simultaneously in the region between C and S (region 2). The remaining gametes are produced by a single crossover in region 1 (the reciprocal pair known as single crossovers in region 1) and a single crossover in region 2 (the reciprocal pair known as single crossovers in region 2).

The numbers shown on the map of this region of the chromosome represent the recombination frequencies in the gene-gene intervals. There are 6 mu between the M and C genes, so 6% (rf = 0.06) of the progeny of this cross will have had a recombination event in region 1. This recombination frequency includes all detectable recombination events between these 2 genes - both SCO in region 1 AND DCO. Likewise, 17% of all the progeny will be the result of a recombination event in region 2. The DCO class is the result of a simultaneous crossover in region 1 and region 2. Recombination in two separate regions of the chromosome should be independent of each other, so we can apply the product rule to calculate the expected frequency of DCOs. Thus frequency of DCO = (0.06) × (0.17) = 0.01. Remember that the recombination frequency between M and C (region 1) includes both SCO 1 and DCO. Thus, 0.06 = SCO 1 + 0.01; solve for SCO 1 = 0.06 - 0.01 = 0.05. The same calculation for region 2 shows that SCO 2 = 0.16. The parental class = 1 - (SCO 1 + SCO 2 + DCO). Also remember that each class of gametes (parental, etc.) is made up of a reciprocal pair. If the frequency of the DCO class is 0.01, then the frequency of the $M c S$ gamete is half of that = 0.005. This is summarized in the table below.

Classes of gametes	Genotype	Frequency of reciprocal pair	Numbers	Frequency of each class
Parental (P)	$M C S$ $m c s$	1 - all else	1 - (0.05 + 0.16 + 0.01) = 0.78	0.39 0.39
SCO 1	$M c s$ $m C S$	rf in region 1 = SCO 1 + DCO	SCO 1 = 0.06 - 0.01 = 0.05	0.025 0.025
SCO 2	$M C s$ $m c S$	rf in region 2 = SCO 2 + DCO	SCO 2 = 0.17 - 0.01 = 0.16	0.8 0.8
DCO	$M c S$ $m C s$	(rf in region 1) x (rf in region 2)	(0.06) x (0.17) = 0.01	0.005 0.005

a. **The proportion of backcross progeny resembling Virginia (parental, $M C S$) = 0.39.**

b. **Progeny resembling $m c s$ (P) = 0.39.**

c. **Progeny with $M c S$ (DCO) = 0.005.**

d. **Progeny with $M C s$ (SCO 2) = 0.8.**

5-18. In foxgloves wild type flower color is red and the mutant color is white; the mutation peloria causes the flowers at the apex of the stem to be very large; normal foxgloves are very tall, and dwarf affects the plant height. When you describe the phenotype of an individual you usually only refer to the non-wild type traits. Diagram the cross:

white flowered × dwarf peloria → F$_1$ white flowered x dwarf peloria → 172 dwarf peloria, 162 white, 56 dwarf peloria white, 48 wild type, 51 dwarf white, 43 peloria, 6 dwarf, 5 peloria white.

a. Because there is only 1 phenotype of F$_1$ plant, the parents must have been homozygous for all three genes. The phenotype of the F$_1$ heterozygote indicates the **dominant alleles: white flowers , tall stems, and normal-sized flowers** (3EQ #2).

b. Designate the alleles for the 3 genes: W = white, w = red; P = normal-sized flowers, p = peloria; T = tall, t = dwarf. The cross is: ***WW PP TT* (white flowered) × *ww dd pp* (dwarf peloria).**

c. Note that all 3 of these genes are genetically linked. There are only 2 classes of parental progeny as defined both phenotypically and numerically. If one (or more) of the genes instead assorts independently of the others, then in a test cross you must see 4 equally frequent group of parental classes (a 1:1:1:1 ratio of the 4 parental types). In order to draw a map of these genes, organize the test cross data, figure out which of the 3 genes is in the middle and calculate the recombination frequencies in regions 1 and 2. By definition, the parentals are the class with the same phenotype as the original parents of the cross, and the DCO class is the least frequent reciprocal pair of progeny. At this point, arbitrarily one of the remaining 2 reciprocal pairs of progeny is SCO 1 and the last remaining pair is SCO 2.

Classes of gametes	Genotype	Numbers
Parental (P)	$W\,P\,T$	162
	$w\,p\,t$	172
SCO 1	$W\,p\,t$	56
	$w\,P\,T$	48
SCO 2	$W\,P\,t$	51
	$w\,p\,T$	43
DCO	$w\,P\,t$	6
	$W\,p\,T$	5

Compare the DCO class with the parental class. Remember that a DCO comes from a meiosis with a simultaneous crossover in regions 1 and 2. As a result, the allele of the gene in the middle switches with respect to the alleles of the genes on the ends. In this data set, take one of the DCO class (for e.g. $W\,p\,T$) and compare it to the most similar parental gamete ($W\,P\,T$). Two alleles out of three in common; the one that differs (p in this case) is the gene in the middle. Thus, the order is $W\,P\,T$ (or $T\,P\,W$).

Once you know the order of the 3 genes, you must calculate the 2 shortest distances - from the end to the center (W to P) and from the center to the other end (P to T). The total number of progeny in this test cross = 543. rf W-P = (56 + 48 + 6 + 5)/543 = 115/543 = 21.2 cM; rf P-T = (51 + 43 + 6 + 5)/543 = 19.3 cM.

The map is:

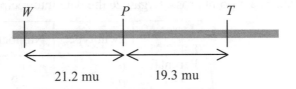

d. Interference (I) = 1 - coefficient of coincidence (coc)

coc = frequency of observed DCO / frequency of expected DCO

As discussed in problem 5-17, the expected percentage of double crossovers is the product of the frequency of recombination in each interval: $(0.193) \times (0.212) = 0.0409$. The observed DCO frequency = 11/543 = 0.0203; **coc = 0.0203/0.0409 = 0.496; I = 1 - 0.496 = 0.504**.

5-19.

a. See problem 5-17 for a detailed explanation of the methodology. Diagram the cross:

$a^+ b^+ c^+ / a b c$ ♀ x $a b c / a b c$ ♂ → ??. Recombination occurs in *Drosophila* females, so the female parent will make the following classes of gametes. Because this is a test cross, these gametes will determine the phenotypes in the progeny.

Classes of gametes	Genotype	Frequency of reciprocal pair	Numbers	Frequency of each class	# of progeny freq x 1,000
Parental (P)	$a^+ b^+ c^+$ $a b c$	1 - all else	1 - (0.18 + 0.08 + 0.02) = 0.72	0.36 0.36	360 360
SCO 1	$a^+ b c$ $a b^+ c^+$	rf in region 1 = SCO 1 + DCO	SCO 1 = 0.2 - 0.02 = 0.18	0.09 0.09	90 90
SCO 2	$a^+ b^+ c$ $a b c^+$	rf in region 2 = SCO 2 + DCO	SCO 2 = 0.1 - 0.02 = 0.08	0.04 0.04	40 40
DCO	$a^+ b c^+$ $a b^+ c$	(rf in region 1) x (rf in region 2)	(0.2) x (0.1) = 0.02	0.01 0.01	10 10

b. Diagram the cross: $a^+ b^+ c^+ / a b c$ ♂ x $a b c / a b c$ ♀ → ??. Here, the heterozygous parent is the male. Recombination DOES NOT OCCUR in male *Drosophila*. Thus, the heterozygous parent will only give 2 types of gametes, the parental types. If you score 1,000 progeny of this cross, you will find **500 $a^+ b^+ c^+$ and 500 $a b c$ progeny**.

5-20.

a. This data is presented as phenotypes of individual spores. Because there is data for 3 genes, this may be analyzed as a 3 point cross. Organize the data into reciprocal pairs of spores.

Classes of gametes	Genotype	Numbers
Parental (P)	$\alpha + +$ $a f g$	31 29
SCO 1	$a + g$ $\alpha f +$	6 6
SCO 2	$\alpha + g$ $a f +$	13 14
DCO	$a + +$ $\alpha f g$	1 1

The $\alpha f g$ DCO spore type is most similar to the $a f g$ parental spore type. Thus, the mating type (a/α) is the gene in the middle. The distances are: f - a/α = (6 + 6 + 1 + 1)/101 = 13.9 cM and a/α to g = (13 + 14 + 1 + 1)/101 = 28.7 cM.

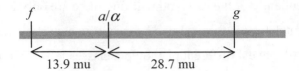

b. This problem says you have an ascus with an $\alpha f g$ spore. This spore is the result of a double crossover (see part a). The reciprocal product would be the $a + +$ spore, but this is not seen in the ascus. Remember that there are 3 different types of double crossovers - 2 strand DCOs, 3 strand DCOs and 4 strand DCOs. Each of these types of DCOs gives a different array of spores in the resulting ascus (**see Figure 5.22**). Draw a meiotic figure of this chromosome and try some different types of DCOs. **A 3 strand DCO gives the desired result**.

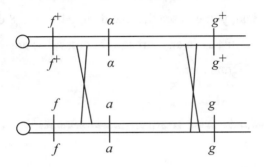

5-21.

a. wild type ♀ × scute echinus crossveinless black ♂ → 16 classes of data! Among the 16 classes there are wild type and sc, ec, cv and bl. This tells you that the parental female was heterozygous for all 4 traits. The fact the parental female is wild type also tells you that the wild-type alleles of all 4 genes are dominant. Remember that if you do a test cross with a female that is heterozygous for **3** linked genes, the data shows a very specific pattern. Because each type of meiosis (no recombination, DCO, etc.) gives a pair of gametes, you will see 8 classes of data that will occur in 4 pairs, both genetically and numerically. Thus, if you begin a cross with a female that is heterozygous for 4 linked genes, you should see 16 classes of progeny in a pattern of 8 genetic and numeric pairs. Although we have 16 classes of data, they do NOT occur in numeric pairs - instead we see numeric groups of 4. This is an earmark of independent assortment, because a 1:1:1:1 ratio is seen when 2 genes assort independently.

Which gene is assorting independently relative to the other 3 genes? To answer this question, list the genotypes of the gametes that came from the heterozygous parent in the largest group of 4. This group should include the parental classes for the 3 linked genes; there are 4 genotypes here to account for the independent assortment of the unlinked gene. Then choose one of the 4 genes, and remove the allele of that gene from all 4 groups. When you do this with the gene that is assorting independently of the rest, you will find there are only 2 reciprocal classes of data left, which are the parental classes for the linked genes. If you choose one of the linked genes to remove, you will still have 4 different phenotypic classes left. Try removing the *b* gene first. You see that when the *b* allele is removed there are only 2 classes left, *s e c* and + + +; when this analysis is repeated removing the allele of the *s* gene there are still 4 different phenotypes left.

Genotype	Numbers	remove *b*	remove *s*
b s e c	653	*s e c*	*b e c*
+ s e c	670	*s e c*	*+ e c*
+ + + +	675	*+ + +*	*+ + +*
b + + +	655	*+ + +*	*b + +*

The *b* gene is therefore assorting independently of the other 3 genes. When the *b* gene is removed from all 16 classes of data, you can see that the data reduces to 8 classes that form 4 genetic and numeric reciprocal pairs, just as in any three-point cross (see answer to part b below). Thus, the genotype of the parental female is:

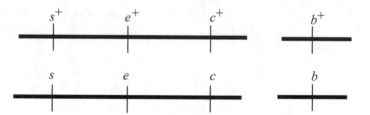

b.   Write the classes out as reciprocal pairs.

Classes of gametes	Genotype	Numbers
Parental (P)	+ + +	1323
	s e c	1330
SCO 1	*s + +*	144
	+ e c	147
SCO 2	*s e +*	171
	+ + c	169
DCO	*s + c*	2
	+ e +	2

Compare the DCO *s + c* to the parental *s e c*; this shows that *e* is in the middle. Calculate rf *s* - *e* = (144 + 147 + 2 + 2)/3288 = 9.0 cM; rf *e* - *c* = (171 + 169 + 2 + 2)/3288 = 10.5 cM.

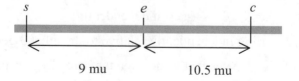

c. Interference = 1 - coefficient of coincidence

coc = observed DCO frequency / expected DCO frequency = (4/3288) / (0.09 × 0.105) = 0.001 / 0.009 = 0.11

I = 1– 0.11 = 0.89

5-22. Diagram the cross:

pink petals, black anthers, long stems × pink petals, black anthers, long stems →

a. The cross is pink × pink → (78 + 6 + 44 + 15) red : (39 + 13 + 204 + 68) pink : (2 + 2 + 117 + 39) white = 143 red : 324 pink : 160 white. The appearance of two new phenotypes (red and white) suggest that flower color shows incomplete dominance. This is confirmed by the 1:2:1 monohybrid ratio seen in the self-cross progeny. **The pink flowered plants are *Pp*, red are *PP* and white are *pp*.**

b. The expected ratio of red: pink : white would be 1:2:1. Calculating for the 650 plants, this equals **162.5 red, 325 pink, and 162.5 white**.

c. The monohybrid ratio of the black and with tan anthers =. (78+26+39+13+5+2) tan : (44 + 15 + 204 + 68 + 117 + 39) black = 163 tan : 487 black = ~ **1 tan : 3 black. Therefore black is dominant to tan**. The monohybrid ratio for the stem length = 487 long stems : 163 short stems. = ~ **3 long :1 short, so long is dominant to short**.

d. Designate the alleles: *P* = red, *Pp* = pink, *p* = white; *B* =black, *b* =tan, *L* =long, *l* =short. Because all 3 monohybrid phenotypic ratios are characteristic of heterozygous crosses, **the genotype of the original plant is *Pp Bb Ll***.

e. If the stem length and anther color genes assort independently, the 9:3:3:1 phenotypic ratio should be seen in the progeny. Totaling all the progeny in each of the classes: 365 long black : 122 short black : 122 long tan : 41 short tan. The observed dihybrid ratio is close to a 9:3:3:1 ratio, so **the genes for anther color and flower color are unlinked**.

   The expected monohybrid ratio for flower color is 1 red : 2 pink :1 white, while that for stem length is 3 long :1 short. If the two genes are unlinked, the expected dihybrid ratio can be calculated using the product rule to give 3/8 long pink : 3/16 long red : 3/16 long white : 1/8 short pink : 1/16 short red : 1/16 short white, or a 6:3:3:2:1:1 ratio. The observed numbers are: 243 pink long : 122 red long : 122 white long : 81 pink short : 41 red short : 41 white short. The ratio is close to that predicted; so **the genes for petal color and stem length are unlinked**.

   The same analysis is done for flower color and anther color. The expected dihybrid ratio here is also 6:3:3:2:2:1. The observed numbers are: 272 pink black : 59 red black : 156 white black :

52 pink tan : 104 red tan : 7 white tan. Because these numbers do not fit a 6:3:3:2:1:1 ratio, we can conclude that **flower color and anther color are linked genes**.

f.   It is clear that the original snapdragon was heterozygous for *Pp* and *Bb*. The genotype of the heterozygous plant could have been either *P B / p b* or it could have been *P b / p B*. If it is the former and the genes are closely linked, then the *pp bb* phenotype will be very frequent (almost 1/4 of the progeny). Instead the *pp bb* class is very <u>infrequent</u>, accounting for only about 1% (7/650) of the progeny. Thus, **the parental genotype must have been *P b / p B***. In this case, the infrequent *pp bb* class received a *p i* recombinant gamete from both parents (**see <u>problem 5-16 parts a & b</u>**). The frequency of the *pp bb* class = (probability of *p b* recombinant)^2. **The frequency of the *p* gametes = rf = 2($\sqrt{(\#pp\ bb/\#\ \text{total progeny})}$) = 2($\sqrt{(7/650)}$) = ~21 map units.**

5-23. In cross #1 the criss-cross inheritance of the recessive alleles for dwarp and rumpled from mother to son tells you that all these genes are X-linked. In cross #2, you see the same pattern of inheritance for pallid and raven, so these genes are X-linked as well. The fact that the F1 females in both crosses were wild type tells you that the wild type allele of all 4 genes is dominant to the mutant allele. Designate alleles: dwp^+ and *dwp* for the dwarp gene, rmp^+ and *rmp* for the rumpled gene, pld^+ and *pld* for the pallid gene, and rv^+ and *rv* for the raven gene. Assign the genes an arbitrary order to write the genotypes. If you keep the order the same throughout the problem, it is sufficient to represent the wild type allele of a gene with +.

Cross 1: *dwp rmp + + / dwp rmp + +* × *+ + pld rv / Y* → *dwp rmp + + / + + pld rv* (wild-type females) and *dwp rmp + + / Y* (dwarp rumpled males).

Cross 2: *+ + pld rv / + + pld rv* × *dwp rmp + + / Y* → *+ + pld rv / dwp rmp + +* (wild-type females) and *+ + pld rv / Y* (pallid raven males).

dwp rmp + + / + + pld rv (cross 1 F$_1$ females) × *dwp rmp pld rv / Y* → 428 *+ + pa ra*, 427 *dw ru + +*, 48 *+ ru pa ra*, 47 *dw + + +*, 23 *+ ru pa +*, 22 *dw + + ra*, 3 *+ + pa +*, 2 *dw ru + ra*.

All 4 genes <u>must</u> be linked, as they are all on the X chromosome. You expect 16 classes of progeny ($2 \times 2 \times 2 \times 2$) in 8 genetic and numeric pairs. Notice that there are only 8 classes of progeny. The data that is seen shows <u>exactly</u> the pattern you expect for a female that is heterozygous for 3 linked genes! If 2 of the 4 genes NEVER recombine then you would expect the pattern of data that is seen. If 2 genes never recombine, they will always show the parental configuration of alleles. Thus, examine the various pairs of genes for the presence or absence of recombinants. Note that you <u>never</u> see recombinants between *pallid* and *dwarp*: all the progeny are either pallid or dwarp, but never pallid and dwarp nor wild type for both traits. This suggests that the two genes are so close together that there is essentially no recombination between the loci. If a much larger number of progeny were examined, you might observe recombinants. Treat *dwp* and *pld* as 2 genes at the same location, so one of them (*dwp*, for instance) can be ignored, and this problem becomes a three-point cross between *pld, rv* and *rmp*.

Classes of gametes	Genotype	Numbers
Parental (P)	*dwp rmp + +*	427
	+ + pld rv	428
SCO 1	*+ rmp pld rv*	48
	dwp + + +	47
SCO 2	*dwp + + rv*	22
	+ rmp pld +	23
DCO	*+ + pld +*	3
	dwp rmp + rv	2

When the *rmp + rv* DCO class is compared to the *rmp + +* parental class, you can see that **rv is in the middle. The *rmp - rv* rf = (48 + 47 + 3 + 2)/1000 = 10 cM; the *pld - rv* rf = (22 + 23 + 3 + 2)/1000 = 5 cM**. I = 1 - coc; coc = observed frequency of DCO/expected frequency of DCO. So coc = (5/1000)/(0.05)(0.1) = 0.005/0.005 = 1; I = 1 - 1 = 0, so **there is no interference**.

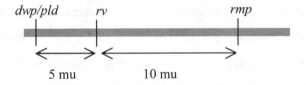

5-24. Use the **Three Easy Rules for Tetrad Analysis** to help you solve this problem.

a. In cross 1, the number of PD (parental ditypes) = NPD (nonparental ditypes) so the *ad* gene and the mating type locus assort independently. In cross 2 the number of PD >> NPD so we can conclude the *p* gend and the mating locus are linked; rf *p - mating type* = (NPD + 1/2T)/ total tetrads = (3) + (1/2)(27)/54 =16.5/54= .31 × 100 = 31 cM between the two genes.

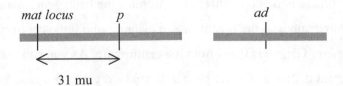

b. **To calculate gene-centromere distances you need information on the order of ascospores in each ascus type**. Only with this information can you calculate gene - centromere distances based on 1/2(# of asci showing MII segregation for the gene)/total asci.

5-25.

a. The number of meioses represented here is the total of the number of asci = **334**. Each ascus contains the 4 products of one meiosis.

b. Diagram the cross: $a + c$ x $+ b +$

To map these genes, use the **Three Easy Rules for Tetrad Analysis**. First designate the type of asci represented. This has to be done for each pair of loci as PD (P), NPD (N) and T refer exclusively to the relationship between two genes. In the table below, the top row shows the designations for all three pairs: the *a-b* comparison is at the lower left, the *b-c* comparison is at the lower right and the *a-c* comparison is at the top of the pyramid.

P	P	T	T	N	P
P P	N N	T P	T N	N P	T T
$a + c$	$a\,b\,c$	$+ + c$	$+ b\,c$	$a\,b +$	$a + c$
$a + c$	$a\,b\,c$	$a + c$	$a\,b\,c$	$a\,b +$	$a\,b\,c$
$+ b +$	$+ + +$	$+ b +$	$+ + +$	$+ + c$	$+ + +$
$+ b +$	$+ + +$	$a\,b +$	$a + +$	$+ + c$	$+ b +$
137	141	26	25	2	3
I I I	I I I	II I I	II I I	I I I	I II I

Rule #1: For genes *a* and *b* PD = NPD, so these two genes are not linked. For genes *b* and *c* PD = NPD, so genes *b* and *c* are not linked. For genes *a* and *c*, PD>>NPD, so the genes are linked. Calculate rf between *a* and *c* = 2 + 1/2(26 + 25)/334 = 8.2 cM (**Rule #2**).

Gene-centromere distances can be calculated in *Neurospora* (**Rule #3**), so analyze the data for MI and MII segregation patterns for the alleles of each gene in each ascus type. This analysis is done <u>separately for each gene</u>, unlike the gene - gene analysis done above. The designation for each gene is presented under that gene at the bottom row of the table (I - MI, II = MII). **Rule #2** shows that the distance between *a* and its centromere = 1/2(26 + 25)/334 = 7.6 mu; the distance between *b* and its centromere = 1/2(3)/334 = 0.4 mu; the distance between *c* and its centromere = 1/2(0)/334 = 0.

Now compile all of these pieces of data into one map. **Rule #1** shows that gene *b* is 0.4 mu from its centromere and is on a different chromosome from genes *a* and *c*. Genes *a* and *c* are on the same chromosome, so the *a*-centromere distance and the *c*-centromere distance refer to the same centromere. Gene *c* is 0 mu from the centromere. As you can see from the map, there are 2 slightly different distances for the gene *a* to gene *c* region - the gene-gene distance is 8.2 mu and the a-centromere distance is 7.6 mu. In this case the longer gene-gene distance is more accurate as it includes the SCOs between *a* and *c* as well as some of the DCOs between *a* and *c* (the 4 strand DCO, **see Figure 5.22**).

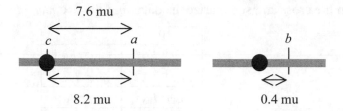

c. Carefully consider the information you have for the different chromosomes in the ascus type chosen (the group with 3 members). The *a* and *c* genes show PD segregation, which can mean either no crossing over between them or a 2 strand DCO. Both genes show MI segregation, which means there haven't been any single crossovers between either gene and the centromere. **Gene *b* shows MII segregation, which means there has been an SCO between the gene and its centromere**. This crossover also means the *a-b* and *c-b* comparisons in this class will show the tetratype pattern (T are due to crossovers between either gene and its centromere when the genes are on separate chromosomes).

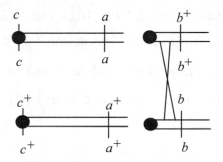

5-26. Genes *a, b,* and *c* are all on different chromosomes. Thus, all crosses between these genes are expected to give an equal number of PD and NPD asci. The fact that the cross involving genes *a* and *b* yields no T indicates that **genes *a* and *b* are both very close to their respective centromeres**. If two genes are on separate chromosomes, tetratypes arise when a crossover occurs between one of the genes and its centromere or between the other gene and its centromere. In the crosses involving genes *a* and *c* or genes *b* and *c,* many T asci are seen. Because genes *a* and *b* are tightly centromere-linked, **gene *c* must be very far from its centromere** in order to generate these T asci.

5-27. Diagram the cross and summarize the data: $met^- \ lys^- \times met^+ \ lys^+$

P	T
$met^+ \ lys^+$	$met^+ \ lys^+$
$met^+ \ lys^+$	$met^- \ lys^+$
$met^- \ lys^-$	$met^+ \ lys^-$
$met^- \ lys^-$	$met^- \ lys^-$
89	11

a. **The two types of cells in the first group of 89 asci are $met^+ \ lys^+$ (could grow on all four types of media) and $met^- \ lys^-$ (require the addition of met and lys to minimal medium).** These asci are parental ditypes (PD). The four types of cells in the second group of 11 asci are $met^+ \ lys^+$; $met^+ \ lys^-$ (grew on min + lys and on min + lys + met); $met^- \ lys^+$ (grew on min + met and on min + lys + met); and $met^- \ lys^-$. These are tetratype (T) asci.

b. Because the number of PD>>> NPD, **the genes are linked.** The distance between them is: (NPD + 1/2(T)) / total tetrads = (0 + 1/2(11))/100 × 100 = **5.5 m.u.**

c. NPD should be seen eventually and would result from four strand double crossovers. There would be two types of spores: $met^- \ lys^+$ (could grow on min + met and on min + met + lys) and $met^+ \ lys^-$ (could grow on min + lys and on min + met + lys).

5-28. Genes c and d are a total of 22 mu apart.

a. If interference = 1, then there are no double crossovers (DCO). Therefore, all the recombinants between the 2 genes are due to single crossovers (SCO). If two genes are linked, then SCOs between the genes give tetratype asci (T). Remember that the rf between C and D is 0.22, DCO between C and D is 0 (so **there can be no NPD tetrads**), and the formula for recombination frequency between 2 genes = NPD + 1/2(T)/total asci. Solve for T: 0.22 = 0 + 1/2(T) so T = 2(0.22) = 0.44. Therefore, **44% of the asci will be tetratypes and the remaining 56% will be parental ditypes.**

b. Interference = 0, so DCOs will occur in the expected frequency (see Problems 5-17 and 5-19). The expected frequency of DCO = (recombination frequency in the C-centromere region) x (recombination frequency in the centromere-D region) = (0.07) × (0.15) = 0.0105. The rf between C-D = SCO + DCO; 0.22 = SCO + 0.0105; SCO = 0.22 - 0.0105 = 0.2095. If SCO frequency between C and D = 0.2095 then T due to SCO = 0.419 (see part a of this problem).

Remember that when analyzing tetrads the three different types of DCOs can be distinguished: 2 strand DCOs, which give PD asci; 3 strand DCOs, which give T asci; and 4 strand DCOs, which give NPD (**see Figure 5.19b-f**). These DCOs occur in the ratio of 1/4 (2 strand DCOs) : 1/2 (3 strand DCOs) : 1/4 (4 strand DCOs). If the total DCO frequency = 0.0105 then 1/4 (0.0105) = 0.003 is the frequency of 4 strand DCOs (NPD), 1/2 (0.0105) = 0.05 is the frequency of 3 strand DCOs (T) and the remaining 1/4 of the DCOs (0.003) are 2 strand DCOs, which will be PD tetrads.

In total, **NPD = 0.003, T = 0.05 (due to DCO) + 0.419 (due to SCO) = 0.469 and the remainder are PD = 0.528**.

c. In *Neurospora* the recombination events that underlie the formation of PD, NPD and T are the same as in yeast. The difference is that the ordered tetrads allow you to distinguish whether a SCO event occurs between *C* and the centromere or between *D* and the centromere. If the SCO occurs between *C* and the centromere then you will see MII segregation for the alleles of the *C* gene and MI segregation for the alleles of the *D* gene. If the SCO occurs between *D* and the centromere then you will see the reverse - MI segregation for *C* and MII segregation for *D*.

Here, I = 1, so there are no DCOs. Therefore, as in part a, T = 0.44 and PD = 0.56. However 7/22 of the SCO events occur between gene *C* and the centromere, while the remaining 15/22 of the SCOs occur between *D* and the centromere. **The expected results are summarized in the table below**.

crossover type and location:	no crossover	SCO *C*-cent.	SCO *D*-cent
ascus type:	PD	T	T
MI or MII gene *C*:	MI	MII	MI
MI or MII gene *D*:	MI	MI	MII
frequency:	0.56	$7/22 \times 0.44 = 0.14$	$15/22 \times 0.44 = 0.30$

d. Here, I = 0, so DCO events are occurring at the expected frequencies. The frequencies of each of the DCO events is as in part b. A DCO represents a single meiosis in which a crossover in the first region (*C* to centromere) happens simultaneously with a crossover in the second region (centromere to *D*). Thus any of the DCO events will show MII segregation for both genes. **The expected results are summarized in the table below**.

crossover type and location:	no crossover	SCO *C*-cent.	SCO *D*-cent	DCO 2 strand	DCO 3 strand	DCO 4 strand
ascus type:	PD	T	T	PD	T	NPD
MI or MII gene *C*:	MI	MII	MI	MII	MII	MII
MI or MII gene *D*:	MI	MI	MII	MII	MII	MII
frequency:	0.528	$7/22 \times 0.419 = 0.14$	$15/22 \times 0.419 = 0.30$	$1/4 \times 0.0105 = 0.003$	$1/2 \times 0.0105 = 0.005$	$1/4 \times 0.0105 = 0.003$

5-29.

a. The genotype of a true breeding wild type diploid strain of *Saccharomyces* can be written $+ / +$. If this diploid undergoes meiosis, **all (100%) of the asci will have 4 viable spores**.

b. The genotype of this strain is $+ / n$, where n = a null activity allele of an essential gene. The diploid cells will be viable, because they have functional enzyme for this essential gene from the $+$ allele. If this diploid undergoes meiosis, then **all (100%) of the tetrads will have 2 + : 2 n spores (that is, only 2 spores in each tetrad will be viable)**.

c. Gene a and gene b are different essential genes; a and b represent temperature-sensitive alleles of these genes. When you diagram a cross you must write complete genotypes for both haploid parents: $a\, b^+$ (strain a) $\times$ $a^+ b$ (strain b) $\rightarrow$ $aa^+ bb^+$. Each haploid parent strain will die when grown under restrictive conditions because they cannot produce a required product, while the diploid cells are viable because they have a wild type allele for both genes.

   If these genes are unlinked, then after meiosis PD asci = NPD asci ('3 Easy Rules for Tetrad Analysis' rule #1). When the genes are unlinked, also remember that T tetrads (asci) arise from SCOs between either gene and its centromere. Because both of these genes are 0 mu from their centromeres, you will not see T asci. Thus, after meiosis, you will have 50% PD (2 $a\, b^+$ spores : 2 $a^+ b$ spores) : 50% NPD (2 $a^+ b^+$: 2 $a\, b$). None of the spores in the PD tetrads are viable under restrictive conditions, while 2 of the 4 spores in the NPD asci are viable ($a^+ b^+$). Thus, **50% of the asci will have 0 viable spores (PD) and 50% of the asci will have 2 viable spores (NPD)**.

d. Again genes a and b are unlinked essential genes, but now gene a-centromere = 0 mu and gene b-centromere = 10 mu. A SCO between gene b and its centromere will happen in 20% of the asci (because only half of the ascospores of an SCO are recombinant) and will give T asci. A T ascus will have a spore ratio of 1 $a^+ b^+$: 1 $a\, b$: 1 $a\, b^+$: 1 $a^+ b$. There is 1 viable spore ($a^+ b^+$) in a T ascus. The remaining 80% of the asci will be equally divided between PD and NPD because the genes are unlinked. Thus, **40% of the asci will have 0 viable spores (PD), 40% of the asci will have 2 viable spores (NPD) and 20% of the asci will have 1 viable spore (T)**.

e. Because both genes are on the same chromosome, you can diagram this cross: $a\, b^+$ (strain a) $\times$ $a^+ b$ (strain b) $\rightarrow$ $a\, b^+ / a^+ b$. When the genes are linked, SCO and 3 strand DCO between the genes give T asci and 4 strand DCO between the genes gives NPD asci. The remainder of the asci are PD (no crossover and 2 strand DCO). If the recombination frequency between gene a and gene b = 0 mu, then the only possible result will be PD asci (2 $a\, b^+$: 2 $a^+ b$) in which none of the spores are viable. Thus, **100% of the asci will have 0 viable spores**.

f. Now the genes are 10 mu apart and there are no DCO events. This means that 20% of the asci produced will be T (see part d) and the remaining 80% will be PD. Therefore, **80% of the asci will have 0 viable spores (PD) and 20% of the asci will have 1 viable spore (T).**

g. A 4 strand DCO gives an NPD ascus. The genotypes of the spores will be 2 $a^+ b^+$: 2 $a\, b$. Thus, **2 of the 4 spores will be viable.**

5-30.

a. Refer to the three strains as trp1, trp2 and trp3. Remember that in *Neurospora* the 4 meiotic products undergo a subsequent mitosis to give 8 spores in the ascus. However, the first 2 spores are identical to each other (they are mitotic products of the same initial spore), the 3d and 4th spores are the same as each other, and so on. For the purposes of discussing the results, assume that the ascus is made up of the 4 original spores prior to this extra mitosis. You cross $trp1^-$ × wild type → diploid → 2 wild type : 2 $trp1^-$. You are seeing a monohybrid ratio of 1:1, which means there is only one gene controlling the trp^- phenotype in strain trp1. Each of the strains gives the same result, so **in each haploid strain a single mutant gene is responsible for the** trp^- **phenotype**.

b. First consider the cross between trp3 and wild type. The diploid is $trp3^-$ / $trp3^+$. After meiosis the first 2 spores were either both $trp3^+$ (could grow on minimal media) or they were both $trp3^-$ (could not grow on minimal media). Thus, all the asci were 2 $trp3^-$: 2 $trp3^+$, and the order of the alleles was either - - + + or + + - -. In other words, all the asci showed MI segregation for the $trp3$ gene! In the case of the crosses with $trp1^-$ x wild type and $trp2^-$ x wild type, the resultant diploids gave some asci that gave a different result - only one spore of the top 2 was trp^+ while the other one was trp^-. In other words, these asci showed MII segregation for the trp gene in question. In summary, **the $trp3$ gene is very closely linked to its centromere (no T = no crossovers between the gene and the centromere), while the $trp1$ and $trp2$ mutations are further from their centromere(s)**.

c. If 2 strains have mutations in the same gene, then the resulting diploid would be unable to grow on minimal media $trpx^-$ / $trpx^-$ and would give asci where all 4 spores were $trpx^-$ (0 spores viable on minimial media). This result is never seen, so each mutant strain must have a mutation in a different gene, for a total of 3 different trp^- genes in the 3 strains. Diagram the cross between **trp1 and trp2**: $trp1^-\ trp2^+$ × $trp1^+\ trp2^-$ → $trp1^-\ trp1^+\ trp2^+\ trp2^-$. When this diploid is allowed to undergo meiosis you see **78 asci** with 0 viable spores (2 $trp1^-\ trp2^+$: 2 $trp1^+\ trp2^-$ =

PD - <u>see Problem 5-29 part f</u>) **and 22 asci** with 2/8 or 1/4 viable spores (1 $trp1^+trp2^+$: 1 $trp1^-$ $trp2^-$: 1 $trp1^-$ $trp2^+$: 1 $trp1^+$ $trp2^-$ = **T**). In the **$trp1$ x $trp3$ cross** you see a new class of asci, those with 2 viable spores (2 $trp1^+trp3^+$: 2 $trp1^-trp2^-$ = NPD). In this cross **there are 46 PD, 48 NPD, and 6 T asci.** In the last cross, **$trp2$ x $trp3$, there are 42 PD, 42 NPD and 16 T asci**.

d. In $trp1 \times trp2$ PD>>NPD (= 0) so the genes are linked. The T asci are caused by SCOs between the 2 genes. The recombination frequency between $trp1$ and $trp2$ = (0 + 1/2(22))/100 = 0.11 = 11 mu. In $trp1 \times trp3$ there are 46 PD = 48 NPD so the genes are unlinked; the 6 T asci arise from SCOs between $trp1$ and its centromere. None of these T asci can arise from crossovers between $trp3$ and its centromere because you found in part b that they were very tightly linked. Thus, $trp1$ is 3 mu from its centromere. In the cross between $trp2 \times trp3$, PD = NPD so the genes are unlinked. There are 16 T asci, so gene 2 must be 8 mu from its centromere. **The map is shown below**.

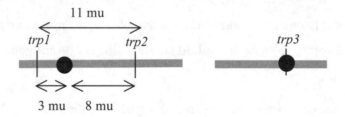

e. Ordered octads show the segregation pattern for each gene separately. In this example, **both mutant genes give the same phenotype (trp⁻). Thus, it is impossible to determine if a trp⁻ spore is + - or -- or - +.** If you cannot distinguish these different types of spores, then you cannot determine the segregation pattern (MI vs MII) for the individual genes.

f. **You can calculate gene-centromere distances because you discovered that one of the genes ($trp3$) is tightly linked to its own centromere.** Therefore, when $trp3^-$ is crossed with either of the other mutants, any T asci must be due to SCO between the other gene and its centromere, and these T asci can then be used to calculate a distance between the other gene and its centromere. This same sort of analysis can also be used in yeast to map gene-centromere distances if a strain is available with a mutation known to be right at a centromere.

5-31. Diagram the yeast cross: $his^- trp^+ \times his^+ trp^- \rightarrow his^-his^+ trp^+trp^- \rightarrow$ 233 PD, 11 NPD, 156 T.

a. The **PD asci must have 2 $his^- trp^+$: 2 $his^+ trp^-$ spores, the NPD asci have 2 $his^+ trp^+$: 2 his^- trp^- spores and the T asci have 1 $his^- trp^+$: 1 $his^+ trp^-$: 1 $his^+ trp^+$: 1 $his^- trp^-$ spores.**

b. **PD>>NPD so the genes are linked; rf between 2 genes = 11 + 1/2(156)/400 = 22.3 mu**.

c. Because the genes are linked, you know that NPD asci are the result of 4 strand DCOs. This sort of DCO is 1/4 of all the DCO events. The 3 strand DCOs (1/2 of all DCOs) give T while 2 strand DCOs (1/4 of all DCOs) give PD. In the data you see 11 NPD tetrads, or 11 meioses that underwent 4 strand DCOs. There are another 22 tetrads that are the result of 3 strand DCOs (and are Ts), and another 11 asci that underwent 2 strand DCOs (and are PDs), for **a total of (11 + 22 + 11 =) 44 asci that underwent 2 crossovers. The remaining (156 - 22 =) 142 T asci underwent a SCO, or 1 crossover, and the remaining (233 - 11=) 222 PD asci underwent 0 recombination events**.

d. There are 44 asci that underwent DCOs for a total of 88 crossover events. There were another 142 asci that underwent SCOs. Thus, there were 142 + 88 = 230 crossover events / 400 meioses = **0.575 crossovers/meiosis**.

e. The equation for recombination frequency between 2 genes used in the textbook and in part b above <u>only</u> includes 4 strand DCO (NPD). **This formula ignores 3 strand and 2 strand DCOs**. This calculation for recombination frequency assumes that all T asci are due to SCOs. Though this is true for the majority of T asci, a small proportion of T asci are due to 3 strand DCO events. Also, it is an oversimplification to assume that all PD asci are non-recombinant, because a very few of them are due to 2 strand DCOs. Remember that the 3 types of DCO events occur in a ratio of 1/4 2 strand DCO : 1/2 3 strand DCO : 1/4 4 strand DCO. With this information we can derive a more accurate formula for recombination frequency between 2 genes = SCO + DCO/total asci: **rf = 1/2(T – 2NPD) + 4(NPD)/total asci** = 1/2(T) + 3(NPD)/total asci. Many yeast geneticists use this more accurate formula in preference to the one in your textbook.

f. rf = 1/2(156 – 22) + 4(11)/400 = 1/2(134) + 44/400 = 111/400 = 0.278 = **27.8 mu**. As you can see, this is somewhat larger than the distance calculated in part b.

5-32 Notice that you are asked for the <u>number of different kinds of phenotypes,</u> not the number of individuals with each of the different phenotypes.

a. **2** (*A*- and *aa*);

b. **3** (*AA*, *Aa* and *aa*);

c. **3** (*AA*, *Aa* and *aa*);

d. **4** (*A*- *B*-, *A*- *bb*, *aa B*-, *aa bb*);

e. **4** (*A*- *B*-, *A*- *bb*, *aa B*-, *aa bb*; because the genes are linked, the frequency of the four classes will be different than that seen in part d);

f. **Nine** phenotypes in total. There are three phenotypes possible for each gene. The total number of combinations of phenotypes is $(3)^2 = 9$ ((AA, Aa and aa) × (BB, Bb and bb)).

g. Normally there are four phenotypic classes, as in parts d and e (A-B-, A-bb, aaB- and $aabb$). In this case, one of the genes is epistatic so two of the classes have the same phenotype, **giving 3 phenotypic classes**.

h. Two genes means four phenotypic classes (A- B-, aa B-, A- bb and aa bb). Because gene function is duplicated the first three classes are all phenotypically equivalent in that they have function, and only the aa bb class will have a different phenotype, being without function. Thus there are only **2 phenotypic classes**.

i. There is 100% linkage between the two genes. The number of phenotypic classes will depend on the arrangement of alleles in the parents. **If the parents are A B / a b × A B / a b, the progeny will be 3/4 A- B- : 1/4 aa bb and there will be two phenotypic classes. If the parents are A b / a B × A b / a B, all of the progeny will be A- B- and there will be one phenotypic class in the offspring**.

5-33.

a. Red sectors arise when one cell in the growing colony becomes $ade2^-$ / $ade2^-$. As all the cells continue to grow, the colony continues to expand and the $ade2^-$ / $ade2^-$ cells form a red sector within the white colony (**see Figure 5.27**). **Red cells of the $ade2^-$ / $ade2^-$ genotype could arise by mitotic recombination (see Figure 5.26), by loss of the entire chromosome containing the $ade2^+$ allele, by a deletion of the portion of the chromsome containing the $ade2^+$ allele, or by spontaneous muation of the $ade2^+$ allele to an $ade2^-$ allele.**

b. The size of the red sectors depends on when in the formation of the colony the event occurred to form the initial $ade2^-$ / $ade2^-$ cell. If the event occurred early in the formation of the colony there will be a larger red sector than if the event occurred near the end of colony formation. All of the events mentioned in part a are rare, so in general they will occur later in colony formation when there are more cells in which they could occur. As a result, **most of the red sectors will be small**.

5-34. After replication in the heterozygous diploid, each chromosome would be composed of a pair of sister chromatids, as shown below. The centromeres on sister chromatids split and segregate from each other during mitosis (**see Figure 5.26**). The centromeres are numbered in the figure below so you can follow them more easily. For example, cen1 and cen2 segregate from each other, so chromatids 1 or 2 and 3 or 4 will end up in a daughter cell. In a normal mitosis, this gives daughter cells that are genotypically $a\ b\ c\ leth\ d\ e\ /\ a^+\ b^+\ c^+\ leth^+\ d^+\ e^+$ and are phenotypically wild type, like the original cell. Mitotic recombination can rarely occur between non-sister chromatids, for example between chromatid 2 and 3. Possible locations for these recombination events are indicated by X-I - X-V in the figure. Assume that no crossovers occur between gene a and the cen since they are so closely linked.

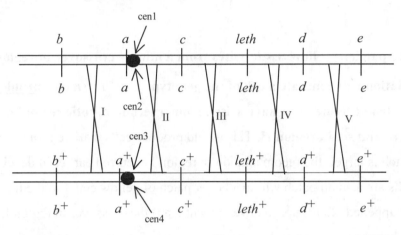

Consider the results of the X-I crossover. In one possible segregation, chromatids 2 and 4 segregate into one daughter cell; this cell will be homozygous for the b^+ allele and heterozygous for the rest of the genes on the chromosome. Thus, this cell will be phenotypically wild type and indistinguishable from the non-recombinant cells surrounding it. The reciprocal product will be the daughter cell with chromatids 1 and 3 whose genotype will be homozygous mutant for gene b and heterozygous wild type for everything else. This cell will continue to divide mitotically and give you a patch of b mutant tissue in a sea of wild type tissue. In effect, the genes that are further away from the centromere than the site of mitotic recombination become homozgous. One segregant (the 2 and 4 chromatids in this example) will be homozygous wild type, and thus indistinguishable from the non-recombinant cells; the other segregant (the 1 and 3 chromatids here) becomes homozygous for the mutant allele of the gene(s) that are further from the centromere than the crossover while those genes that are closer to the centromere than the recombination event (or on the other side of the centromere, i.e. those genes unaffected by the mitotic recombination event) remain heterozygous. Next consider crossover X-II. In this case the same segregations will give you a wild type cell (chromatids 2 and 4) while the reciprocal daughter cell will be $b^+\ a^+\ c\ leth\ d\ e$ (lethal, chromatids 1 and 3). Crossover X-III will also give a lethal recombinant product. Crossover X-IV will give a $d\ e$ patch of mutant tissue, and crossover X-V will give a patch of e mutant tissue. Thus, **the only phenotypes that will be found in sectors as a result of mitotic recombination will be b, e, and $d\ e$ (#2, 5, and 9)**.

5-35. The genotype of the female fly is $y^+ sn^+ / y\ sn$:

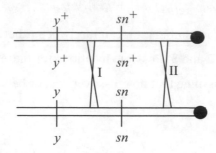

a.   **The larger patch of yellow tissue arises from a mitotic crossover in region I. This mitotic recombination will generate a cell of the genotype $y\ sn^+ / y\ sn$ (<u>see problem 5-34</u>). The smaller patch of tissue must have arisen from a second mitotic recombination between the centromere and sn, in region II.** This would produce cells that are $y\ sn / y\ sn$. Because the yellow patch is larger, the recombination in region I occurred earlier in development. The yellow, singed cells are a small patch within a larger patch of yellow cells, so the mitotic crossover in region II happened after (later in development than) the crossover in region I, and this second crossover happened in a recombinant daughter cell of the first recombination event. No wonder these patches within patches are rare!

b.   If the genotype of the female was $y^+ sn / y\ sn^+$, then **a recombination event in the region I would give you a detectably recombinant cell with the genotype $y\ sn / y\ sn^+$. A subsequent second mitotic crossover in region II in one of these originally recombinant $y\ sn / y\ sn^+$ cells will give you a patch of $y\ sn$ tissue inside a patch of yellow tissue, as in part a.**

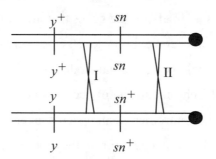

There is one mitotic recombination event which will give a different result in a $y^+ sn^+ / y\ sn$ female (part a) than a $y^+ sn / y\ sn^+$ female (part b). After a recombination event in region II the female from part a will show a $y\ sn$ patch of tissue. The other recombinant product will be homozygous $y^+ sn^+$, and thus indistinguishable from non-recombinant cells. The same sort of recombination event in region II in the female from part b will give one daughter cell of genotype $y^+ sn / y^+ sn$ which will give a patch of sn tissue. The other product of the mitotic recombination will be $y\ sn^+$ / $y\ sn^+$ and will give an adjacent patch of yellow tissue. This phenomenon is called twin spots.

5-36. List the phenotypes seen in the normal tissue and the 20 tumors and their frequency of occurrence:

- normal tissue = $A^F A^S\ B^F B^S\ C^F C^S\ D^F D^S$

- tumor type 1 = $A^F\ B^F B^S\ C^F C^S\ D^F$ = 12 tumors

- tumor type 2 = $A^F\ B^F B^S\ C^F C^S\ D^F D^S$ = 6 tumors

- tumor type 3 = $A^F\ B^S\ C^F C^S\ D^F$ = 2 tumors

Remember that all of the tumor tissues were also homozygous for $NF1^-$ while the normal tissue is heterozygous $NF1^- NF1^+$.

a. **Mitotic recombination** could have caused all 3 types of tumors.

b. Remember that mitotic recombination causes genes further from the centromere to become homozygous, while genes between the recombination event and the centromere, or those on the other side of the centromere, remain heterozygous (review problem 5-34). Notice that all 3 types of tumor cells are homozygous for $NF1^-$, so by definition all 20 mitotic recombination events that gave rise to the 20 tumors occurred between the centromere and the $NF1$ gene. Notice also that all 20 tumors are homozygous for gene A, specifically the A^F allele. Because gene A is affected by all of the mitotic recombination events, it must be further from the centromere than the $NF1$ gene. Because these tumors become homozygous for the A^F allele, that allele must be on the same homolog with the $NF1^-$ allele. There are 6 tumors that are homozygous for these 2 genes (tumor type 2). As you work your way along the chromosome from $NF1$ toward the centromere the next gene to also become homozygous is the D gene (tumor type 1), specifically the D^F allele, then the B gene (tumor type 3), the B^S allele. Gene C never becomes homozygous, yet we are told it is on this chromosome as well. Thus it could either be on the other side of the centromere from $NF1$, A, D and B or it could be on the same side of the centromere as the rest of the genes but very closely linked to the centromere (like the a gene in problem 5-34). **The order of the genes and coupling of the alleles is shown below:**

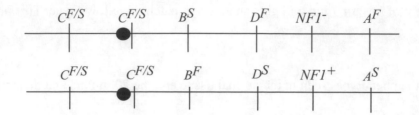

It is possible to use the **mitotic** recombination frequencies with which the various genotypes of tumors arose as a rough, relative approximation of the distances between the genes. These numbers are NOT to be confused with the meiotic recombination frequency we have been calculating in the other problems in this and other chapters! In tumor type 2 the mitotic recombination event happened between gene *NF1* and gene *D* (*NF1⁻* is homozygous, and so further from the centromere than the recombination event while gene *D* is still heterozygous), and this occurred in 6/20 tumors = 0.3, so **the relative recombination frequency between NF1 gene D is 0.3**. In tumor type 1 the mitotic recombination event happened between gene *B* and gene *D* (*B* is still heterozygous in these tumors while *D* is homozygous D^F), and this happened in 12/20 tumors = 0.6, so **the relative recombination frequency between genes B and D is 0.6**. In tumor type 3 the mitotic recombination event happened between the centromere and gene *B* (or between gene *C* and gene *B* if you place gene *C* on the same side of the centromere as the NF1 gene). This event happened in 2/10 tumors = 0.1 so **the relative recombination frequency between the centromere and gene B is 0.1**.

c.   In this mechanism an entire homolog of one chromosome is lost. If the lost homolog was the one with the $NF1^+$ allele, then **the resulting cell would be hemizygous for the NF1⁻ allele, and would develop into a tumor. This did NOT occur here, because ALL 3 genotypes of tumors are still heterozygous for at least the C gene**.

d.   Yes, **deletions of portions of the $NF1^+$ homolog which cause loss of the $NF1^+$ allele** could cause tumors to develop in the resulting *NF1⁻* cells. If these deletions extended to the neighboring genes *A, D,* or *B,* the tumor cells could show the same protein variants as in the problem.

Chapter 6   DNA: How the Molecule of Heredity Carries, Replicates, and Recombines Information

Synopsis:

The statement "DNA's genetic functions flow directly from its molecular structure" is a good focus for reviewing DNA structure. By focusing on function, the beauty of the structure will become more evident. Make sure you really understand the structure and get a good mental image of the DNA molecule and its construction. Understanding where hydrogen and covalent bonds are found, the polarity of the strands of DNA, and why complementarity is important will provide a good basis for understanding many of the cellular processes (for example, transcription, translation and recombination) and the manipulations of recombinant DNA technology, from cloning to genetic screening.

DNA is the nearly universal genetic material. Experiments showing that DNA causes bacterial transformation (**see Figures 6.3 and 6.4**) and that DNA is the agent of virus production in phage-infected bacteria (**see Figure 6.5**) demonstrated this fact.

According to the Watson-Crick model the DNA molecule is a double helix composted of two antiparallel strands of nucleotides; each nucleotide consists of one of four nitrogenous bases (A, T, G or C), a deoxyribose sugar and a phosphate. An A on one strand pairs with a T on the other, and a G pairs with a C. DNA carries information in the sequence of its bases, which may follow one another in any order.

The DNA molecule reproduces by semiconservative replication. In this type of replication the two DNA strands separate and the cellular machinery then synthesizes a complementary strain for each (**see Feature Figure 6.17**). By producing exact copies of the base sequence information in DNA, semiconservative replication allows life to reproduce itself.

Recombination arises from a highly accurate cellular mechanism that includes the base pairing of homologous strands of nonsister chromatids (**see Figure 6.19 and Feature Figure 6.22**). Recombination generates new combinations of alleles.

Significant Elements:

After reading the chapter and thinking about the concepts, you should:

♦ Become familiar with the evidence that replication is semiconservative, and understand the general processes of initiation and elongation.

♦ Think about why special enzymes (such as topoisomerases and telomerases) are needed for replication of the genome.

♦ In the process of recombination, understand how heteroduplex regions are formed and what effect they have on recombination outcomes.

♦ Work your way through the recombination process by drawing it out for yourself.

♦ Assign the 5' and 3' designations to backbone strands of DNA and RNA.

♦ Indicate complementary bases to an RNA or DNA template.

♦ Explain the importance of structural features of DNA for function in copying (replication) in recombining information.

♦ Think about experimental design.

Problem Solving Tips:

♦ The nature of the problems changes with this material. The problems still require that you have a basic understanding of the concepts and use that knowledge, but the problems are more based on experimental design. These problems can be viewed as more creative synthesis problems.

♦ The experimental type of question may cause you to go back and refresh your understanding of some techniques.

♦ A good way to approach these problems is to identify concepts relevant to the problem and review your knowledge of the topic and any relevant experimentation.

♦ One technique that comes up frequently is tagging a molecule with a label so the molecule can be followed. This is often done using radioactivity. Radioactive label can be incorporated into protein or DNA if a cell or organisms is grown in or fed a radioactive precursor that goes specifically into the type of molecule you want to follow. In designing experiments using radioactive labeling, be sure to consider how you can get a unique label into the molecule of interest.

Solutions to Problems:

6-1. a. **6**; b. **11**; c. **9**; d. **2**; e. **4**; f. **8**; g. **10**; h. **12**; i. **3**; j. **13**; k. **5**; l. **1**; m. **7**.

6-2. (<u>See Figure 6.4</u>). The proof that DNA was the transforming principle was the treatment of the transforming extract with an enzyme (DNase) that degrades DNA. After this treatment, the extract was no longer able to transform rough, nonvirulent strains of *Streptococcus pneumoniae* bacteria into smooth, virulent cells that could kill host mice. Avery, MacCleod and McCarty also showed that treatments with RNase and proteinase did not abolish the transforming activity of their extracts, indicating that the transforming principle was not RNA or protein. These experiments using enzymatic treatments were important because it could be argued that the purified "transforming principle" these investigators isolated as DNA might have contained proteins or other molecules. Even with this extensive evidence, many scientists still remained unconvinced the DNA carries genes for almost 20 more years.

6-3. In DNA transformation, the DNA that enters the cell is in the form of randomly sized fragments, usually generated by mechanical forces that shear the DNA while it is prepared from the bacterial cell. Therefore, 2 genes that are closer together on the chromosome will end up on the same fragment more often than genes that are far apart. A high cotransformation frequency between 2 genes indicates that they are close together. **Gene *a* is closer to *c* than it is to gene *b* because there are many instances when *a* and *c* were cotransformed but only a few instances when *a* and *b* were cotransformed.**

6-4. Sulfur is found only in proteins, never in DNA, while phosphorus is a major constituent of the backbone of the DNA molecule and is found only very rarely in proteins (none of the amino acids in proteins contain phosphorous, though as we will see in Chapter 8 phosphorous can sometimes be added to certain proteins at certain times). Nitrogen and carbon, on the other hand, are found in both proteins and DNA. Hershey and Chase needed to differentiate between protein and DNA, so they needed to be able to specifically label the proteins and not the DNA and vice-versa. **If they had used labeled nitrogen or carbon there would be no way to differentiate protein and nucleic acid.**

6-5. (**See Figure 6.9b for an overview of DNA structure**) In **Tube #1** all the sugar phosphate bonds are broken. You would see **individual pairs of complementary nucleotides held together by hydrogen-bonds and attached to a sugar** with no the phosphate group and free phosphates (**see Figure 6.8**). In reality, the hydrogen bonds that hold together individual complementary nucleotides (2 for A-T pairs and 3 for G-C pairs) are not very stable. These hydrogen bonds would be disrupted by the thermal forces working at room temperature. You usually need at least 4 nucleotide pairs in order to have DNA that is stably double stranded at room temperature. In **Tube #2** the bonds that attach the bases to the sugars are broken. You would see **base pairs** (similar to **Figure 6.8** without the 'sugar') **and sugar phosphate chains without the bases** (similar to **Figure 6.7c** without the 'bases'). **Tube #3 would contain single strands of DNA** since the hydrogen bonds between bases were broken (**see Figure 6.7c**).

6-6. X-ray diffraction studies yielded a crosswise pattern of spots, indicating that **DNA is a helix** containing repeating units spaced every 3.4 Å. One complete turn of the helix occurs every 34 Å. The diameter of the molecule is 20 Å, indicating that DNA must be **composed of more than one polynucleotide chain**. (**See Figure 6.7c**). The key X-ray diffraction pictures were taken by Rosalind Franklin and Maurice Wilkins in 1951-1952; James Watson and Francis Crick then built models based on the known chemistry of the nucleotide building blocks to fit the X-ray data.

6-7.
a.   Human DNA is double stranded. If 30% of the bases are A, and A pairs with an equal amount of T (30%), that leaves 40% to be C + G. Thus, there will be **20% C**.

b.   **30% T**.

c.   **20% G** (must equal the amount of C).

6-8. Remember that in double stranded DNA the amount of A = T (purines) and the amount of G = C (pyrimidines). The true statements are **a, b and e**. It is useful to keep in mind the fact that statement c is false; therefore A + T does not equal G + C. In fact, the DNA of different species can vary a great deal in the proportions of A-T base pairs relative to C-G base pairs. Moreover, this ratio can be very different in different regions of the same chromosome. In most organisms, the regions between genes have a higher proportion of A-T base pairs (they are "A-T rich") than the genes themselves.

6-9. Double-stranded DNA contains equal amounts of complementary bases while single stranded DNA does not. To distinguish the two forms of DNA, **determine the percentage of each base. If A = T and C = G, the DNA is double-stranded**. There are some other options as well. For example, you can treat the DNA with restriction enzymes, which were discussed in the **Fast Forward box** in this chapter, see **Figure A**. Double stranded DNA is digested by restriction enzymes while single-stranded DNA is not. Another possibility is to examine the DNA in the electron microscope: as shown in **Figure 6.11**, double stranded DNA looks relatively smooth and wide while single stranded DNA looks kinky and narrow.

6-10.

a. **The A-T base pairs have only two hydrogen bonds, so it takes less heat energy to denature these base pairs.** G-C base pairs have three hydrogen bonds holding them together. It thus takes more energy to break the bonds between Cs and Gs. Remember that the DNA of different species can vary a great deal in the proportions of A-T base pairs relative to C-G base pairs. Moreover, this ratio can be very different in different regions of the same chromosome. In most organisms, the regions between genes have a higher proportion of A-T base pairs (they are "A-T rich") than the genes themselves. In the early stages research on genomes scientists sometimes tried to locate genes by looking for regions of DNA that were more resistant to heat denaturation, and thus had a higher G-C content.

b. **The denatured single stranded DNA must contain stretches of nucleotides that are complementary to a nearby sequence but in an inverted orientation.** These 'stem-and-loop' structures are regions where the a single strand of DNA formed a double-stranded region. The loops and the strings holding the stems together are still single-stranded DNA.

6-11. The complementary DNA would have the complementary sequence with the opposite polarity. Note also the presence of T in DNA in contrast with U in RNA.

3' GGGAACCTTGATGTTTCGGCTCTAATT 5'

6-12. RNA from virus type 1 was mixed with protein from virus type 2 to reconstitute a "hybrid" virus. In a parallel experiment, RNA from virus type 2 was mixed with protein from virus type 1. When these reconstituted hybrid viruses were used to infect cells, **the progeny viruses in each case had the protein that corresponded to the type of RNA in the parent hybrid virus.** The protein in the progeny did <u>not</u> correspond to the protein in the parent hybrid virus.

6-13. Primers for DNA synthesis are RNA molecules that are made by the "primase" enzyme. The primer is complementary to the DNA sequence shown and has the opposite polarity:

5' UAUACGAAUU 3'

6-14. After one S phase, the label would be in one strand of each DNA double helix (call this the round 1 helix), so each chromatid (a double-stranded molecule) contains label on one of its two strands. The labeled ^{3}H-thymidine was removed before the next S phase, so the next set of new strands are not labeled. When the unlabeled strand of each round 1 chromatid is used as a template in round 2, the resulting double stranded chromatid will be unlabeled. When the labeled strand in the round 1 helix is replicated, the resultant chromatid contains this labeled strand and an unlabelled, newly synthesized complementary strand. Thus, after this second round of replication is complete, every chromosome would have one labeled chromatid (but the label is only on one strand) and one unlabeled chromatid. Therefore, for each homologous chromosome pair, **one chromatid of each of the two chromosomes would contain label. This is option c**.

6-15. It is best to consider the individual strands of DNA when calculating the amount of DNA of different densities. Meselson and Stahl started with H:H double stranded DNA (with nucleotides only containing ^{15}N). After one generation in 14N media (with L or light ^{14}N nitrogen), this original molecule becomes two molecules - H:L and H:L, both of which have an intermediate density. After a second generation these 2 molecules will become 4 - H:L, L:L and H:L, L:L. Thus there will be equal amounts of the intermediate band and the light band, as stated in the problem. After another round of DNA replication in the light media (round 3) these 4 molecules will become 8 - H:L, L:L, L:L, L:L and H:L, L:L, L:L, L:L. Thus **after 3 rounds 1/4 of the total DNA will be intermediate density and 3/4 will be light density**. After the next round of replication (round 4) the 8 molecules will become 16 - H:L, L:L, L:L, L:L, L:L, L:L, L:L, L:L and H:L, L:L, L:L, L:L, L:L, L:L, L:L, L:L. **At this point (after round 4), 1/8 of the double-stranded DNA molecules would be of the intermediate density and the remaining 7/8 would be light density**.

6-16. A bubble is formed by 2 replication forks proceeding in opposite directions form a single origin of replication.

a.   There are 3 bubbles in this figure, so there must be **3 origins of replication** in this DNA molecule.

b.   There are **6 replication forks,** one at each of the two ends of each of the 3 replication bubbles.

c.   If all replication forks move at the same rate, then the largest bubble was the first one activated. The **smallest bubble (the one in the middle) was the last origin of replication to be activated**.

6-17.

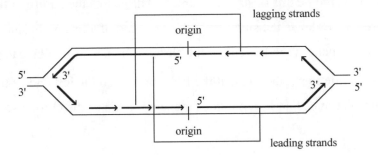

6-18.

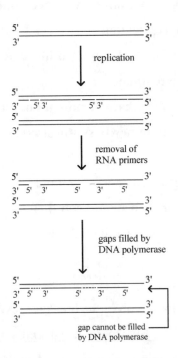

Normally, after an RNA primer is removed from the 5' end of an Okazaki fragment, the "lost"
information in the strand being synthesized can be replaced. This occurs when DNA polymerase
extends the 3' end of the preceding Okazaki fragment by copying the template strand exposed by
primer removal. However, this cannot occur at the 5' ends of the new DNA strands, because there is
no preceding Okazaki fragment that could be extended. In other words, **when the primer is
removed from the very 5' end of a newly synthesized strand in a linear chromosome, bases are
exposed at the 3' end of the (old) template strand. There is no way to synthesize complementary
DNA for those exposed nucleotides. As a result, you would expect that information equal to the
length of the removed primer will be lost from the 5' end of the new strand at each end of the
chromosome** if there is not some sort of alternate replication methodology at the chromosome ends.
This leads to a successive shortening of the chromosomes in each generation of cells. It turns out that

linear chromosomes in eukaryotic organisms have special structures called "telomeres" at their ends that allow them to overcome this obstacle. Chapter 12 discusses the nature of these telomeres in detail. The figure above shows the situation at the right end of a linear chromosome. The same loss of DNA sequence happens at the left end of the chromosome as well. You can picture this by rotating this diagram 180º while remembering that the RNA primer on the left end of the chromosome will be found on the pair of strands that is continuous at the right end of the chromosome.

6-19.

a. **Topoisomerase relieves the stress of the overwound DNA ahead of the replication fork.**

b. **Helicase unwinds the DNA by breaking hydrogen bonds between base pairs to expose the single strand templates for replication.**

c. **Primase synthesizes a short RNA oligonucleotide, which DNA polymerase requires as a primer in order to copy the template.**

d. **DNA ligase joins the sugar phosphate backbones of adjacent Okazaki fragments in order to construct a continuous strand of newly synthesized DNA.**

6-20. In order to unlock the intertwined helices, **you must break both strands of one of the intertwined molecules. The unbroken strand can then pass through the break in the other DNA molecule.** The broken helix must then be rejoined. This breakage/rejoining is mediated by topoisomerase.

6-21. The process of recombination involves the breakage, exchange, and rejoining of strands to make the Holliday intermediate (**see Figure 6-22, especially steps 6-8**). This is followed by a final resolution involving cutting and rejoining the strands. **If the same two strands of the DNA molecule that were cut during the initiating event are cut during the resolution, there will be no crossing over.** (Crossing over is defined as the genetic recombination of markers outside of the gene conversion region.) **There is an equal likelihood that the other strands will get cut and genetic recombination is the result.** Regardless of which strands are cut during resolution, those resulting in crossing-over or those not leading to crossing-over, **mismatches within the heteroduplex region that are generated early in the process can be corrected to the same allele, resulting in gene conversion** (also **see Figure 6.20**). Therefore, gene conversion occurs equally as frequently as recombination of genetic markers.

6-22. The numbers of *B* and *b* alleles are not in the 2:2 ratio predicted from reciprocal exchange during recombination. The 3:1 ratios indicate that gene conversion occurred. **The same type of recombination event occurred in the formation of the spores in the two tetrads. Both cases involved the formation of a heteroduplex region including the *B* gene, but correction of the mismatch in the region of heteroduplex was different (see Figures 6.20 and 5.21).** In the first case (*A B C, A B C, a B c* and *a b c*) the correction of mismatches in both duplexes was to *B* and there was no crossing over between the flanking markers (**see Figure 6.22 steps 6-8**). In the second case (*A B C, A b c, a b C* and *a b c*) the correction in one duplex was to *b* and in the other correction was to *B*. In tetrad II there was also crossing-over between *A* and *C* (**see Figure 6.22 step 8**).

6-23. A mutant *recA E. coli* strain will not be able to undergo recombination. Any event requiring recombination would not occur normally. (**See Figure 6.22, step 3**).

6-24.

a. **The crossover initiated either between *e* and *f* or between *f* and *g*.**

b. **If you said the initiating cut was between *e* and *f* then the resolving cut occurred between *f* and *g*. If you said the initiating cut was between *f* and *g* then the resolving cut occurred between *e* and *f*.** In either case, gene *f* must have been included in the heteroduplex region to get the resulting octad (because gene *f* does not segregate 4:4).

c. **The chromosome that ended up with the f^+ - f^- mismatch in the heteroduplex region was corrected by mismatch repair to $f^- f^-$. The *e* gene was not within the heteroduplex region and therefore it was not affected by the mismatch repair, so the alleles segregated 4 e^+ : 4 e^-.**

Chapter 7   Anatomy and Function of a Gene: Dissection Through Mutation

Synopsis:

This chapter is about MUTATIONS! They are the heart and soul of genetics - the basis of genetic variation, the raw material for evolution, the basic tool of genetic analysis. Now that you know the structure of DNA, you can understand the molecular nature of different kinds of mutations, how errors arise that can result in mutation (**see Figure 7.12**), and how errors can be corrected. You can also understand some of the implications of various mutations on protein structure and function.

The functions of proteins produced within an organism determine phenotype. The connection between genes and what the gene products do is apparent in the consequences of a defect (mutation) in a gene encoding a protein that is needed in the pathway. The order of genes acting in an enzymatic pathway is determined based on the ordering of intermediate compounds and the enzymes that catalyze the conversion of one compound to another.

Two of the most important tools for genetic analysis are complementation and deletion mapping. Complementation analysis is used to determine if mutations that result in the same phenotype are in the same gene. In this way, the number of genes required for a particular process can be determined. In complementation analysis, gametes containing chromosomes with two different mutations are combined. If a gamete containing a mutation in gene A, but not gene B is combined with a gamete having a mutation in gene B but not A, wild type gene products for both A and B will be made in the resulting zygote and the zygote will be wild type, not mutant. Thus, these mutations complement each other. But if the mutations are in the same gene, there is no good copy of the gene available so there is no functional gene product and the cell is still mutant. Thus, the mutations do not complement.

Deletion analysis is used to determine the location and order of genes. It is a quicker way to determine map order than doing linkage analysis on each pair of mutations (or genes) along a chromosome.

Recombination can occur within a gene, even between adjacent nucleotide pairs.

Significant Elements:

After reading the chapter and thinking about the concepts, you should be able to:

♦ Identify different types of mutations that can occur in DNA: frameshift, transition, transversion, deletion and inversion.

♦ Know how to set up a complementation test; a genotype with the mutations in trans is a complementation test; a genotype with a mutation in cis is a test for dominance/recessiveness.

$$\frac{a^+ \quad b^+}{a \quad b}$$ $$\frac{a^+ \quad b}{a \quad b^+}$$

cis genotype   trans genotype

♦ Determine the number of genes represented by the results a of complementation test.

♦ Understand the difference between complementation and recombination analysis. Complementation is a test for function and does not require any interaction between the DNA molecules. Recombination requires physical exchange between 2 homologous chromosomes. This occurs during meiosis in eukaryotes and during DNA replication in prokaryotes and bacteriophage. Make sure you understand which process is being analyzed.

♦ Use deletion mapping to position mutants on a map (see Problem 7-17).

♦ Use fine structure analysis to order mutations and calculate recombination frequencies.

♦ Determine the order of intermediate compounds in an enzymatic pathway using data on the ability of mutants to grow on intermediate compounds (See Problem 7-24).

♦ Determine the order that genes act in an pathway composed of several enzymatic steps (see Problem 7-24).

♦ Understand how to use mutations to dissect a complex biosynthetic process like the assembly of the bacteriophage T4 (see Figure A in the Fast Forward box).

Problem Solving Tips:

♦ Many of the problems in this chapter involve cross schemes - remember to apply the 3 Essential Questions to help with assigning genotypes - see Chapter 5 'Problem Solving - How to Begin' on page 61 of this Student Guide.

♦ A complementation test is a test for function. Lack of complementation means the mutations are in the same gene. If a wild type phenotype is seen in all cells then the mutations are in different genes (are complementing).

♦ Group the negatives (i.e. mutations that are in the same gene) when analyzing complementation data.

♦ Remember that dominant mutations cannot be mapped to a single gene based on complementation data.

♦ Recombination between the mutant chromosomes and reversion of the mutation on one of the mutant chromosomes are both rare events that regenerate a wild-type gene. Therefore, a small number of wild-type cells when most of the cells in the complementation test are mutant are due to recombination or reversion.

♦ Deletions do not revert and can be recognized by this characteristic. Deletions do not complement mutations that are located in the deleted region.

♦ Deletions can also be defined by their behavior in fine structure recombination mapping. A deletion is a mutation that does not recombine with 2 other mutations which do recombine with each other (consider deletion #3 in Problem 7-17).

♦ Fine structure recombination analysis can be used to order closely linked mutations. When 2 closely linked mutations are in trans, the rare single recombination events between these mutations will give one product that is wild type for the phenotype. By examining the genotypes of flanking markers it is possible to order the original mutations (see Problem 7-18). Recombination frequency = 2(# wild type recombinant progeny) / total progeny.

♦ An enzymatic pathway is a series of steps each catalyzed by a gene product (enzyme). To order the compounds, work from the final product toward the beginning of the pathway. Remember that the final product is the compound on which all the mutants will grow. If there is a mutation early in the pathway, providing later intermediate compounds will allow growth, because the subsequent enzymes are available. The block point has effectively been bypassed in these cases.

Solutions to Problems:

7-1. This answer has been grouped 2 different ways: as the terms (numbered column) that apply to each mutational change (lettered column) and vice versa.

a. **1, 5**; b. **1, 2, 5**; c. **2**; d. **8**; e. **9**; f. **3, 6, 9**; g. **4, 7**; h. **1, 5**; i. **6, 8, 9**; j. **3, 4**; k. **6, 7**.

Presented alternately:

1. **a, b, h**; 2. **b, c**; 3. **f, j**; 4. **g, j**; 5. **b, h**; 6. **f, i, k**; 7. **g, k**; 8. **d, i**; 9. **e, g, i**.

7-2. Each independently derived mutation will be caused by a different single base change. When you find a base that differs in only one of the sequences, it is the mutation. Determine the wild-type sequence by finding the base that is present at that position in the other two sequences. The wild-type sequence is therefore:

5' ACCGTAGTCGACTGGTAAACTTTGCGCG

7-3. Of the achondroplasia births observed, 23/27 are due to new mutations because there is no family history of dwarfism. Achondroplasia is an autosomal dominant trait, so it will be expressed in the child that receives the mutant gamete. There were 120,000 births registered, so there were 240,000 parents in which the mutation could have occurred during meiosis, leading to a mutant gamete. **The mutation rate = 23 mutant gametes/240,000 gametes = 9.5×10^{-5}. This rate is higher than 2 to 12×10^{-6} mutations per gene per generation which is the average mutation rate for humans**.

7-4. Dominant mutations can be detected immediately in the heterozygous progeny who receives the mutant gamete (see Problem 7-3). **Recessive mutations can only be detected when they are homozygous. To detect the appearance of new recessive alleles you must test cross with a recessive homozygote.** This can be done in mice, where the researcher can control the mating, but it cannot be done in humans!

7-5. Kim's hypothesis is that the bacteriophage is able to induce resistance in ~1 in every 10^4 bacteria. If she is right, then several (~10 if there are 10^5 bacteria on each plate) of the colonies on each of the replica plates should contain resistant cells that will continue to grow after exposure to phage. **These resistant cells and the surviving colonies that grow from them will be randomly distributed over the three plates**. **Maria**'s hypothesis is that the resistant cells are already present in the population of cells they plated on the original plate. If she is right then some of the colonies on the original plate (about 10 of them) should have contained resistant bacteria that will give rise to colonies after treatment with the phage. **These resistant colonies will appear at the same locations on all three of the replica plates. (See Figure 7.4e)**.

7-6. To ensure that the mutants you isolate are independent you should follow procedure #2. If you follow procedure #1, a mutation causing resistance to the phage could have arisen several generations before the time when you spread the culture and several of the colonies you isolate could be clones of the same cell and thus have the identical mutation. Different mutations in the same gene often give different information about the role of the gene product. Therefore, geneticists generally strive to find many independent and thus different mutations in the same gene in order to understand as much about the gene as possible.

7-7. Diagram the crosses (InX^{Bar} = X chromosome with inversions and the mutant allele of Bar eyes, X* = mutagenized X chromosome:

X / InX^{Bar} ♀ × X* / Y → F_1 InX^{Bar} / X* ♀ × X / Y → 1/2 InX^{Bar} / Y (Bar ♂) : 1/2 X* / Y (wild type ♂). These F_1 females must also have had daughters: 1/2 InX^{Bar} / X (Bar ♀) : 1/2 X* / X (normal ♀). However, there were 3 F_1 females who did NOT give these ratios of sons: ♀A, ♀B and ♀C.

F_1 InX^{Bar} / X^A* (♀A) x X / Y → 1/2 ♀ : 1/4 *Bar* ♂ : 1/4 white ♂

F_1 InX^{Bar} / X^B* (♀B) x X / Y → 2/3 ♀ : 1/3 *Bar* ♂

F_1 InX^{Bar} / X^C* (♀C) x X / Y → 4/7 ♀ : 2/7 *Bar* ♂ : 1/7 normal ♂

Remember that each F_1 daughter inherited a different mutagenized X chromosome from her mutagenized male parents. Thus, ♀A, ♀B and ♀C all are each heterozygous for a different mutagenized X chromosome. Any unusual results obtained with these females is due to some alteration of the X that underwent the mutagenesis. **The mutagenized X chromosome carried by ♀A must have a recessive white-eye mutation**. There is no effect on viability, but the sons who inherited mutant (non-*Bar*) X chromosome (1/2 of the total sons) have white eyes. In ♀B, there is a reduction in the number of males by half. This indicates that **♀B is heterozygous for a mutagenized X chromosome with a new recessive lethal mutation**, and the sons that inherit that X chromosome die. Female C produced fewer sons than expected among those that inherited the mutagenized X chromosome; however, some of the sons that inherited that chromosome survived. Thus, **the mutagenized X chromosome from ♀C may contain a recessive mutation that causes lethality but is not completely penetrant, or she could be a mosaic** where some of her cells are of one genotype, while other cells are a different genotype. In the case of mosaicism, ♀C inherited a mutagenized X chromosome from her father in which one strand of DNA contained a lethal mutation and the other strand had the wild-type sequence. This scenario implies that DNA replication or repair mechanisms have not yet acted to make sure both strands were completely complementary in sequence. As a result, some of her tissues will have the wild type chromosome from her father and others will have the mutant homolog. All of the cells in ♀C will have a second X chromosome, the InX^{Bar} chromosome that she inherited from her mother. Inheritance of these chromosomes could cause the germ line of ♀C to be a mosaic that contains a mix of gametes: InX^{Bar}, wild-type and the lethal mutation.

7-8. Diagram the cross; * denotes the mutagenized fly:

   wild type ♀ × wild type* ♂ → F$_1$ ♀ × *y cv ct sn m* → 1/3 wild type ♀ : 1/3 *ct sn* ♀ : 1/3

   wild type ♂

a.   X-rays cause breaks in DNA. The *ct sn* ♀ in the second generation suggests that **the mutagenized wild type X chromosome is missing the *ct*, *sn* region**. Thus this mutagenized chromosome has a deletion that removes both *ct*$^+$ and *sn*$^+$ and therefore uncovers the recessive alleles on the other homolog.

b.

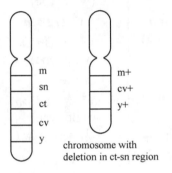

chromosome with
deletion in ct-sn region

c.   *ct sn* ♀ (from 2nd generation of cross above) × wild type ♂. The genotype of this female is *y cv ct sn m / y*$^+$ *cv*$^+$ (del) *m*$^+$. Therefore, **this female will make a variety of non-recombinant and recombinant gametes**. The table shows the reciprocal pairs of parental gametes and the SCO classes: region 1 between *y* and *cv*; region 2 between *cv* and one end of the deleted region and region 3 between the other end of the deleted region and *m*):

female gametes	male gamete $y^+ cv^+ ct^+ sn^+ m^+$	male gamete Y
$y\ cv\ ct\ sn\ m$ (parental)	$y\ cv\ ct\ sn\ m\ /$ $y^+ cv^+ ct^+ sn^+ m^+$ (wild type)	$y\ cv\ ct\ sn\ m\ /\ Y$ (y cv ct sn m ♂)
$y^+ cv^+$ (del) m^+ (parental)	$y^+ cv^+$ (del) m^+ / $y^+ cv^+ ct^+ sn^+ m^+$ (wild type)	$y^+ cv^+$ (del) m^+ / Y (dead)
$y\ cv^+$ (del) m^+ (SCO region 1)	$y\ cv^+$ (del) m^+ / $y^+ cv^+ ct^+ sn^+ m^+$ (wild type)	$y\ cv^+$ (del) m^+ / Y (dead)
$y^+ cv\ ct\ sn\ m$ (SCO region 1)	$y^+ cv\ ct\ sn\ m$ / $y^+ cv^+ ct^+ sn^+ m^+$ (wild type)	$y^+ cv\ ct\ sn\ m$ / Y (cv ct sn m ♂)
$y^+ cv^+ ct\ sn\ m$ (SCO region 2)	$y^+ cv^+ ct\ sn\ m$ / $y^+ cv^+ ct^+ sn^+ m^+$ (wild type)	$y^+ cv^+ ct\ sn\ m$ / Y (ct sn m ♂)
$y\ cv$ (del) m^+ (SCO region 2)	$y\ cv$ (del) m^+ / $y^+ cv^+ ct^+ sn^+ m^+$ (wild type)	$y\ cv$ (del) m^+ / Y (dead)
$y^+ cv^+$ (del) m (SCO region 3)	$y^+ cv^+$ (del) m / $y^+ cv^+ ct^+ sn^+ m^+$ (wild type)	$y^+ cv^+$ (del) m / Y (dead)
$y\ cv\ ct\ sn\ m^+$ (SCO region 3)	$y\ cv\ ct\ sn\ m^+$ / $y^+ cv^+ ct^+ sn^+ m^+$ (wild type)	$y\ cv\ ct\ sn\ m^+$ / Y (y cv ct sn ♂)

The female will also make rarer DCO classes of recombinants. Note that half of all the sons inherit the deletion, which is lethal when hemizygous because many essential genes are missing.

7-9. Yes. The rat liver supernatant contains enzymes that convert substance X to a mutagen, and *his*$^+$ revertants occur. Our livers contain similar enzymes that process substances, converting them into other forms that cause mutation and can lead to cancer.

7-10. The mutagen initially mutates somatic cells, not a gamete-producing cell. These somatic cell mutations give rise to the tumor cells. **When the tumor cells are injected into a new mouse, they will divide in an uncontrolled manner and cause a tumor to devolop. The somatic mutation that caused the original cell to become cancerous are not present in the germ cells of the mouse.** Thus, the cancer phenotype cannot be inherited in a Mendelian fashion.

7-11. Cross each of the reversed (reverted) colonies to a wild type haploid to generate a heterozygous wild type diploid. Then sporulate the diploid and examine the phenotype of the spores. **If the met^+ phenotype is due to a true reversion, then the cross was:** $met^- \times met^+ \rightarrow met^+ / met^- \rightarrow 2$ $met^+ : 2\ met^-$. If there is an unlinked suppressor mutation in another gene, the suppressor (su^-) and original met^- mutation should assort from each other during meiosis: $met^-\ su^-$ (phenotypically met^+) $\times met^+\ su^+$ (wild type) $\rightarrow met^- / met^+ ; su^- / su^+ \rightarrow 1/4\ met^-\ su^-$ (met^+) : $1/4\ met^-\ su^+$ (met^-) : $1/4\ met^+\ su^-$ (met^+) : $1/4\ met^+\ su^+$ (met^+) $= 3/4\ met^+ : 1/4\ met^-$. If the two genes are linked, you will still get the four classes but the proportions of the recombinant gametes ($met^-\ su^+$ being the one you can detect as it is the only one with a met^- phenotype) will increase proportionate to the distance between the 2 genes.

7-12. Do a complementation test by mating the two mice. If the mutations in each mouse are in the same gene all the progeny will be mutant (albino). If the mutations causing albininsm in each mouse are in different genes, all the progeny will be wild-type (the genes will complement).

7-13. Mutations 5 and 6 do not revert, so they are deletions. Deletions are not included in the complementation groups shown in part a below. All the rest of the mutations do revert, so they are point mutations.

a. When analyzing complementation data, group the mutants that do NOT complement as these are mutations of the same gene. Mutation 1 does not complement mutation 8, but these 2 mutations do complement all the other point mutations. Therefore, mutations 1 and 8 make up one complementation group. Mutation 2 complements every other mutation, therefore it is the sole mutation in another complementation group. Mutation 3 does not complement 4 nor 7, so these form a third complementation group. **In total there are three complementation groups: (1, 8); (2); and (3, 4, 7).**

b. The diploid cells from part a are allowed to undergo meiosis. **If the diploid is heterozygous for 2 mutations on separate chromosomes, then independent assortment will occur**: $2 / 2^+ ; 1 / 1^+ \rightarrow 1/4\ 2\ 1$ (lys$^-$) : $1/4\ 2\ 1^+$ (lys$^-$) : $1/4\ 2^+\ 1$ (lys$^-$) : $1/4\ 2^+\ 1^+$ (lys$^+$) $= $ **3/4 lys$^-$: 1/4 lys$^+$**. If the 2 mutations in the heterozygous diploid are in genes on the same chromosome, or if the 2 mutations are in the same gene, **recombination can occur between and within genes, producing prototrophic (lys$^+$) spores**. The example below shows mutations 1 and 8, which are in the same complementation group. After the recombination event shown, the 4 spores will be $1/4\ 1^+\ 8$ (lys$^-$) : $1/4\ 1^+\ 8^+$ (lys$^+$) : $1/4\ 1\ 8$ (lys$^-$) : $1/4\ 1\ 8^+$ (lys$^-$) $= $ **3/4 lys$^-$: 1/4 lys$^+$**. In the case of genetic linkage, the numbers of tetrads showing the 3/4 lys⁻ : 1/4 lys⁺ ratio will depend on the distance between the mutations.

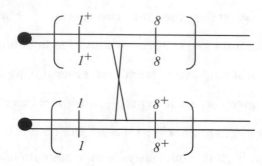

c.   A '−' result in the recombination data in part b means that the 2 mutations in the diploid cannot recombine. Therefore, they must affect the same nucleotide, since recombination occurs between adjacent nucleotides (**see Figure 6.22 step 1**). This is expected for the diploids that were generated by mating the same mutation (e.g. *1 × 1*), as by definition these occupy the same position. All point mutations recombine with all other point mutations, so no two point mutations affect the same nucleotide. Note that mutations *5* and *6*, which are known deletions (do not revert) do not recombine with various of the point mutants. In the complementation data, deletion *5* acts like any of the point mutations in the complementation group (*4, 3, 7*). However, it acts differently in the recombination data. It does not recombine with (overlaps) point mutations *4* and *3*, but it does recombine with *7*. Thus it is possible to genetically define mutation 5 as a deletion = a mutation that does not recombine with 2 other mutations that DO recombine with each other. If two point mutations recombine, they must affect different nucleotides, and because the deletion fails two recombine with the two mutations, it must remove more than one nucleotide. Deletion 6 does not recombine with mutations *1* nor *4*. Because *1* and *4* are in different genes (complementation groups), this deletion must span the distance between these 2 genes as well as 'uncover' the point mutations themselves. Deletion *6* does recombine with mutation *8*, which places *8* on the far side of its gene from the gene containing mutation *4*. (**See Figure 7.18a**). Combining this information with the complementation data from part a allows you to **draw the map below**:

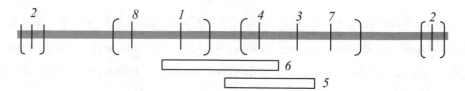

The location of gene 2 (to the right or left side) cannot be determined from this data.

7-14.

a. **The starting tube (call it tube A) contains 5 ml of bacteriophage at a concentration of 1.5×10^{10}; take 1 µl (0.001 ml) of tube A, corresponding to 1.5×10^7 phage, to 999 µl (in** practice, 1 ml) of diluent in tube B. This step is a 10^{-3} dilution. Repeat this step with 1 µl of tube B (1.5×10^4 phage) and mix it with 999 µl of diluent in tube C (10^{-6} dilution). Next take 1 µl of tube C (1.5×10^1 phage) and mix it with bacteria (about 100x more cells than phage = a low multiplicity of infection (MOI)). Allow the phage to infect the cells, then add to a top agar and pour on an agar plate. Repeat the infection/top agar step with 10 µl of tube C (1.5×10^2 phage) and plate. There should be 15 plaques on the first plate and about 150 plaques on the second plate.** This describes only one of many possible protocols: other dilution steps, such as 10^{-2} dilutions or 10^{-1} dilutions, could also be employed.

b. **To figure out the total number of phage, you need to look at a particular dilution in the electron microscope and count all the phage particles. The ratio of plaques to total phage is the plating efficiency. In part a, it is fair to assume that only one phage initiated each plaque because of the very low MOI. Because there were many more bacterial cells than phage, the chances are very high that any individual bacterial cell could have been infected only by a single phage.**

7-15.

a. Deletion mutations can be identified in a couple of ways. **Deletions are mutations that never revert to wild type. Deletions are also mutations that don't recombine with 2 other mutations that do recombine with each other**. For example, mutations a and b <u>do</u> recombine with each other, but mutation c does not recombine with mutation a nor mutation c; this implies that mutation c is a deletion (<u>see problem 7-13c above</u>).

b. The length of the T4 chromosome in micrometers predicts the number of nucleotide pairs because the physical size of the molecule is known from the Watson and Crick model. Recombination analysis with many mutants distributed over the T4 chromosome suggests the total map units in the T4 genome. Thus, **Benzer could estimate the number of map units/nucleotide pair. He then compared this to the smallest distance he could measure between 2 mutations by recombination.**

c. *rII⁻* **mutants in the same nucleotide pair cannot recombine with each other to produce *rII⁺* phage** (again, <u>see problem 7-13c above</u>).

7-16.

a.   **There are two complementation groups and therefore two genes**.

b.   **The complementation groups are (1, 4) and (2, 3, 5)**.

7-17.

a.   Based on the complementation data (the first table), mutation 6 is almost certainly a deletion, because it doesn't complement with any of the other mutations. The second table gives the recombination results. A '+' in this table is the result of recombination that occurred in the *E. coli* B host to generate wild-type phage that can now grow in *E. coli* K (λ). The '-' designation indicates that a few phage were able to form plaques. These are revertants. **Remember that point mutations can revert to wild type, while deletion mutations cannot**. Mutants 3, 6, and 7, did not form any plaques here (as designated by '0'), so these three must be deletions. **Thus, mutants 3, 6, and 7 are deletion mutants (non-reverting)**.

b.   The first table, based on the coinfection of *rII* mutants into *E. coli* K(λ), gives the results of complementation analysis and lets us place mutations in the two *rII* complementation groups. Notice that deletion 6 does not complement any of the mutants so it must delete at least part of each of the two *rII* genes The two complementation groups are (1, 2, 5) and (4, 8, 9). The second complementation group is the *rIIB* gene as defined by the problem. Next use the recombination data to order the mutations with respect to the deletions. Deletion 6 does not recombine with mutation 1 (*rIIA*) nor with mutations 8 and 4 (*rIIB*), so these mutations must be near each other because deletion 6 spans both genes. Deletion 3 allows you to order 8 (outside deletion 3), then mutation 4 (overlaps both deletions 6 and 3), then mutation 9 (only overlaps deletion 3). Mutations 2 and 5 cannot be ordered relative to each other, except to say that they do not overlap deletion 6, and so they are surrounded by {}. **The map is:**

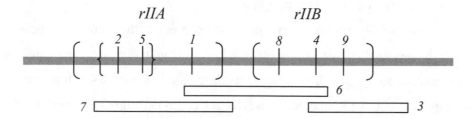

c.   **The order of mutations 2 and 5 in *rIIA* cannot be determined form this data. To determine the order, you would need to use other deletions that occur in *rIIA* in recombination testing, or cross mutants 2 and 5 and test for linkage to appropriate genetic markers in genes that lie to either side of the *rIIA/B* loci.**

7-18.

a. Diagram the cross:

$Ly^+ ry^{41} Sb / Ly ry^{564} Sb^+ ♀$ x $ry^{41} / ry^{41} ♂$ → 8 $Ly ry^+ Sb$ and lots of ry progeny

Recombination within the *rosy* gene, between the two *ry* mutations in the heterozygous female, generates the eight offspring with wild type eyes. Such a recombination event will give one recombinant gamete with the wild type sequence for both mutations, thus giving a ry^+ phenotype in the progeny of this cross. The reciprocal recombinant gamete will be a double mutant, ry^{41} ry^{564}, which will yield *ry* progeny indistinguishable from the parental type *ry* progeny. We assume the $ry^{41} ry^{564}$ progeny are found in equal numbers to the ry^+ recombinants, so the **recombination frequency = (8 + 8) / 100,000 = 0.0016%. The distance between ry^{41} and ry^{564} = .0016 mu**.

b. The wing and bristle phenotypes of those eight recombinant offspring are a consequence of the order of the ry^{41} and ry^{564} with respect to the flanking markers, the *Ly* and *Sb* genes. Try both orientations of the two *ry* mutations to see which order produces wild-type eyes together with Lyra wings and Stubble bristles as the result of a crossover between them.

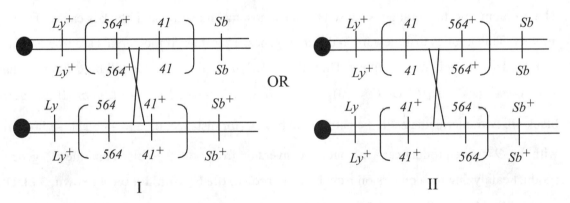

Orientation II produces the $Ly ry^+ Sb$ recombinants obtained so the order must be $Ly ry^{41}$ $ry^{564} Sb$.

7-19.

a. Diagram the cross, assuming the genes are unlinked (**see Figure 5.17a-c**):

$argE^- argH^+$ × $argE^+ argH^-$ → $argE^-/argE^+$; $argH^-/argH^+$

When this diploid is sporulated, the PD asci are: 2 $argE^- argH^+$ (arg⁻) : 2 $argE^+ argH^-$ (arg⁻), and the NPD asci are: 2 $argE^+ argH^+$ (arg⁺) : 2 $argE^- argH^-$ (arg⁻). The frequency of PD spores = frequency of NPD. Next, diagram the cross assuming the genes are linked (**see Figure**

<u>**5.18 a and f**</u>): $argE^-\ argH^+$ x $argE^+\ argH^-$ $\rightarrow$ $argE^-\ argH^+$ / $argE^+\ argH^-$. In this case, the PD ascus is the same as above: **2 $argE^-\ argH^+$ (arg⁻) : 2 $argE^+\ argH^-$ (arg⁻). The NPD asci are the same as above also: 2 $argE^+\ argH^+$ (arg⁺) : 2 $argE^-\ argH^-$ (arg⁻). In this case there would be many more PD asci than NPD asci.** In all cases the distribution of spores will show MI segregation for both genes (<u>**see Figure 5.23**</u>). In other words, the center of the ascus corresponds to the ':' in the ratios above.

b. **The $argE^-\ argH^+$ spores of the PD will grow when you supplement the media with ornithine, citrulline, arginosuccinate or arginine. For the 2 $argE^+\ argH^-$ spores, only arginine itself in the media allows growth. In the case of the 2 $argE^-\ arg\ H^-$ spores of the NPD asci, only arginine allows growth. The two 2 $argE^+\ argH^+$ spores are prototrophs that grow on minimal medium without supplementation.**

7-20.

a. Diagram the cross:

orange $\times$ black $\rightarrow$ F_1 brown

The problem says that orange is caused by one autosomal mutation and black is caused by another. This implies that they are in different genes. Therefore, if the parents are true breeding then the F1 is doubly heterozygous. Thus, the underlying ratio in the F2 will be; *BB oo* (orange) $\times$ *bb OO* (black) F_1 *Bb Oo* (brown) $\rightarrow$ F_2 9 *B- O-* : 3 *B- oo* : 3 *bb O-* : 1 *oo bb*. If orange and black are two intermediates in the pathway to brown (for instance, orange $\rightarrow$ black $\rightarrow$ brown), with the O^+ gene product carrying out the conversion from orange to black and the B^+ gene product catalyzing the conversion from black to brown, **the F$_2$ would have 9 brown, 3 black, 4 orange**. In other words, epistasis would be seen. (Note: We don't know the order of orange and black in this pathway. **If the order were black $\rightarrow$ orange $\rightarrow$ brown, a 9 brown : 3 orange : 4 black ratio would be seen.**)

b. If there are only two pathways, one producing orange and the other black, then there would be four different phenotypes in the F$_2$ generation: **9 brown (*O- B-*) : 3 black (*oo B-*) : 3 orange (*O- bb*) : 1 nonpigmented (*oo bb*).**

7-21. Designate the genes and alleles. The *W* gene product converts a colorless (white) pigment to green. The *G* gene product converts green to blue flowers, the mutant allele is *g*. Either of two gene products *B* or *L* can convert blue to purple flowers; *b* and *l* are the mutant alleles. Diagram the cross.

Note that both parents are *WW*. All progeny will be *WW*, and it will not affect the array of phenotypes in the progeny. For this reason, it is not considered in this cross:

gg BB LL (green) × *GG bb ll* (blue) → F$_1$ *GgBbLl* → F$_2$:

3/4 *G-* × 3/4 *B-* × 3/4 *L-* = 27/64 *G- B- L-* (purple);

3/4 *G-* × 1/4 *bb* × 3/4 *L-* = 9/64 *G -bb L-* (purple);

3/4 *G-* × 3/4 *B-* × 1/4 *ll* = 9/64 *G- B- ll* (purple);

3/4 *G-* × 1/4 *bb* × 1/4 *ll* = 3/64 *G- bb ll* (blue);

1/4 *gg* × 3/4 *B-* × 3/4 *L-* = 9/64 *gg B- L-* (green);

1/4 *gg* × 1/4 *bb* × 3/4 *L-* = 3/64 *gg bb L-* (green);

1/4 *gg* × 3/4 *B-* × 1/4 *ll* = 3/64 *gg B- ll* (green);

1/4 *gg* × 1/4 *bb* × 1/4 *ll* = 1/64 *gg bb ll* (green)

The ratio is 45 purple : 16 green : 3 blue. You can see why the problem specified that the green parent was mutant in only a single gene, as *gg bb LL* or *gg BB ll* plants would still be green yet would yield a very different ratio of phenotypes in the F$_2$.

7-22.

a.   The deleted homolog has a known null activity allele for all the genes within the deletion (because the genes are not present at all). Therefore, **if a *mutant / deletion* genotype has the same mutant phenotype as the *mutant / mutant* genotype, this suggests that the mutant allele has the same level of activity as an allele with known zero activity (the deletion). One limitation of this assumption is that some phenotypes have a threshold level of enzyme activity. In other words, the mutant phenotype is seen as long as the level of enzyme activity is below some critical threshold.** Once the level of enzyme activity rises above this level, the phenotype becomes wild type. For example, imagine a situation where any individual has the mutant phenotype if the enzyme activity is <30%. Suppose allele *m* of an autosomal gene produces an enzyme with 20% of the activity as the enzyme encoded by the wild-type allele. Thus, an *m / m* individual would have 20% the enzyme activity of wild-type homozygotes, and an *m / deletion* heterozygote would have only 10% of normal enzyme activity. Because of the threshold, both of these individuals would have the same mutant phenotype, yet the *m* allele is clearly not null.

b.   You can determine that a mutant allele of a gene is a true null activity allele **if an assay for enzyme activity of the encoded protein shows none present in individuals homozygous for that allele (or in *mutant / deletion* heterozygotes).** In some cases, antibodies can be used to determine the amount of gene product present in individuals of these genotypes. Thus, **if the antibody does not detect any protein, you can safely assume there is no enzyme activity = null allele**. Remember that the converse of this statement is not true: the presence of protein as detected with an antibody does NOT mean there is enzyme activity, as even the change of a single amino acid in a large protein can completely abolish its function.

7-23.

a.   Diagram the crosses:

1. blue × white → purple → 9 purple : 4 white : 3 blue

2. white × white → purple → 9 purple : 7 white

3. red × blue → purple → 9 purple : 3 red : 3 blue : 1 white

4. purple × purple → purple → 15 purple : 1 white

3EQ #1 - there are 2 genes controlling the phenotypes in each cross, because all four crosses show epistatic modifications of the 9:3:3:1 ratios; 3EQ #2 - in all 4 crosses the purple phenotype corresponds to the "A- B-" class; 3EQ #3 - none of the genes are X-linked. Assign genotypes in all the crosses:

Cross 1. *AA bb* (blue) × *aa BB* (white) → *Aa Bb* (purple) → 9 *A- B-* (purple) : 4 *aa --* (white) : 3 *A- bb* (blue)

Cross 2. *AA bb* (white) × *aa BB* (white) → *Aa Bb* (purple) → 9 *A- B-* (purple) : 7 *aa --* + *-- bb* (white)

Cross 3. *AA bb* (red) × *aa BB* (blue) → *Aa Bb* (purple) → 9 *A- B-* (purple) : 3 *A- bb* (red) : 3 *aa B-* (blue) : 1 *aa bb* (white)

Cross 4. *AA bb* (purple) × *aa BB* (purple) → *Aa Bb* (purple) → 15 *A- --* + *-- B-* (purple) : 1 *aa bb* (white)

b.   1. colorless $\xrightarrow{\text{A}}$ blue $\xrightarrow{\text{B}}$ purple

2. colorless1 $\xrightarrow{\text{A}}$ colorless2 $\xrightarrow{\text{B}}$ purple

3. colorless $\xrightarrow{\text{A}}$ red   (red pigment + blue pigment = purple)

   colorless $\xrightarrow{\text{B}}$ blue

4. colorless $\searrow^{\text{A}}$

                  purple

   colorless $\nearrow_{\text{B}}$

c. **Cross #2** is compatible with a single-step pathway in which genes *A* and *B* encode two different subunits of a multimeric enzyme that catalyzes the step. In such a case, enzyme activity would result only if at least one allele of each gene were the dominant allele specifying subunit production.

d. Assuming that "tightly linked" means the distance between the *A* and *B* genes is 0 mu, we can rewrite all four of the crosses in the form: *A b / A b* × *a B / a B* → *A b / a B* (selfed) → 1/4 *A b / A b* : 1/4 *A b / a B* : 1/4 *a B / A b* : 1/4 *a B / a B*. The phenotypes for each cross:

Cross 1. 2 purple (*A b / a B* + *a B / A b*) : 1 blue (*A b / A b*) : 1 white (*a B / a B*)

Cross 2. 1 purple (*A b / a B* + *a B / A b* + *A b / A b*) : 1 white (*a B / a B*)

Cross 3. 2 purple (*A b / a B* + *a B / A b*) : 1 red (*A b / A b*) : 1 blue (*a B / a B*)

Cross 4. all purple (*A b / a B* + *a B / A b* + *A b / A b* + *a B / a B*)

Note that if there is any recombination between the two genes, then recombinant *a b* gametes will be produced, allowing the emergence of white F$_2$ plants (*a b / a b*) in crosses 3 and 4. The greater the distance between the two genes, the more the ratios will resemble those in part a.

7-24. First, order the compounds from final product to first one in the pathway. The final compound is the one on which all of the mutants in the pathway will grow. The compound before that (E in this example) is the one that allows all the mutants except one class to grow. Continue working toward the beginning of the pathway in this manner. Next, order the mutants. Again, you can do this by working backwards from the final product through the intermediates, look for the mutant which grows only when supplied with G. In this problem it is mutant 2. The mutation must be in the gene encoding the enzyme catalyzing the last step synthesizing compound G. Then look for the mutant that grows only when supplied with G or one other intermediate. Mutant 7 can grow only when supplied with intermediate E or with G. This verifies our earlier assignment of E as the intermediate that precedes G, and it also tells us that the gene in which mutation 7 is located encodes the enzyme that allows the synthesis of E. In this way, continue working back through the pathway to get the answer.

$$X \xrightarrow{6} F \xrightarrow{1} D \xrightarrow{5} A \xrightarrow{3} C \xrightarrow{4} B \xrightarrow{7} E \xrightarrow{2} G$$

7-25.

a. This problem is worked in the same way as problem 7-24.

$$X \xrightarrow{18} D \xrightarrow{14} B \xrightarrow{9} A \xrightarrow{10} C \xrightarrow{21} \text{thymidine}$$

b.   **Double mutant 9 and 10 blocks at the B to A step first, and so it accumulates intermediate B. Similar reasoning predicts that double mutant 10 and 14 would accumulate intermediate D.**

7-26. To solve this problem, consider first only those mutants that are only defective in biosynthesis of one amino acid. The mutants defective in only the proline pathway are those that grow when given proline in the media but not when given glutamine. Mutants 2, 6, and 1 are of this type. There is no intermediate that allows the growth of mutant 2 so the defect must be in the final enzyme that produces proline. Working backward from this point in the pathway, mutant 6 grows when supplied with intermediate A, so A is the final intermediate and mutant 6 is blocked in the step that leads to A. Mutant 1 grows when supplied with intermediates E or A, indicating that E is prior to A in the proline pathway. Now conduct the same analysis for glutamine. Mutants 7 and 4 are only defective in glutamine biosynthesis. Mutant 4 grows only on glutamine, while mutant 7 grows when supplied with B or glutamine, indicating that mutant 7 is blocked in the production of B, and mutant 4 cannot convert intermediate B to glutamine. Now look at the mutants that area defective in both glutamine *and* proline biosynthesis. Mutants 5 and 3 are of this type. Mutant 3 grows only if given intermediate C, so it must be blocked just prior to this step. Mutant 5 grows if given C or D, so it is blocked prior to the D intermediate. This represents the first part of the pathway that is used both in proline and glutamine biosynthesis. Putting all of this information together, we have the following branching pathway:

$$X \xrightarrow{5} D \xrightarrow{3} C \xrightarrow{7} B \xrightarrow{4} gln$$
$$\xrightarrow{1} E \xrightarrow{6} A \xrightarrow{2} pro$$

7-27. This problem is similar to an ordered enzymatic pathway except the gene products are a series of nonenzymatic proteins that make up a structure that is assembled in a particular order. The loss of one protein due to mutation will prevent all the subsequent proteins from being added. The loss of the first protein at the surface would prevent all others from being at the cell surface. The mutant that fits this description is E. Mutants A and C have a similar pattern in which only E and C or A respectively are at the surface, so genes A and C encode the two proteins that form the dimer structure shown second from the embryo surface. The logic is continued to place the remaining three gene products in their order.

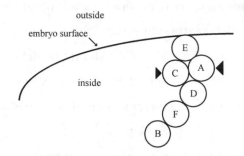

7-28. The data represent complementation experiments done at the biochemical level in the test tube. The results suggest that **there are 2 different X-linked genes that cause hemophila when mutant**. Individuals 1 and 2 are mutant in one gene (call it gene A) and individuals 3 and 4 are mutant in the other gene (gene B). Individuals 1 and 2 thus lack the function of one factor needed for clotting, while 3 and 4 lack a different factor. When the two kinds of blood complement (as in the mixture of blood from individuals 1 and 3), clotting occurs because the blood of each patient supplies the factor lacking in the other patient.

The results exclude a pathway in which the product of one of these genes (say the *A* gene) is required for the synthesis of the protein encoded by other gene (*B*). In such a case, the blood of a patient mutant for gene *A* would have neither protein A nor protein B, so the mixture of mutant bloods would have no source of compound B. Another excluded scenario is one in which protein A is a substrate for a reaction catalyzed by enzyme B, and this reaction could not take place in the test tube (for example, if protein A were rapidly degraded if not immediately converted into something else by the action of enzyme B). Many other linear, convergent, or divergent pathways are still consistent with the results.

As an interesting historical sidelight, the cited article in the *British Medical Journal* was published in the December 27th (Christmas) issue of 1952. The first patient whose blood could complement that of most other hemophiliacs in the test tube (thus indicating the existence of two different kinds of X-linked hemophilia) was a 5 year-old boy whose family name was Christmas. Because of these facts, the rarer form of the disease, usually called hemophilia B, is still often called Christmas disease.

7-29. Remember that the von Willebrand factor is necessary at fairly high levels to stabilize factor VIII.

a.   This should be a **successful treatment** because the normal plasma contains both vWF and factor VIII. The **effect should be immediate** because both factors are present **and prolonged** because vWF stabilizes factor VIII.

b. This treatment should **not be successful** because vWD plasma has neither vWF nor factor VIII.

c. This should be a **successful treatment** because the hemophilia A plasma contains vWF even though it does not have factor VIII. The **effect should be delayed** because the patient has no factor VIII and the added vWF will only stabilize factor VIII newly synthesized by blood cells in the patient, and this takes time. The **effect should also be prolonged** because vWF stabilizes factor VIII.

d. This should be a **successful treatment** because the normal blood contains both vWF and factor VIII. The **effect should be immediate** because both factors are present **and prolonged** because factor VIII in the transfused plasma is already stabilized.

e. This **treatment will not be successful**. The vWD plasma has neither vWF nor factor VIII, and the patient cannot synthesize any factor VIII.

f. This **treatment should not be successful**. Neither the patient's blood nor the transfused plasma has any factor VIII.

g. This treatment **should be successful, but only after a delay** to allow the patient's blood cells to synthesize enough factor VIII that can be stabilized by the injected vWF. The **effects should be prolonged** because of the stabilization.

h. This treatment should be **unsuccessful** because no factor VIII can be made.

i. This treatment **should be successful immediately** because you are injecting factor VIII, but the **effects will be only very short-term** because the injected factor VIII will be degraded in the absence of vWF.

j. This treatment **should be successful immediately, and the effects will be prolonged** because the patient's blood has vWF which can stabilize the injected factor VIII.

7-30.

a. Two loci are needed, one for the α globin polypeptides and one for the β globin polypeptides.

b. Assuming that both alleles of both genes are expressed at the same levels, then you would see 1/2 α1 : 1/2 α2 for the 2 forms of the hemoglobin α subunit and 1/2 β1 : 1/2 β2 for the forms of the β subunits. The α subunits will form the following sorts of dimers: 1/4 α1α1 : 1/2 α1α2 : 1/4 α2α2. The β subunits will assemble into dimers in the same way, giving a genotypic monohybrid ratio of: 1/4 β1β1 : 1/2 β1β2 : 1/4 β2β2. In order to figure out the types of hetero-tetramers and their frequencies, apply the product rule to the 2 monohybrid ratios: **1/16 α1α1 β1β1 1/8 α1α2 β1β1 : 1/16 α2α2 β1β1 : 1/8 α1α1 β1β2 : 1/4 α1α2 β1β2 : 1/8 α1α2 β2β2 : 1/16 α1α1 β2β2 : 1/8 α1α2 β2β2 : 1/16 α2α2 β2β2**.

7-31. In unequal crossing-over the homologous chromosomes align out of register, and instead use homology between related but different genes. **The result of crossing-over is a homolog with a duplication, β δ/β δ, and a homolog with a deletion, β/δ.** The genes with a slash indicate hybrid genes. Even though these genes are very similar, they do have differences. One of the more important differences between the δ and β genes is the time of expression. In these hybrid genes the regulatory region of the β gene may have been replaced with the regulatory information of the δ gene, or vice-versa, so the hybrid gene may be expressed inappropriately (that is, during the wrong time in development). In addition, the polypeptides formed from the hybrid genes may affect the affinity of hemoglobin for oxygen in unpredictable ways.

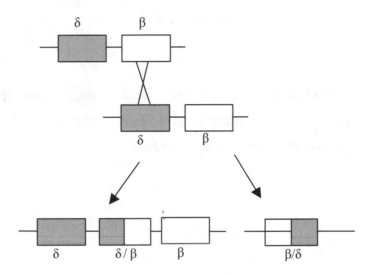

7-32. <u>See Figures 7.25, 7.26 and 7.27</u>.

a. **Null mutations have no functional gene product. Hypomorphic mutations have a lower level of protein activity while hypermorphic mutations have a higher level of activity. Dominant negative mutations interfere with the functioning of the normal polypeptide made by the other allele or they interfere with the functioning of other proteins that interact with the gene product. Neomorphic mutations exhibit a new phenotype.**

b. **Null and hypomorphic mutations would usually be recessive** to wild-type, unless the phenotype is particularly sensitive to decreases in the amount of a gene product (as in the cases of incomplete dominance or of haploinsufficiency). **Hypermorphic and neomorphic mutations would probably be dominant,** but it is possible that in rare cases, the phenotype might not be sensitive to the extra or new function unless the mutant allele were present in two doses. **Dominant negative mutations are dominant by definition.**

7-33. The problem states that the contributions of each allele are additive. The 100% level must be two wild type alleles, so each β^+ allele contributes 50% activity.

a. & b. Bill has a 70% level of function, so he has a normal, β^+, allele and a mutant allele that contributes 20% (β^{20}): $\boldsymbol{\beta^+/\beta^{20}}$. Carol, with 50% activity, is heterozygous for a normal allele and a null allele: $\boldsymbol{\beta^+\beta^0}$.

c. There are four possible genotypes of a child of these parents: 1/4 $\beta^+\beta^+$ (100% activity); 1/4 β^+ β^{20} (70% activity), 1/4 $\beta^0 \beta^+$ (50% activity), 1/4 $\beta^0\beta^{20}$ (20% activity). Only the last genotype leads to clinical thalassemia, so the **probability = 1/4.**

d. **The probability of $\beta^+\beta^+$ (100% activity) = 1/4.**

7-34.

a. Most mammals, including New World primates, are dichromats, while Old World primates are trichromats. Primates diverged from mammals 65 million years ago (65 Myr), while Old World primates diverged from each other about 35 Myr ago.

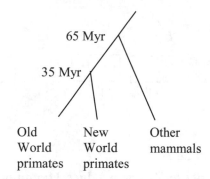

Compare this with **<u>Figure 7.28d</u>**. You can see that the final gene duplication event – the one eventually giving rise to different red and green photoreceptor genes – most likely occurred in the lineage leading to the trichromatic Old World primates (like humans) but not in the lineage leading to the dichromatic New World primates (like marmosets). This dates the final gene duplication event to some time subsequent to 35 Myr. The information also lets you conclude that the two previous gene duplication events shown in Figure 7.28 occurred at some time prior to 65 Myr because all mammals have rhodopsin plus at least two color photoreceptors.

b. Assuming there is only one allele for the autosomal receptor gene, female monkeys will have 2 alleles of the X-linked receptor while males will only have one. **Males will be dichromats**, as they are hemizygous for the X-linked gene. However, **there will be 3 classes of males that see colors in different ways. Females can be dichromats** if they are homozygous for an allele of the X-linked gene, again there will be 3 classes. **Females can also be trichromats** if they are heterozygous for this gene, and there are 3 different combinations: 1-2, 1-3 and 2-3.

c.   The fact that 95% of our receptors work best in low light conditions suggests that **early mammals were active in low light situations**, for example in forests at night!

7-35. See Chapter 7, Figure A.

a.   You will see: **i) the base plate** - this is the structure just to the left of the number 19 in the Figure; **ii) a completely formed head filled with DNA; and iii) completely formed tail fibers**. None of these should be attached to each other.

b.   You will see: **i) an immature head, not filled with DNA, ii) replicated phage genomes not inserted into heads; iii) completely formed tail in which the sheath is on top of the base plate** (the structure to the right of numbers 3 and 15); and **iv) completely formed tail fibers**. None of these should be attached to each other.

c.   Because these two mutations are in different genes, the two strains should complement each other and you should see **completely formed new progeny T4 bacteriophage** in the electron microscope.

d.   You will see: **i) nearly complete tail assemblies** (the figure just to the left of numbers 3 and 15); **ii) mature heads filled with phage DNA; iii) and iv) two separate halves of the tail fiber** (notice that the final tail fibers have a kink between the 2 halves). All four parts should be separate from each other.

Chapter 8   Gene Expression: The Flow of Genetic Information from DNA via RNA to Protein

Synopsis:

This chapter describes how the information in DNA is converted into usable machinery (proteins) in the cell via the processes of transcription and translation. This flow of information is part of the central dogma of genetics. You need to become very comfortable with using the terms transcription and translation <u>accurately</u>. In <u>transcription</u>, DNA information is converted into RNA information. In <u>translation</u>, RNA information is converted into protein information. Work on developing some mental pictures for yourself so you can see the process occurring when you speak the words. The three letter DNA code and the correspondence between DNA sequence and protein sequence is described in this chapter. Tie together your knowledge of transcription/translation and the genetic code. This chapter contains many new vocabulary terms. The best way to know you have a good grasp on the terms is use the terms while pretending you are describing transcription, translation and the genetic code to another person.

Begin to introduce more inquiry into your learning process. For example, think about the components involved in transcription, RNA processing, and translation. How could they be affected by alterations (mutations) in any one of the components? Start thinking about how we know what we know and what evidence supports a particular view of how a process occurs.

Significant Elements:

After reading the chapter and thinking about the concepts, you should be able to:

♦ Identify open reading frames in a DNA sequence.

♦ Assign 5' and 3' end designations to DNA or RNA sequences.

♦ Describe the parameters for transcription - describe the enzymes (RNA polymerase) and proteins that play a role and what information in the DNA (promoter and terminator sequence) is important for their ability to function (**see Feature Figure 8.11**).

♦ Identify the mRNA-like strand in the double-stranded DNA sequence either by knowing the sequence of the mRNA transcript or by looking for open reading frames.

♦ Describe the parameters for translation - describe the RNAs (rRNAs, tRNAs), proteins (aminoacyl synthetases) and RNA-protein complexes (ribosomes with the P and A sites) that are important and the information in the mRNA (ribosome binding site, start codon, stop codon) that is important for their ability to function (**see Feature Figure 8.24**).

- Understand that the nearly universal genetic code consists of 64 codons. Of these, 61 specify amino acids, while the other 3 (5' UAA, 5' UAG and 5' UGA) are nonsense or stop codons.

- Use a codon chart (**see Figure 8.3**) to do a virtual translation of DNA into protein sequence or reverse translate protein into DNA sequence.

- Explain the steps in processing a eukaroytic primary transcript into an mRNA: addition of 5' methyl CAP (**see Figure 8.13**), addition of 3' poly-A tail (**see Figure 8.14**) and splicing out of introns (**see Figure 8.15**).

- Answer questions that require you to know the roles of the nucleic acids and proteins in transcription, translation, and RNA processing.

- Understand the ways that different types of mutations affect gene expression (**see Figure 8.27 for an overview**): silent mutations, missense mutations, nonsense mutations (and nonsense suppressors, **see Figure 8.28**), frameshift mutations (**see Figures 8.5 and 8.6**) and mutations outside of coding sequences that alter signals required for transcription, processing of the mRNA or translation.

Problem Solving Tips:

- The convention is to write DNA sequences with the top strand running 5' to 3' left to right.

- Transcription occurs in a 5' to 3' direction along the template DNA strand. The complementary strand of DNA is the mRNA-like strand. The mRNA-like DNA strand has the same polarity and sequence (with Ts instead of Us) as the mRNA.

- Ribonucleotides are added to a 3' end of the growing mRNA strand.

- Information in the 5' portion of a coding region of the mRNA will be information in the NH_2 terminal portion of the protein.

- There is mRNA sequence at the 5' and 3' ends of the transcript that is NOT translated into protein. These sequences are called the 5' and 3' untranslated regions (UTRs).

- By definition an open reading frame (ORF) is an RNA sequence with no stop (nonsense) codons in the adjacent groups of 3 nucleotides.

    If you are given an mRNA sequence, there are 3 possible reading frames to examine to see if they are open or closed (have an in-frame stop codon). Remember that the ribosome reads the mRNA from 5' to 3', so you check for ORFs by starting at the 5' end of the mRNA sequence. The first reading frame begins with the first nucleotide, the second frame begins with the second nucleotide and the third frame begins with the third nucleotide. When you begin with the fourth nucleotide you are back in the first reading frame.

If you are given a DNA sequence to examine, there are 6 possible reading frames! Remember that the template strand of DNA is transcribed 3' to 5' to synthesize a complementary mRNA that is 5' to 3'. Use the complementary mRNA-like strand of DNA as a stand-in for the mRNA. Read the mRNA-like strand of DNA from its 5' end, checking the 3 possible reading frames. Then assume the other strand of DNA is the template, and repeat the process.

♦ Codons are found in mRNAs and anticodons are found in tRNAs. To avoid confusion, be consistent in how you write them. In the Study Guide the codons are written 5' to 3' and the anticodons are written 3' to 5'.

Solutions to Problems:

8-1. a. **5**; b. **10**; c. **8**; d. **12**; e. **6**; f. **2**; g. **9**; h. **14**; i. **3**; j. **13**; k. **1**; l. **7**; m. **15**; n. **11**; o. **4**; p. **16**.

8-2. a. **4**; b. **6**; c. **1**; d. **2**; e. **3**; f. **5**.

8-3.

a. **GU GU GU GU GU or UG UG UG UG UG.**

b. **GU UG GU UG GU UG GU UG GU.**

c. **If you start with the first base: GUG UGU GUG. If you start with the second base: UGU GUG UGU.**

d. **GUG UGU GUG UGU GUG UGU GUG UGU.** This is the result of reading the first 3 nucleotides then going back to the second nucleotide and reading a codon, etc. **There are other possibilities**, such as reading codons starting on 1, 2, 4, 5, etc. Overlapping codes will always give more coding information for the same number of bases compared to the non-overlapping code.

e. **GUGU GUGU or UGUG UGUG.**

8-4.

a. **Comparing the mutant to the wild-type sequence you can see where insertions, corresponding to + mutations, and deletions, corresponding to - mutations, occurred**, see part b.

b. The amino acids in the wild-type and mutant protein are shown:

```
                    Lys Ser Pro Ser Leu Asn Ala
wild-type:   5'  AAA AGT CCA TCA CTT AAT GCC 3'
                    (-)                 (+)
mutant:      5'  AAA GTC CAT CAC TTA ATG GCC 3'
                    Lys Val His His Leu Met Ala
```

The five amino acids in between the - and the + mutations are different from wild-type.

c. **The substitutions of amino acids between the - and + mutations in the mutant must not alter the structure of the protein significantly enough to alter protein function.**

8-5. Glutamic acid can be encoded by either GAA or GAG. In sickle cell anemia this amino acid is changed to valine by a single base change. Valine is encoded by GUN with N representing any of the four bases. Therefore, the second base of the triplet was altered from A to U in the Hb^S allele. In Hb^C the glutamic acid codon (GAA or GAG) is changed to a lysine (AAA or AAG). The change here is in the first base of the codon. **The mutation causing the Hb^C allele therefore precedes the Hb^S mutation in the sequence of β-globin gene** when reading in the 5' - 3' direction that the RNA polymerase travels along the gene.

8-6. Note that the nucleotide sequences below are written as mRNA sequences. From this you can easily convert to the DNA sequence of the gene. Remember that proflavin causes frameshift mutations (single base insertions and deletions) in the DNA. **All mutagens work at the level of the DNA, even though the changes are often written at the level of the mRNA!** Notice that there are several ambiguous bases in the wild-type sequence (any one of four bases possible is indicated by N; other amino acids are encoded by two different codons). Line up the invariant bases in the mutant with the wild-type and it is clear that a single base insertion occurred in the mutant. Knowing the amino acid sequence of the mutant and therefore the nucleotide sequence, all but one of the third base ambiguities in the wild type can be resolved. The mRNA sequence of the wild-type gene would be:

```
                  Gly   Ala    Pro    Arg    Lys
wild-type mRNA:  5' GGN  GCN    CCN    AGA/G  AAA/G 3'
                                       CGN
mutant mRNA:     5' GGN  CAU/C  CAA/G  GGN    AAA/G 3'
                  Gly   His    Gln    Gly    Lys
```

After comparing the wild type and mutant sequences you can see that the first nucleotide of the second codon was deleted to make the mutant sequence. The deduced DNA sequence of wild-type is:

5' GGN GCA CCA AGG AAA 3'

8-7. Nierenberg and Leder used an *in vitro* translation system to determine that 5' CUC is the leucine codon and 5' UCU codes for serine. **The basis of the assay is that the combination of a synthetic triplet RNA codon, matching charged tRNA, and ribosome bound together would be too large to pass through a filter.** They set up 20 reactions, each containing 5' CUC, one radioactive amino acid attached to its tRNA, and the other 19 non-radioactive amino acids attached to their tRNAs. **In the mixture containing the radioactive amino acid leucine that corresponds to the codon 5' CUC, the radioactivity would be trapped on the filter.** The same experiment was done for the 5' UCU triplet; in this case, serine was the radioactive amino acid that was trapped on the filter.

8-8.

a. **If the Asn6 (5' AAC) is changed to a Tyr residue, the nucleotide change is to a UAC. In protein B this means that the Gln (5' CAA) at position 3 becomes a Leu (5' CUA).**

b. **Leu (5' CUA) at position 8 is changed to Pro (5' CCA). In protein B the Thr (5' ACU) at position 5 is still an Thr residue even though the codon is different (5' ACC).**

c. **When Gln (5' CAA) at position 8 in protein B is changed to a Leu (5' CUA), the Lys codon (5' AAG) at position 11 in protein A is changed to a stop codon (a5' UAG).** This would cause the production of a truncated form of protein B only 10 amino acids long.

d. This is a thought question that involves some speculation; the following are two reasonable possibilities. (1) As parts a-c have shown, a mutation in the region of overlap has a high probability of causing alterations in both of the proteins simultaneously. Any change in this DNA sequence has the potential to affect two proteins instead of just one, and an organism would be less likely to tolerate mutations affecting the production of two proteins than mutations affecting the production of a single protein. There would thus be strong evolutionary selection against overlapping reading frames. (2) If a region of DNA evolved so as to encode a protein with a stable three-dimensional conformation, it is very unlikely that a stable protein could be produced by the sequence shifted by one nucleotide. This is because the alternate reading frame is likely to have a stop codon every 3 codons out of 64. This means that in reality, among the very few examples of overlapping reading frames that exist, either one or both of the proteins is very small, composed of only a few amino acids.

8-9. DNA sequences are generally written with the 5' to 3' strand on top and the 3' to 5' strand on bottom. If the protein coding sequence for **gene F** is read from left (N terminus) to right (C terminus), then top strand is the RNA-like strand which is read from 5' to 3', and the **template strand**

must be the bottom strand of DNA. The template for gene G is the opposite since the coding sequence is read in the opposite direction (right to left). **The template strand for gene G is the top strand.** Note that this means the enzyme RNA polymerase moves from left-to-right along the DNA in transcribing gene F, and from right-to-left in transcribing gene G.

8-10. <u>See Figures 8.19 for tRNA structure, 8.21 for codon/anticodon pairing, 8.23 for ribosome structure and 8.24 for translation</u>.

Assume that the complete gene sequence is shown:

For part e. only one codon is labeled, but any of the 9 codons in this gene could have been labeled. Part q. is only present if this is figure represents an mRNA from a eukaryotic cell. **The following items are not found in this figure: a, d, g, i, j, n and t.**

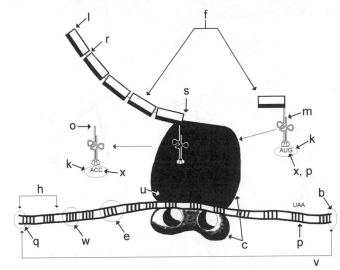

8-11. Refer to the <u>figure in the answer to Problem 8-10 above</u>.

a.   This figure represents **translation**.

b.   **The next anticodon is 5' GUA which is complementary to the codon 5' UAC which codes for tyrosine. The tyrosine is added to the carboxy-terminal (C-terminal) end of the growing polypeptide chain. The protein will be 9 amino acids long when completed**.

c.   There are 2 other building blocks with known identities. **The amino acid just placed on the carboxy-terminus of the growing polypeptide chain is tryptophan** (anticodon of departing tRNA is 5' CCA which is complementary to the codon 5' UGG which codes for tryptophan). This would be the fourth amino acid from the protein's N terminus. There is also **an assumed building block with a known identity – the initiation codon (5' AUG) that starts off protein synthesis, placing the amino acid methionine at the amino-terminal (N) end of the protein.**

d.   **The first amino acid at the N terminus would be f-Met in a prokaryotic cell and Met in a eukaryotic cell. The mRNA would have a cap at its 5' end and a poly(A) tail at its 3' end in a eukaryotic cell but not in a prokaryotic cell. If the mRNA were sufficiently long, it might encode several proteins in a prokaryote but not in a eukaryote.**

8-12. The heavier lines in the figure below represent mRNAs from 2 different genes, I and II; two genes are portrayed to show that the results would be slightly different depending upon whether or not the gene has introns. In general the DNA strands form a double stranded DNA structure, except where their pairing is interrupted by the presence of the mRNA/DNA heteroduplex. Gene II pairs with the template strand of the DNA, forcing the other strand to loop out. Gene I has an intron which has been processed out of the mature mRNA. When the mRNA pairs with the template strand of DNA there is a loop-out of the DNA in the region corresponding to the intron.

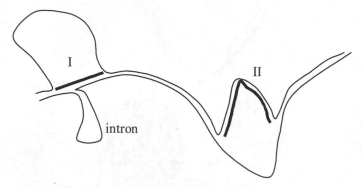

8-13. Both strands of a double-helical DNA molecule are possible template strands (see for example problem 8-9 above). There are three possible reading frames on both the top and bottom strands. If a stop codon is found in a frame, then it is not an open reading frame (that is, the entire sequence cannot not code for a protein). Assume first that the bottom strand is the template strand, so that the top strand is the mRNA-like strand. Treat it as the equivalent to an mRNA, which means it would be translated in the 5' to the 3' direction and that you would substitute T for U to allow translation from the genetic code. Thus, the first potential reading frame begins with the first nucleotide, the second reading frame with the second nucleotide, etc. In the following sequence, an 'x' above the nucleotide means the reading frame is closed (that is, it contains a stop codon), while an 'o' means it is open. The stop codons in the top and bottom strands are shown in bold. Scan the sequence looking for stop codons (or their direct DNA equivalents: TAA, TAG and TGA). Repeat this process assuming the top strand of the DNA is the template strand and the bottom strand is the mRNA-like strand, with 3 possible reading frames beginning at the 5' end of the sequence. **When you analyze this DNA sequence you find that there are two open reading frames in the top strand (reading from left to right) and one in the bottom strand (reading from right to left).**

```
            X  OO
   5' CTTACAGTTTATTGATACGGAGAAGG 3'
   3' GAATGTCAAATAACTATGCCTCTTCC 5'
                            O XX
```

8-14. Eukaryotic genes contain intervening sequences (introns) that do not code for proteins.
Because they are spliced out of the primary transcript and thus are not included in the mature mRNA, introns do not have to contain open reading frames. In fact, introns almost always contain stop codons that would halt all possible reading frames. Thus, the reading frames of almost all eukaryotic genes are interrupted by introns that contain stop codons in that frame.

8-15.

a. The mutant nucleotide is marked in bold. The corresponding change to the amino acid sequence is shown below the RNA sequence.

wild type	5' AUG	ACA	CAU	CGA	GGG	GUG	GUA	AAC	CCU	AAG
	Met	Thr	His	Arg	Gly	Val	Val	Asn	Pro	Lys
mutant 1	5' AUG	ACA	CAU	CCA	GGG	GUG	GUA	AAC	CCU	AAG
transversion				Pro						
mutant 2	5' AUG	ACA	CAU	CGA	GGG	UGG	UAA	ACC	CUA	AG
deletion						Trp	STOP			
mutant 3	5' AUG	ACG	CAU	CGA	GGG	GUG	GUA	AAC	CCU	AAG
transition		Thr (no change)								
mutant 4	5' AUG	ACA	CAU	CGA	GGG	GUU	GGU	AAA	CCC	UAA G
insertion						Val	Gly	Lys	Pro	STOP
mutant 5	5' AUG	ACA	CAU	UGA	GGG	GUG	GUA	AAC	CCU	AAG
transition				STP						
mutant 6	5' AUG	ACA	UUU	ACC	ACC	CCU	CGA	UGC	CCU	AAG
inversion			Phe	Thr	Thr	Pro	Arg	Cys	Pro	Lys

b. **The frameshift mutations 2 and 4 (single base insertions and deletions) can be reverted with proflavin**. Mutations 1, 3 and 5 are single nucleotide substitutions, either transitions (changes from one purine to the other or from one pyrimidine to the other) or transversions (changes a C-G base pair to a G-C base pair). **EMS causes transitions (see Figure 7.12b), so it can revert mutations 3 and 5 (transition mutations) back to the original DNA sequences**. Mutation 1 is a transversion, and EMS can not change the G-C back to C-G. **Often "reversion" is used in a more general sense, meaning simply a restoration of wild type protein function.** In this case, it is possible that another, non-wild type amino acid at the mutant position will restore function. An EMS-induced transition in the mutant codon could give a functional protein. With this second meaning of "reversion", **all three point mutations (1, 3, and 5) could be reverted by EMS.**

8-16. First look at the sequence to determine where the Met Tyr Arg Gly Ala amino acids are encoded. The top strand clearly does not encode these amino acids. On the bottom strand, there are Met Tyr codons on the far right (reading 5' to 3') and Arg Gly Ala much farther down the same strand. Why aren't these codons adjacent? There could be an intron in the DNA sequence which is spliced out of the primary transcript. Thus, the mature mRNA would encode this short protein.

a. **The bottom strand is the RNA-like strand, so the top strand is the template. The RNA polymerase moves right to left along the template**.

b. The sequence of the processed nucleotides in the mature mRNA is shown below. The junction between exons is marked by a vertical line:

5' CCC AUG UAC A<u>G</u>|G GGG GCA UAG GGG 3'

The sequence of this mRNA contains nucleotides prior to the AUG initiation codon and subsequent to the UAG stop codon to emphasize that the mRNA does not begin and end with these codons: remember that mRNAs contain both 5' and 3'- untranslated regions (UTRs).

In fact, this problem has a simplified DNA sequence to allow the sequence to fit on the page and to facilitate your analysis of the sequence in a reasonable amount of time. If you look very carefully at the sequence, you can see that there are canonical splice donor and splice acceptor sequences at the borders between the intron and the two exons that flank it. However, the intron does not contain a canonical branch site (**see Figure 8.16b** to review the nature of the three sequences needed for splicing, and then try to verify these statements yourself). In reality, the shortest introns are about 50 bp long, instead of the 24 bp in the problem, and must contain branch sites.

c. **A Thr residue at this position could occur if the G base on the bottom strand that just precedes the junction between the intron and the first exon (underlined in the answer to part b above) was mutated to a C. This base change would also alter the splice donor site, so splicing does not occur. The next codon after the ACG for Thr is a UAA stop codon, so the polypeptide encoded by the unspliced RNA is only three amino acids long**.

8-17. In transcription, complementary base pairing is required to add the appropriate ribonucleotide to a growing RNA chain. In translation, complementary base pairing between the codon in the mRNA and the anticodon in the tRNA is responsible for aligning the tRNA that carries the appropriate amino acid to be added to the polypeptide chain.

8-18.

a. **The minimum length of the coding region is 477 amino acids x 3 bases/codon = 1431 base pairs** (not counting the stop codon). The gene could be longer if it contained introns.

b. Look at both strands of this sequence for the open reading frame (remember that the given sequence is part of a protein-coding exon, so there must be an open reading frame). It occurs on the bottom strand, starting with the second base from the right (x = closed reading frames, o = open reading frames). The direction of the protein is N-terminal to C-terminal going from right to left in the coding sequence on the RNA-like strand (that is, the bottom strand of the DNA).

```
                                xxx
   template strand     5' GTAAGTTAACTTTCGACTAGTCCAGGGT 3'
   mRNA-like strand    3' CATTCAATTGAAAGGTGATCAGGTCCCA 5'
                                                 xo x
```

mRNA   5' ACCCUGGACUAGUCGAAAGUUAACUUAC 3'

c. The amino acid sequence of this part of the mitotic spindle protein is: **N...Pro Trp Thr Ser Gly Lys Leu Thr Tyr...C**. Notice that the open reading frame begins with the second nucleotide. You cannot determine the amino acid corresponding to the A nucleotide at the 5' end of the mRNA because you don't know the first two nucleotides of the codon.

8-19.

a. The protein would terminate after the His codon due to a nonsense mutation. **The Trp codon (UGG) could have been changed to a either a UGA or a UAG codon**. These stop codons result from changing the second or third base of the Trp codon to an A.

b. Reverse translate the amino acid sequence:

```
             N Ala   Pro   His    Trp   Arg     Lys  Gly  Val  Thr C
   mRNA      5' GCN   CCN   CAU/C  UGG   CGN     AAA  GGN  GUN  ACN
                                         AGA/G
```

If you restrict the possibilities to mutations that substitute one base pair for another, there are four possible ways to generate a nonsense mutation from this sequence. (i) If the second base of the Trp codon UGG changes to A, a UAG stop codon will result. (ii) If the third base of the Trp codon UGG changes to A, a UGA stop codon will result. (iii) If the Lys codon was AAA, and there is an A to T substitution at the first position, a UAA stop codon would be produced. (iv) If the Gly codon is GGA, a mutation of G to T at the first position will generate a UGA nonsense codon.

There are many other changes that could cause the premature termination of the protein encoded by this sequence, for instance, the insertion or deletion of a single base pair causing a frameshift mutation. As just one example, if the Arg codon is CGU, a single base insertion in the DNA before or within this codon would lead to a UAA codon in the mRNA, with the U would come from the former Arg codon and AA from the Lys codon.

8-20. The extra amino acids could come from an intron that is not spliced out due to a mutation in a splice site. The genomic DNA sequence in normal cells should contain this sequence. An alternative is that the extra amino acids could come from the insertion of DNA from some other part of the genome, such as the insertion of a small transposable element. In this case, the normal allele of the gene will not have this sequence. Note that the defect in this case could not be caused by a mutation that changes the stop codon at the end of the open reading frame to an amino-acid-specifying codon. If this were the case, the extra amino acids would be found at the C-terminus of the longer protein, not in the middle.

8-21.

a.   The anticodon is complementary to the codon. **The anticodon in this nonsense suppressor tRNA is 3' AUC 5'.** Because the 5'-most nucleotide in the anticodon is C, the wobble rules predict that it can pair only with a 5' UAG 3' stop codon and not with any other triplet.

b.   The wild-type tRNA^Gln recognizes either a 5' CAA 3' or a 5' CAG 3' codon. Only a single nucleotide was changed to turn the wild type tRNA into the nonsense suppressor. The suppressor tRNA recognizes a 5' UAG 3' codon, so the wild type tRNA must have recognized the 5' CAG 3' codon if only one base change occurred. The anticodon sequence in the wild-type tRNA is therefore 3' GUC 5'. The template strand employed to produce the tRNA is complementary to the tRNA sequence itself, so **the sequence of the template strand of DNA is 5' CAG 3'**.

c.   In any wild type cell of any species there is a minimum of one tRNA^Gln gene. This one has to be able to recognize both of the normal Gln codons, 5' CAA 3' and 5' CAG 3'. The wobble rules say that a tRNA with a 3' GUU 5' anticodon can recognize both Gln codons, so this would be the minimum tRNA gene in any organism (**see Figure 8.22**). However, *B. adonis* is a species that can harbor the Gln nonsense suppressing tRNA discussed in part a. **Therefore there must be two tRNA^Gln genes in a wild-type *B. adonis* cell. One would code for the tRNA (anticodon of 3' GUC 5') that was changed into a nonsense suppressor, and the other gene would code for the tRNA (with an anticodon of 3' GUU 5') that recognizes both of the normal Gln codons.**

8-22.

a.   There is a progression of mutations from Pro (5'CCN, wild type) to Ser (5'UCN) to Trp (5'UGG, strain B) codons. **The original wild type proline codon must have been 5' CCG, so the sequence of the DNA in this region must have been: 5' CCG**

3' GGC

b.   The wild type amino acid is Pro (proline). The amino acid at the same position in strain B is Trp (tryptophan). Although these are not the same amino acids, the phenotype of strain B is wild type. This means that **Trp at position 5 is compatible with the function of the enzyme encoded by the gene.** This implies that the change from Pro to Trp is a conservative substitution. In fact, if you look at **Figure 7.21**, you can see that Pro and Trp are both amino acids with nonpolar R groups. The original mutation changed Pro to Ser (serine). This mutant is non-functional, so the **Ser at position 5 is not compatible with enzyme function - it is a non-conservative substitution.** **Figure 7.21** shows that Ser has an uncharged polar R group, so its chemical properties are likely to be very different than those of Pro.

c.   **Strain C does not have any detectable protein, so it is likely to be a nonsense mutation that stops translation after only a few amino acids have been added.** One way in which thus could have happened is that the codon for Ser in the mutant (5' UCG) could have been changed by mutation to a 5' UAG. However, the nonsense mutation could also have occurred at other locations in the gene's coding region.

d.   Strain C-1 is either a same-site revertant (5' UAG to sense codon) or a second site revertant (nonsense suppressing tRNA mutation). **Because the reversion mutation does not map at the enzyme locus, it must be a nonsense suppressing tRNA.**

8-23. A missense suppressing tRNA has a mutation in its anticodon so that it recognizes a different codon and inserts an inappropriate amino acid. Problem 8-21 dealt with the change of the tRNA^{Gly} anticodon from 3' GUC 5' to 3' AUC 5', making a nonsense suppressor tRNA. Imagine instead that the tRNA^{Gln} anticodon had mutated to 3' GCC 5'. This mutant tRNA will respond to 5' CGG 3' codons, thus putting Gln into a protein in place of Arg. This is a missense suppressor.

a.   Consider here the effect of the presence of a missense or nonsense tRNA on the normal proteins in a cell that are encoded by wild-type genes without missense or nonsense mutations. **A missense suppressing tRNA has the potential to change the identity of a particular amino acid found in many places in many normal proteins.** Using the example above of a missense suppressor that replaces Arg with Gln, most proteins have many Arg amino acids, so a missense suppressor could potentially substitute Gln for Arg at many sites in any single protein. **In** contrast, a nonsense suppressing mutation can only affect a single location in the expression of any wild-type gene (that is, the stop codon that terminates translation), making a normal protein longer. The longer protein will still have all of the amino acids comprising its normal counterpart. Together, these considerations mean that the presence of a missense suppressing tRNA has more likelihood of damaging proteins synthesized in the cell than the presence of a nonsense suppressing tRNA.

b.   In addition to the situation described above in which a mutation in a tRNA gene would change the anticodon to recognize a different codon, there are other possible ways to generate missense suppression, including: **(i) a mutation in a tRNA gene in a region other than that encoding the anticodon itself, so that the wrong aminoacyl-tRNA synthetase would sometimes recognize the tRNA and charge it with the wrong amino acid; (ii) a mutation in an aminoacyl-tRNA synthetase gene, making an enzyme that would sometimes put the wrong amino acid on a tRNA; (iii) a mutation in a gene encoding either a ribosomal protein, a ribosomal RNA or a translation factor that would make the ribosome more error-prone, inserting the wrong amino acid in the polypeptide; (iv) a mutation in a gene encoding a subunit of RNA polymerase that would sometimes cause the enzyme to transcribe the sequence incorrectly**.

8-24. If a tRNA was suppressing +1 frameshift mutations, then **it must have an anticodon that is complementary to 4 bases, instead of 3**.

8-25. Use the wobble rules (**see Figure 8.22**) to help solve this problem.

a.   The nonsense codons that differ only at the 3' end are 5' UAG and 5' UAA. **A tRNA with the anticodon 3' AUU 5' could recognize both of these nonsense codons** because the U at the 5' end of the anticodon could pair with G or A at the 3' end of the nonsense codons.

b.   **This nonsense suppressing tRNA would suppress 5' UAA 3' and 5' UAG 3'**.

c.   This question asks which wild type tRNAs could have their anticodons changed to 3' AUU 5' with a single nucleotide change. It is easier to answer this question from the perspective of the mRNA sequences. In other words, what sense codons could have mutated to become 5' UAA/G? **These are: 5' CAA/G = Gln, 5' GAA/G = Glu, 5' AAA/G = Lys, 5' UCA/G = Ser, 5' UUA/G = Leu, 5' UAU/C = Tyr**, and 5' UGG = Trp. However, there is an additional consideration that excludes Trp as an answer. There is only one possible anticodon for a Trp-carrying tRNA. This anticodon is 3' ACC 5'. A normal Trp tRNA could not have an anticodon of 3' ACU 5' because wobble would allow such a hypothetical tRNA to recognize a 5' UGA 3' stop codon. Thus two mutational changes would be required to change the only possible normal tRNA^{Trp} with a 3' ACC 5' anticodon to a nonsense suppressing tRNA with a 3' AUU 5' anticodon.

8-26. In the second bacterial species where the isolation of nonsense suppressors was not possible, there must be only a single tRNA^{Tyr} gene and a single tRNA^{Gln} gene. Thus, if either gene mutated to a nonsense suppressor, it would be lethal to the cell, as there would not be any tRNA that could put Tyr or Gln where they belong. In this scenario, the single tRNA^{Tyr} would have to have an anticodon of 3' AUG 5' to recognize the two Tyr codons of 5' UAU 3' and 5' UAC 3' based on the wobble rules. The single tRNA^{Gln} would have to have an anticodon of 3' GUU 5' to recognize both 5' CAG 3' and 5'CAA 3' Gln codons.

8-27. Nonsense or frameshift mutations that affect codons for amino acids near the C-terminus are usually less severe than nonsense or frameshift mutations in codons for amino acids near the N-terminus because less of the protein that will be affected.

a.   **very severe effect** as there will be no functional protein

b.   **probably mild effect** if none of the last few amino acids are important for enzyme function

c.   **very severe effect**, see part a

d.   **probably mild effect**, see part b

e.   **no effect**, by definition a silent mutation maintains the same amino acid

f.   **mild to no effect** as the replacement is with an amino acid with similar chemical properties

g.   **severe effect** as the mutation is likely to destroy the enzyme activity

h.   **could be severe** if it affects the protein structure enough to hinder the action of the protein, **or could be mild** if the enzyme's function can tolerate the substitution

8-28. Mitochondria do not use the same genetic code! In yeast mitochondria, the codon 5' CUA 3' codes for Thr, not Leu as it does in yeast or human nuclear genes. If you want to ensure that the correct protein will be made by the yeast cell, **you should mutate all the 5' CUA 3' codons in the mitochondrial gene to 5' ACN 3' before putting the gene into a chromosome in the yeast nucleus**. This ensures that the cellular translation machinery will put a Thr at all positions it is required in the protein.

8.29. The size of the protein is 2532 amino acids. Thus the mRNA must be 2532 x 3 = 7.6 kb. To be detectable, any changes must be more than 1% of normal size or amount. The answers below are presented in the order of mRNA size; mRNA amount; protein size; protein amount. A '+' means there will be a >1% change, while a '–' means there will be no change. We assume in the answers below (excepting part j) that mutant mRNAs or proteins have normal stability in the cell, though this is not always true in practice.

a.   – – – –. This is a non-conservative amino acid substitution changing the identity of only a single nucleotide in the mRNA and a single amino acid in the protein, so it will probably affect the ability of the protein to function normally, but it will not affect the size or amount of mRNA or protein.

b.   – – – –. This is a conservative amino acid change, so it probably won't affect the function of the protein either.

c.   – – – –. This is a silent change, so it won't affect any detectable parameters.

d.   – – + –. This is a nonsense mutation, so it won't affect mRNA size or amount, nor amount of protein. It will affect the size of the protein.

e.   – – + – or +. Met1Arg could mean that no protein is made, as the ribosome can't initiate translation at an Arg codon. However, it is possible that a protein could be made if there is another downstream, in frame 5'AUG codon that can be used to initiate translation. This second possibility would result in a smaller protein.

f.   – + – +. A mutation in the promoter would most likely make it a weaker promoter, so fewer RNA polymerase molecules would bind. With less mRNA, less protein would be produced.

g.   – – + –. This change would not be detectable in the mRNA, but it will cause a frameshift mutation in the protein which will obviously affect the size of the protein.

h.   – – – –. A deletion of 3 bases (a codon) would only remove 1 amino acid / 2532 amino acids; this is a <1% change in size of the mRNA or protein.

i.   + – + –. This mutation causes alternate splicing to occur, removing exon 19. Depending on the size of exon 19, this could easily have a >1% affect on the size of the mRNA and protein.

j.   + + – +. Lack of a poly(A) tail could make the mRNA noticeably smaller. It could also decrease the stability of the mRNA. If the mRNA is degraded faster then less protein will be translated.

k.   – – + or – –. This substitution in the 5' Untranslated Region will not affect the mRNA. However, it could have an affect on the amount of protein made if it affects ribosome binding, for example. Conversely, it may have absolutely no affect on how well the protein is translated.

l.   – – – –. If this insertion is into an intron, then it should be spliced out of the primary transcript, producing a wild type mature mRNA.

8-30.

a. Null mutants are those with no enzyme activity. **The following changes could lead to loss of enzyme activity: a (if this amino acid substitution blocks protein activity); b (if this amino acid substitution blocks protein activity); d (would destroy most of the protein activity); e (if there is no translation initiation); f (if promoter does not function); g (frameshift would alter most of the amino acid sequence of the protein); h (it is possible that the deletion of 1 amino acid blocks protein function or alters the tertiary structure of the protein); i (will produce an altered protein that might be unable to function); j (could make the mRNA completely unstable); and k (losing exon 19 could inactivate the protein).** In practice, mutations like j and k seem to cause less severe reductions in gene expression.

b. Any mutation that makes an altered protein with impaired function or lower levels of otherwise functional protein is likely to be recessive to wild type. Most null or hypomorphic alleles are in fact recessive, loss-of-function mutations. Thus, **any of the mutations in the list (except c and l) could be recessive. Those listed in the answer to part a are also the most likely to be recessive.**

c. **Any of the mutations, other than c and l, could potentially be dominant to wild type.** For most of these, if the mutation was null or strongly hypomorphic there could be a **dominant effect from haploinsufficiency**, where one wild type allele of the gene does not express enough gene product for a wild type phenotype. This is true for any mutation that alters the protein product in size or amount. Any of the mutations that change the size of the protein, as well as a, b and h (which alter the protein to a lesser degree) could potentially have **dominant negative or neomorphic dominant effects** that would depend on the protein (for example, whether it is multimeric) and the exact consequences of the mutation. Finally, it is possible that a promoter mutation, as in f, could turn on a gene in the wrong tissue or at the wrong time, leading to an **ectopic dominant phenotype**. However, you should remember that most of the scenarios leading to dominant phenotypes are generally much more rare than recessive effects due to loss of function.

8-31. The order of these elements in the gene are: **c; e; i; f; a; k; h; d; b; j; g**.

8-32.

a. **All of these elements are in the RNA (that is, the terms are abbreviated) except the promoter (c) and the transcription terminator (g). These 2 are the only structures in this list** that the RNA polymerase enzyme recognizes on the DNA. RNA processing and translation occur post-transcriptionally.

b. **a, e, f and i are found partly or completely in the first exon; a, h and k are found partly or completely in the intron** (note that the splice donor site shown on the splicing figure includes nucleotides in both the upstream exon and the intron); and **b, d, j and g are found partly or completely in the second exon**.

Chapter 9   Deconstructing the Genome: DNA at High Resolution

Synopsis:

This chapter introduces you to many of the recombinant DNA techniques that have provided a powerful new approach for studying the mechanisms of inheritance and functions of specific genes. Restriction enzymes, cloning DNA, making libraries, identifying clones of interest, DNA sequencing, PCR amplification are now just a part of the toolkit that all biologists (not just geneticists) use. These techniques will be referred to over and over throughout this textbook (and probably in your other biology courses as well) so it is worthwhile to get a solid understanding of these techniques from this chapter.

As you read about the various techniques and apply them to solve problems, try to keep in mind which techniques are done in solutions in test tubes (restriction enzyme digests, ligating fragments together, PCR, DNA sequencing, making cDNA) and which techniques involve analyzing or manipulating DNA in cells (transformations, screening libraries, preparing large amounts of cloned DNA, total genomic DNA or cellular RNA). This should help your understanding of the techniques and their uses. Hybridization of nucleic acids is central to many techniques but is often challenging to understand. The basis of hybridization is *complementarity* of bases in forming double stranded nucleic acids. A probe DNA or RNA molecule is used to locate a specific sequence (on a nitrocellulose or membrane based blot after electrophoresis in a gel, as a clone inside a cell, or in a chromosome squash) based on hybridization. A probe contains a recognizable radioactive or fluorescent tag that makes it possible to identify the place where the probe found a complementary sequence.

Significant Elements:

After reading the chapter and thinking about the concepts, you should be able to:

♦ Describe the essential steps in cloning.

♦ Describe the basic components and uses of different types of cloning vectors.

♦ Make a map of restriction enzyme sites.

♦ Read and interpret DNA sequencing gels.

♦ Design PCR primers.

♦ Determine which technique(s) you must use to achieve a desired goal. There is often more than one way to reach a goal. However, there is usually one most efficient, preferred way to solve a problem.

- The technique used determines what is being examined and limits the interpretation of the data. For instance, probing a genomic library will give you a clone that is homolgous to the probe, but this clone probably won't be transcribed and translated in *E. coli*. Probing a cDNA library will give you a clone which can be translated and transcribed in *E. coli*.

Problem Solving Tips:

Essential Steps in Cloning:

Basically a straightforward process that has lots of options and variations that can be used depending on what is desired. Basic components are insert DNA and vector. There are relatively few sources for the insert DNAs. However there are many, many types of vectors that have been developed for various purposes.

Types of insert DNA -

- cDNAs contain only the regions of genes that are present in processed (spliced) transcripts synthesized in the cell from which they were isolated (**see Figure 9.2**).

- genomic DNAs are digested fragments of the genomic DNA of an organism, and so contain all of the DNA (genes and non-coding regions) from the cells.

Basic vector criteria -

- vectors must have an origin of replication so they can be replicated in the host organism, usually *E. coli*.

- vectors must have a selectable marker(s) so you can determine that they are present in the host organism; the selectable marker is often an antibiotic resistance.

- vectors also often have multiple cloning sites with known restriction sites, ways to detect the presence of an insert DNA after cloning. One example of this is the β-galactosidase/ X-gal detection system. Insertion of a fragment into the middle of the *lacZ* gene inactivates the gene. Cells carrying an insert within the *lacZ* gene are unable to cleave a lactose-like substrate (X-gal) and are phenotypically Lac⁻. They are recognized as white colonies while colonies that received intact copies of the vector (no insert interrupting the *lacZ* gene) can cleave the substrate, turning the cells blue.

Types of vectors/purpose of cloning (**see Table 9.2**)-

- library vectors; often used with partially digested insert DNAs so the sequences of inserted DNAs will overlap.

- expression vectors (**see Figure 9.12**) to allow transcription and translation of cloned genes; must be used with cDNA inserts. Use your knowledge of the requirements for transcription and translation when considering if genes cloned into expression vectors will be expressed in the host cell.

Cloning -

♦ after restriction enzyme digestion, mix insert and vector DNAs and ligate together sticky ends that have complementary overhanging single-stranded bases can be. It may be helpful to draw out the 5' and 3' ends generated (including the individual bases of the recognition site) when a double stranded DNA is cut by a restriction enzyme (**see Figure 9.2**).

♦ transform the ligation mix into the host cells, usually *E. coli*.

♦ select for presence of vector (may also be able to isolate those vectors that you know have an insert).

♦ grow up lots of the clone(s).

Identifying the desired clone -

♦ often you must identify a particular desired clone from a large variety of different inserts; this often involves probing, or hybridization with a labeled DNA.

Other Techniques -

♦ gel electrophoresis separates DNA fragments according to their size (**see Figure 9.5**).

♦ blotting is the process of transferring the material in the gel to a nitrocellulose filter or a nylon membrane and covalently binding the material from the gel to the filter or membrane. A Southern blot has DNA on the membrane (a genomic Southern has genomic DNA), a Northern blot has mRNA on the membrane and a Western blot has protein on the membrane (**see Figure 9.15**).

♦ Restriction mapping is part science and part art, like putting together a jigsaw puzzle. Use a pencil and an eraser. Be patient. The first step is usually ascertaining if you began with a linear or a circular piece of DNA. Usually this is gotten out of context - a plasmid clone is circular, for instance. Begin the map by examining a single digestion lane on the gel and determining the total size of the DNA (the sum of all the fragments) and the number of restriction sites for that enzyme (2 fragments when you digested a circular piece of DNA means there were 2 restriction sites; 2 fragments when you digested a linear piece of DNA means there was only 1 restriction site). Next, look at the double digestion lane. Determine which bands from the single digestion are left undigested in the double enzyme digestion. The fragments from the single enzyme digestion that disappear in the double digestion must have a restriction site for the second enzyme within them. Figure out which smaller fragments they have been broken into, then begin mixing and matching various combinations of bands until you find one that gives you an order that will give the correct pattern of bands when you digest the DNA with the second restriction enzyme alone (see

Problems 9-15, 9-16 and 9-17). Make sure the final sites you put on a map are consistent with results from all digests.

♦ DNA sequencing provides the ultimate description of a cloned fragment of DNA. Make sure you can explain the Sanger sequencing method (dideoxy sequencing) to a frien (**see Figure 9.17**).

♦ PCR rapidly purifies and amplifies a single DNA fragment from a complex mixture (**see Feature Figure 9.16**). In order to do PCR you must know something about the DNA sequence of 2 short stretches of the DNA to be amplified. The DNA fragment to be amplified is defined by a pair of oligonucleotide primers that are each complementary to one of the strands of the DNA template. These primers are extended at their 3' ends. The size of the final product of the PCR reaction is determined by the distance between the 5' ends of the primer pair.

Solutions to Problems:

9-1. a. **10**; b. **1**; c. **9**; d. **7**; e. **6**; f. **2**; g. **8**; h. **3**; i. **5**; j. **4**.

9-2.

a. *Sau*3A recognition sites are 4 bases long and are expected to occur randomly every 4^4 or 256 bases. The human genome contains about 3×10^9 bases, one would expect $3 \times 10^9/256 = 1.2 \times 10^7$ **~12,000,000 fragments**.

b. *Bam*HI recognition sites are 6 bases long and would be expected every 4^6 or 4096 bases;. $3 \times 10^9/4,100 = 7.3 \times 10^5$ **~700,000 fragments are expected**.

c. The *Sfi*I recognition site is 8 specific bases. The N indicates that any of the four bases is possible at that site and therefore does not enter into the calculations. Recognition sites would be expected every 4^8 or 65,536 bases; $3 \times 10^9/65,500 = 4.6 \times 10^4$ **~46,000 fragments are expected**.

9-3. Selectable markers in vectors provide a **means of determining which cells in the transformation mix take up the vector**. These markers are often drug resistance genes so a drug can be added to the media and only those cells that have received and maintained the vector will grow.

9-4. An artificial chromosome is **a vector which can accept very large pieces of foreign DNA** – up to 1Mb in the case of yeast artificial chromosomes. Commonly used artificial chromosome vectors include BACs (bacterial artificial chromosomes) from *E. coli* and YACs (yeast artificial chromosomes) from yeast. **Such vectors make it possible to clone large pieces of genomic DNA**, so they are useful for making genomic libraries of organisms with large genomes, like humans. This sort of vector must be able to act as a chromosome in the host cell (*E. coli* for BACs and yeast for YACs) so they must include an origin of replication, a selectable marker and an origin of replication. YACs must also include a centromere and telomeres at the end of the vector.

9-5. First, work through the digestion and ligation of the DNA fragments and the vector. The vector is cut with *Bam*HI, leaving the following ends:

```
5' —G          GATCC—
3' —CCTAG          G—
```

The insert DNA is cut with *Mbo*I, leaving the following sticky ends:

```
5' —            GATC—
3' —CTAG            —
```

The ligation of an *Mbo*I fragment to a *Bam*HI sticky end will only occasionally create a sequence that can be digested by *Bam*HI. It depends on the exact base sequence at the ends of the *Mbo*I fragment. The 'X' in the sequence below indicates this ambiguity. In all cases the following sequence will be found: The sequences from the inserted *Mbo*I fragment are in bold.

```
5' —GGATCX————————XGATCC—
3' —CCTAGX ————————XCTAGG—
```

a. **100%** of the junctions can be digested with *Mbo*I

b. A junction that can be digested with *Bam*HI must have a C at the 3' end of the *Mbo*I recognition sequence. This would occur 1/4 or **25% of the time**.

c. **None** of the junctions will be cleavable by *Xor*II.

d. The first five bases fit the recognition site for *Eco*RII. The final position must be a pyrimidine (C or T). There is a **1/2 chance** that the junction will contain an *Eco*RII site.

e. For the restriction site to be a *Bam*HI site in the human genome it must have had a G at the 5' end. This G was in the vector sequence in the clones created. The chance that the 5' end was NOT a G=**3/4**.

9-6.

a. The **genomic library** is based on the most inclusive and complex starting material, so it would consist of the greatest number of different clones.

b.   **All of these libraries would overlap each other to some extent**. The genomic library contains all the DNA sequences, while the other libraries are made up of subsets of the genomic sequences. All cells express a common subset of genes (housekeeping genes). These genes would result in some overlap of clones, although the cDNA libraries will each contain some unique sequences. Although introns often have repeated DNA, the transcribed and translated portions of sequences are usually unique, so the library of unique genomic sequences will overlap with the cDNA libraries as well.

c.   **The genomic library uses the total chromosomal DNA inside the cell. The repetitive sequences in the genomic DNA would have to be removed to create the unique DNA library. The cDNA libraries all start with the mRNA present in the cells and thus represent therefore the expressed genes in these cells**.

9-7.

a.   You need **4-5 genome equivalents** to reach a 95% confidence level that you will find a particular unique DNA sequence.

b.   The number of clones needed depends on the total size of the genome of your research organism and the average insert size in the vector. BAC inserts can be 500kb while plasmid vectors normally have inserts smaller than 15 kb. **Divide the number of base pairs in the genome by the average insert size then multiply by five** to get the number of clones in five genome equivalents.

9-8. <u>See Table 9.2</u>

a.   An intact copy of the whole gene would be on a fragment larger than 140 kb and would therefore have to be cloned into a **YAC vector**.

b.   The entire coding sequence of 38.7 kb could be cloned into a **cosmid vector** (30-45 kb inserts) as a cDNA copy of the gene.

c.   Exons are usually small enough to clone into a **plasmid vector** (<15 kb inserts). If this exon was larger than 15 kb it could be cloned into a **lambda vector**.

9-9. One order is: **c, j, f, a, k, i, g, b, e, h d**. However, there is some flexibility in the order. Steps c and j can be switched, but they must be done before j; steps c, j and f must be completed before a; c, f, j and a must occur before k and all of these must happen before i. The next step must be the g, b, and e group, although these 3 can be changed among each other, then step h and lastly step d.

9-10.

a. **Such a vector is able to be transformed, selected, and maintained in animal cells and** *E. coli* **must contain a selectable marker in** *E. coli,* **a selectable marker in animal cells and replication origins for each type of cell**.

b. The **human gene must be a cDNA** - no introns. For a human protein to be expressed in *E. coli* the insert **must be next to regulatory sequences for bacterial transcription and translation** so the protein will be expressed in *E. coli*.

9-11. See Figure 9.5 and 'Gel Electrophoresis Distinguishes DNA Fragments According to Size.' The rate at which a piece of DNA moves through a gel is dependent on the strength of the electric field, the gel composition, the charge density and the physical size of the molecule. When electrophoresing DNA the only variable is the size of the molecule - all the rest of the variables are the same for each molecule. **Longer DNA molecules take up more volume and therefore bump into the gel matrix, slowing down the molecule's movement**. Shorter molecules can easily slip through many pore sizes in the gel matrix.

9-12. When you digest a circular DNA one fragment indicates that the DNA has 1 restriction site for the enzyme. Thus, *Bam*HI and *Eco*RI each cut the plasmid once. The double digest gives information about the relative positions of these two sites. The 2 restriction sites are at two different positions on the plasmid. The *Eco*RI site is 3 kb away from the *Bam*HI site and it is 6 kb around the rest of the circle back to the *Eco*RI.

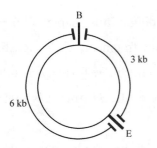

9-13. When the vector (pWR590) is digested with *Eco*RI **you get one 2.4 kb fragment. When the vector is digested with** *Mbo*I **there are 3 fragments - 0.3, 0.5 and 1.6 kb**. The somatostatin insert was cloned into the vector at the *Eco*RI site. There is also an *Eco*RI site very near one end of the insert DNA. Therefore, after digestion of the recombinant plasmid with *Eco*RI, a small *Eco*RI insert fragment of 49 bp and the vector fragments of 900, 500, 300 and 700 bp (2400 bp total length) will be generated. Next, consider the *Mbo*I restriction pattern. The insert fragment contains an *Mbo*I site 5 bp from one end. The insert fragment could ligate into the vector in either of 2 possible orientations. **In one orientation the *Mbo*I site in the insert is nearest the 700 bp *Mbo*I vector fragment, so digestion with *Mbo*I produces 705, 300, 500 and 944 (formed from the 900 bp vector fragment + the rest of the insert) bp fragments. In the other orientation, the *Mbo*I digest produces 905, 50, 300 and 744 bp fragments.**

9-14. Draw the recombinant plasmid to help you determine the fragment sizes before sketching the gel.

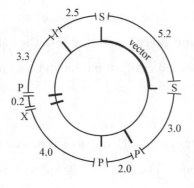

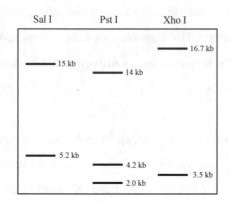

9-15.

a.   Remember the Problem Solving Tips at the beginning of this chapter! If there is one restriction site then digesting a circular molecule results in one fragment, while digesting a linear molecule generates two fragments. Digestion of a circular molecule will always result in one fewer restriction fragments than the digest of a linear molecule. **Sample A is therefore the circular form** of the bacteriophage DNA.

b.   The length of the linear molecule is determined by adding the lengths of the fragments from one digest. 5.0+3.0+2.0 kb = **10.0 kb**. (This size is not realistic - λ DNA is, in fact, about 50 kb in length.)

c.   The circular form is the same length - **10.0 kb**.

d.   Comparison of the circular and linear maps gives you information on which fragments contain the ends of the linear molecule. The 5.0 kb *Eco*RI fragment is present in the circular but not the linear digest so the 4.0 and 1.0 kb fragments must be joined in the circular map while they are at either end of the linear molecule. Begin drawing a picture of the molecule for yourself at this point. The same logic applies to the 2.7 kb *Bam*HI fragment – it is present in the circular but not the linear digest so the 2.2kb and 0.5 kb pieces must be at the ends of the linear molecule. If the 0.5 kb *Bam*HI fragment was at the end where the *Eco*RI 1.0 kb fragment is, the 1.0 kb *Eco*RI fragment would have been cut by *Bam*HI in the double digest. However, the 1.0 kb fragment is still in the double digest, so the 0.5 kb fragment must be within the 4.0 kb *Eco*RI fragment. The remaining *Eco*RI site is placed based on the double digests. The 2.0 kb *Eco*RI fragment is not cut by *Bam*HI but the 3.0 kb fragment is, so place the site within the 3.0 kb. Now double check that all the *Bam*HI+*Eco*RI fragment sizes are as seen in the different double digests.

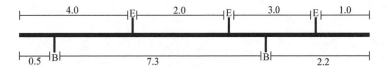

9-16. Plasmids are circular pieces of DNA, thus the *Eco*RI and *Sal*I digests indicate that there is one site for each of these enzymes. *Hin*dIII, in contrast, cuts the molecule at three sites. Draw a circle showing the three *Hin*dIII sites. In the *Sal*I+*Hin*dIII digest the 4.0 kb *Hin*dIII fragment is cut into 2.5 and 1.5 kb fragments. The *Sal*I site is therefore 1.5 kb from one end or the other in the 4.0 kb *Hin*dIII fragment. Similarly the *Eco*RI+*Hin*dIII double digest splits the 1.0 kb *Hin*dIII fragment into 0.6 and 0.4 kb fragments, but the orientation of the *Eco*RI site within the 1.0 kb *Hin*dIII is ambiguous. Try placing the *Eco*RI site in the two different positions in the 1 kb *Hin*dIII fragment In each case see how this fits with the *Eco*RI+*Sal*I digestion results. The orientation that works places the 0.4 kb *Hin*dIII-*Eco*RI fragment adjacent to the 2.5 kb *Sal*I-*Hin*dIII fragment.

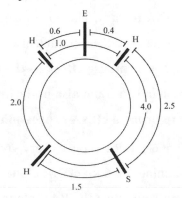

9-17. You start with circular DNA, so *Eco*RI, *Hin*dIII and *Pst*I each cut the plasmid once while *Bam*HI cuts twice. Start your map by placing a *Bam*HI sites to create the 4.5 and 2.5 kb fragments.

The 4.5 kb *Bam*HI fragment is missing in the *Bam*HI+*Eco*RI double digest and therefore is cut by *Eco*RI into 1.5 and 3 kb pieces. The 2.5 kb fragment *Bam*HI fragment is cut into 2 and 0.5 kb the *Bam*HI+*Hin*dIII double digest. Try placing the *Eco*RI and *Hin*dIII sites within the *Bam*HI fragments so that the sites are consistent with the double digest results. The 2.0 kb *Hin*dIII-*Bam*HI fragment must be adjacent to the 3.0 kb *Bam*HI-*Eco*RI fragment to generate a 5.0 kb *Hin*dIII-*Eco*RI fragment in the double digest. *Pst*I must be right next to *Eco*RI because there is only one band seen in the *Pst*I + *Eco*RI double digest and that band is the same size as in the *Pst*I or *Eco*RI digests alone. Consistent with this is the fact that the *Pst*I+*Bam*HI digest looks just like the *Eco*RI+*Bam*HI digest.

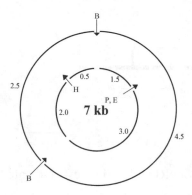

9-18.

a. You could do **tryptic digests** of the SWT protein and **get partial amino acid sequences** for the SWT peptides. Once these partial protein sequences of the SWT protein are known, the **amino acid sequences can be 'reverse translated'** into potential DNA sequences that could encode parts of that protein. This mixture of degenerate **oligonucleotides can be used as a probe to hybridize to a library of genomic or cDNA library**. The homologous clones would contain parts of the SWT gene. Alternately, these degenerate **oligonucleotides can be used as PCR primers using either mouse genomic DNA or mouse cDNA as the template**. Because you have unlimited quantities of the SWT protein you could **also make antibodies to the protein in rabbits and then use these antibodies to probe a cDNA expression library** of mouse cDNAs.

b. *Not*I will digest approximately every 4^8 nucleotides or once every ~65.5 kb. Thus after digestion, electrophoresis and ethidium bromide staining **the 6 kb cDNA clone with its 2 kb insert will have a single band because there are no *Not*I sites; the BAC clone with its 150 kb genomic insert will have 2 or 3 *Not*I fragments; and the purified mouse genomic DNA will have**

many, many *Not*I fragments which will look a smear on the gel. These *Not*I fragments must be separated using Pulse Field Gel Electrophoresis (PFGE) because are so large.

c.   *Rsa*I sites occur every 4^4 nucleotides on average, or once every ~256 bp. Thus **the cDNA clone will have 6000 bp/256 bp = roughly 23 bands; the BAC clone will have 150 kb/0.256 kb = about 590 bands** which will appear as a smear; and **the liver genomic DNA will also look like a smear**, though the average fragment size in the smear will be much smaller than it was in the *Not*I digestion.

d.   When the cDNA insert is used as a probe any DNA fragments that are homologous to the cDNA will hybridize and be seen as bands on the X-ray film. In mice, 2 kb genes usually span about 20 kb of genomic DNA. Thus, **when the *Not*I Southern is probed the single band in the cDNA lane will hybridize**. Most of the BAC clone is made up of mouse DNA so all the bands have mouse genomic DNA in them. Although only 2 kb of the 150 kb is actually homologous to the probe, this 2 kb is spread over about 20 kb of genomic DNA separated by introns. **One, or perhaps 2, of the bands in the BAC clone will hybridize and 2 or 3 of the bands in the liver genomic lane will hybridize.** In the liver genomic DNA lanes **the 1 or 2 *Not*I fragments** that correspond to the DNA found in the BAC clone will hybricize. A very different pattern will be seen when the ***Rsa*I digested** Southern blot is probed with the cDNA insert. **About 1/3 (2 kb insert/6 kb total size) of the bands in the cDNA** clone will hybridize because the rest are vector sequences. Most of the bands in the BAC clone contain only intron sequences and so will NOT hybridize to the cDNA probe. Only roughly 2000 bp of exons/256 bp fragments = **~8 fragments in the BAC clone will hybridize. The same bands will be seen in the genomic DNA** sample because the *Rsa*I sites will occur in the same places with respect to the introns and exons.

9-19.

a.   The full length is represented by the largest fragment in the partial digestion, as this **1.83 kb** fragment is the one that has not been digested.

b.   After digestion only one of the products is visualized on the autoradiograph. Because the fragment is only labeled at the *Eco*RI end, the *Bam*HI end 'disappears.' Thus you see a set of nested bands with the longest one representing digestion at the enzyme site furthest from the *Eco*RI end, the next longest digested at the next site closer to the *Eco*RI end, etc. If you start with the biggest fragment and subtract each subsequent fragment size from the preceding one you will generate a map the begins at the unlabeled or *Bam*HI end of the fragment. For *Hha*I (from left to right): 0.42, .09, 0.93, 0.26, 0.13; for *Sal*I: 0.26, 0.64, 0.35, 0.10, 0.24, 0.24.

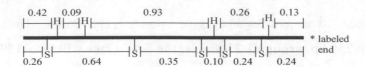

c.   The complete double digest will yield the following fragments: 0.26, 0.16, 0.09, 0.39, 0.35, 0.10, 0.09, 0.15, 0.11, 0.13. All of these fragments will be seen on an ethidium bromide stained gel. Only the fragment that includes the labeled *Eco*RI end will be seen on the autoradiograph. The 0.13 kb fragment contains the labeled end. It is rightmost on the map.

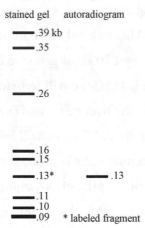

9-20.

a.   **(1) 3.1, 6.9 kb; (2) 4.3, 4.0, 1.7 kb; (3) 1.5, 0.6, 1.0, 6.9 kb; (4) 4.3, 2.1, 1.9, 1.7 kb; (5) 3.1, 1.2, 4.0, 1.7 kb.**

b.   The **6.9 kb fragment in the *Eco*RI+*Hin*dIII digest; the 2.1 and 1.9 kb fragments in the *Bam*HI+*Pst*I, and the 4.0 kb fragment in the *Eco*RI+*Bam*HI digest** will hybridize with the 4.0 kb probe.

9-21. If the gene has been isolated from other organisms, you could use **a probe of the gene from another organism**. Because the protein sequence of the ozonase enzyme is known, the **amino acid sequence could be 'reverse translated' into potential DNA** sequences that could encode that protein. A mixture of **oligonucleotides that could encode that peptide sequence could be used as a probe to hybridize to a library of genomic or cDNA clones**.

9-22.

a. The newly synthesized strand is read from the gel beginning with the smallest band which corresponds to the 5' end of this strand. This newly synthesized strand is complementary to the template strand. Reading the sequence from the gel:

newly synthesized strand: **5' TAGCTAGGCTAGCCCTTTATCG 3'**
template strand: **3' ATCGATCCGATCGGGAAATAGC 5'**

b. Any DNA strand has 3 possible reading frames, which begin at the 5' end with the first nucleotide, the second nucleotide and the third nucleotide. If the template strand is the mRNA-like strand, then one of the possible reading frames should not contain stop codons (should be an open reading frame or ORF). **There are stop codons in each frame so it is unlikely that this is an exon sequence of a coding region**.

9-23. First read the DNA sequence of the mutant from the gel. From the shortest to longest fragment seen, the sequence of the synthesized strand is:

5' TAGATAAGGAATGTAAGATAT<u>A</u>ACTGAGATTTAAC

The corresponding mRNA sequence is:

5' UAGAUAAGGAAUGUAAGAUAU<u>A</u>ACUGAGAUUUAAC

Compare the mRNA sequence of the mutant to the wild type mRNA sequence given in the problem. **There was a single nucleotide change from a T in the wild type sequence to an A in the mutant sequence**. The single base change in the gene (and therefore in the mRNA) is underlined.

9-24.

a. Synthesis occurs in the 5' to 3' direction, so the smallest fragment would contain the 5'' T added to the primer and the next sized product would incorporate the C.

b. First write out the sequence of both strands and scan each strand for stop codons. **The newly synthesized strand has stop codons in all three frames (underlined) and therefore would**

not be the coding (exon) sequence. On the DNA sequencing template strand the reading frame that starts with the first nucleotide does not contain a stop codon and therefore is the ORF in this RNA-like strand.

Synthesized strand: 5' TCTAGCCTGAACTAATGC 3'
DNA sequencing template: 3' AGATCGGACTTGATTACG 5'

c.   The peptide sequence begins with the amino terminal end which corresponds to the 5' end of the mRNA-like DNA sequence (the DNA sequencing template) is **N Ala-Leu-Val-Gln-Ala-Arg**.

9-25. Primers have to be 5' to 3' and have the 3' end toward the center so DNA polymerase can extend into the sequence being amplified. Only **set b.** satisfies these criteria.

Chapter 10   Reconstructing the Genome Through Genetic and Molecular Analysis

Synopsis:

This chapter describes the tools and goals of genome analysis, including the construction of both large scale, <u>high density linkage or genetic maps</u> (**see Figure 10.8**) and <u>long range physical maps</u> (**see Figure 10.7**).

The linkage maps are ultimately used to locate genes and explore their function. These maps have been developed by computer analysis of the raw DNA sequencing data show the relative positions of closely spaced DNA markers. High-throughput platforms have been developed for the rapid analysis of large numbers of genetic markers (millions/day). Two types of commonly used genetic markers are single nucleotide polymorphisms (SNPs) and simple sequence repeats (SSRs or microsatellites).

The physical maps chart the features of the chromosomes using various techniques to localize DNA sequences to specific locations. The DNA clones have been ordered into overlapping clusters that span the chromosomes. This can be done by either using the markers of a high-density linkage map and sequence tagged sites (STSs) as hybridization probes to obtain overlapping, large-insert clones that span the chromosomes. Gaps between the clusters (contigs) are filled by surveying new large insert libraries with STS markers from either contig end. The second approach does not rely on linkage mapping and genetic markers. Instead, physical analysis begins with the complete set of individual clones in a whole-genome library. Fingerprints are determined for each clone and a computer search determines the overlaps in the corresponding clones. With a large enough number of clones it is possible to form contigs for every chromosome (**see Figure 10.9**).

The contigs are then characterized at a more detailed level using restriction mapping and hybridization with probes. Computers combine the information for individual clones into detailed, whole chromosome physical maps.

Long-range DNA sequence maps compiled from the sequences of subclones provide a complete sequence of the entire genome. The subclones may be derived either from previously mapped large insert clones (heirarchical shotgun approach) or directly from the genome (whole-genome shotgun approach).

The linkage, physical and sequence maps can all be integrated because the markers for the linkage and physical maps (SSRs, SNPs and STSs) can be uniquely and specifically localized on the genome using PCR.

The Human Genome Project has also led to paradigm changes in our approaches to biology and medicine. For the first time it is becoming possible to take a systems approach where all the elements of a system are studied.

This comparative genome work has lead to a variety of major insights of human and model organism genomes:

♦ there are surprisingly few genes in the human genome - 40,000 to 60,000

♦ genes fall into 2 major classes - noncoding RNA genes and protein coding genes

♦ the collection of proteins in a particular cell is a proteome and the collection of mRNAs is the transcriptome. These are both dynamic and change throughout the time spans of development and physiological responses

♦ repeat sequences constitute >50% of the human genome, with some having evolved to become genes or control sequences

♦ evolution can occur by the lateral transfer of genes from one organism to another

♦ different human races have very few, if any, uniquely distinguishing genes

♦ the sequences of microbes, plants and animals all employ the same genetic code and show a remarkable similarity

♦ the Human Genome Project has catalyzed the development of high-throughput equipment for the studies of genomics and proteomics

Significant Elements:

The problems in this chapter use many of the techniques that were introduced in chapter 9, including restriction mapping, cloning, probing, PCR and dideoxy (Sanger) sequencing (**see Figure 10.23**). After reading the chapter and thinking about the concepts, you should be able to:

♦ Describe fluorescent *in situ* hybridization (FISH, **see Figure 10.6**) which locates a cloned locus to a particular band on a particular chromosome.

♦ Explain microarrays (**see Figures 10.24 and 10.25**).

Solutions to Problems:

10-1. a. 9; b. 4; c. 2; d. 6; e. 3; f. 8; g. 1; h. 5; i. 7.

10-2. The **human genome contains a large amount of repeated DNA** and intervening sequences **(introns) can be quite large**. The **human genome also has many duplicated genes** (gene families). These factors make the total DNA to protein-coding DNA ratio higher than that in bacteria.

10-3.

a. The clones you isolated will contain microsatellite DNA sequences since that was how they were identified. Do they contain unique sequences also? To determine this, you would have to cut the clones into small fragments, **subclone the small fragments and use each as a probe in a hybridization with genomic DNA**. If only a single band was seen after hybridization with one of these subfragments, you would have identified unique sequence within your clone.

b. To determine if the microsatellite associated with an STS is polymorphic in the population, the **oligonucleotides used to detect the STSs surrounding a microsatellite should be used to amplify the DNA from a large sample of individuals in the population**.

10-4. The FISH protocol for locating a gene on a chromosome **involves a single hybridization** with your labeled gene sequence **to a chromosome spread**. This is especially useful in organisms where the large numbers of matings required for linkage analysis are expensive (e.g., mouse) or not feasible (e.g., humans) and the cytology is good (chromosomes are easily distinguished from one another). **FISH gives results quickly and can be used with any cloned piece of DNA**, while polymorphic alleles are required for linkage analysis. One disadvantage of FISH is the low resolving power.

10-5. Each pair of PCR primers detects a different locus (e.g. the alpha primer pairs detect a 100 bp piece of DNA). Each locus can have 2 alleles – if both primers of the pair anneal then there will be a band (+ band) on the PCR gel OR if one of the primers does NOT anneal because of a single base mismatch with the DNA then there is no band (- band). If the man is homozygous for one allele detected by a given primer pair then all of his sperm would have the same pattern.

a. Because the 200 bp fragment is never amplified, the man may be **homozygous for a mismatched sequence to which the Beta primers anneal**, preventing the fragment from being amplified. (He could also have 1 mismatch on one homolog and a different mismatch on the other homolog.)

b. He is heterozygous for the sequences to which the alpha (a) and delta (d) primers bind because the sequence is amplified in about half of the sperm. If the sequences to which the alpha and delta primers anneal are unlinked the following sperm will be 1/4 a^+d^+ : 1/4 a^+d^- : 1/4 a^-d^+ : 1/4 a^-d^-. The data shows there are 10 a^+d^+ sperm, 11 a^-d^- sperm, 2 a^+d^- sperm and 2 a^-d^+ sperm, showing that the loci are linked. The first 2 classes are the Parentals, so the man's genotype is a^+d^+ / a^-d^-. The last 2 classes are the recombinants, so rf = 4/25 = 16 cM. Thus **the regions to which the alpha and delta primer pairs anneal are 16 cM apart**.

10-6.

a. Linkage distances, expressed in centimorgans, are based on recombination frequencies. The difference in genetic distances between the sexes can be explained by **a difference in the amount of recombination in males and females**.

b   **No**, the physical distances between the 2 sexes are the same. **The same chromosomes are passed from generation to generation** (e.g., grandmother to father to daughter).

10-7.

a. The question asks you to map the genomic insert The data includes the plasmid, so make a map for the sites on the entire clone - indicate which sites are in the vector and which are in the genomic DNA. The insert was cloned into the *Bam*HI site. The *Not*I and *Sal*I sites are immediately adjacent to the *Bam*HI site, with one on either side of the insert. The small fragments between the *Not*I, *Bam*HI and *Sal*I sites can be ignored. The vector is 2.5 kb in length. The Southern blot tells you which fragment in each lane contains vector sequences - 4.2 kb *Not*I, 3.0 kb *Sal*I, 2.5 kb *Not*I+*Sal*I. Remember that when you digest a circular molecule the # bands = # sites. Thus 2 *Not*I fragments = 2 NotI digestion sites. You already know that one of these sites is at one end of the vector and the other one must be in the insert. There are 3 *Sal*I sites and again, one is at one end of the vector and the other 2 are in the insert. Comparing the *Sal*I digest with the *Sal*I+*Not*I lane you can see that the 6.8 kb *Sal*I fragment is not digested by *Not*I. The 3.0 kb *Sal*I fragment contains the vector (2.5 kb) + 0.5 kb of DNA attached to the end of the vector with the *Not*I site. Try placing the 6.8 kb *Sal*I fragment next to the 0.5 kb *Not*I-*Sal*I fragment. The 4.2 kb *Not*I fragment contains the 2.5 kb *Not*I-*Sal*I vector fragment and then a 1.7 kb *Sal*I-*Not*I fragment, so place it on the other side of the 0.5 kb *Not*I-*Sal*I fragment. The second *Not*I site cannot be in the 6.8 kb *Sal*I fragment, so the larger *Not*I fragment includes the 0.5 kb *Not*I-*Sal*I fragment + the 6.8 kb *Sal*I fragment + the 2.3 kb *Sal*I-*Not*I fragment (0.5 + 6.8 + 2.3 = 9.6 kb *Not*I fragment). From the map you can see the *Sal*I sites that are present in the genomic region. Remember that the *Not*I and *Sal*I sites next to the ends of the vector are not present in the chromosome and the map is actually circular with the left and right ends joined into 1 *Not*I site.

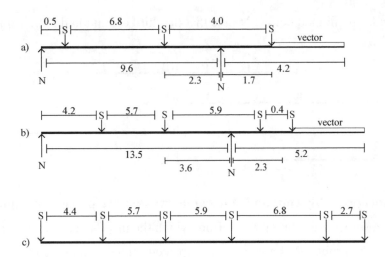

b. The Southern blot hybridization here indicates the overlap of sequences between clone 1 and 2. This includes the vector sequences and the insert sequences in common. Compare the double digestion lanes in clones 1 and 2. Notice that only the 2.3 kb *Not*I-*Sal*I genomic fragment seen in clone 1 is also present in clone 2. Clone 2 has two *Not*I sites (one at the end of the vector) and four Sal I sites (one at the other end of the vector). The 5.7 and 0.4 kb *Sal*I fragments are not digested by *Not*I. The 6.7 and 5.9 kb *Sal*I fragments are cut into 2.5 (vector) + 4.2 and 3.6 + 2.3 kb fragments. The 6.7 kb *Sal*I fragment has the vector, so there is a *Sal*I 4.2 kb beyond the *Not*I vector site (at left end of the figure below). The second *Not*I site has to be located 13.5 kb away from the *Not*I site at the end of the vector to accommodate the *Sal*I site and the fragments present in the double digest. What *Not*I-*Sal*I fragments can add up to this fragment? We know the 4.2 kb *Not*I + *Sal*I fragment is part of this *Not*I fragment; that leaves 9.3 kb. If the 5.7 kb *Sal*I fragment were within this 13.5 kb fragment, that would leave an additional 3.6 kb *Not*I + *Sal*I fragment. But how do we know the order of the 5.7 and 3.6 kb fragments? Looking at the hybridization, the 3.6 kb fragment hybridizes with the clone 1 probe but the 5.7 does not. Clone 1 and 2 overlap in the 2.3 kb region and parts of adjacent fragments. The 3.6 kb should therefore be next to the 2.3 kb *Not*I-*Sal*I fragment. This also helps us place the next restriction fragments- 2.3 and 0.4 kb on our map.

c. From the maps of clones 1 and 2 we can put together an overlapping map of the *Sal*I sites. We know from previous hybridizations that there is overlap through the 2.3 kb *Sal*I-*Not*I region, and we know that the 2.3 kb *Not*I-*Sal*I fragment is part of a 5.9 kb *Sal*I fragment. The fragments flanking this are 5.7 kb (from clone 2) and 6.8 kb (from clone 1). The fragment next to the 5.7kb has to be the 4.4 kb fragment, the only one not already accounted for in the clone 2 hybridization to genomic DNA. Next to the 6.8 kb fragment is the 2.7 kb fragment, the only fragment not accounted for in the clone 1 hybridization to genomic DNA.

10-8. The four cosmids are aligned below. Notice that the third cosmid had to be flipped over for the alignment.

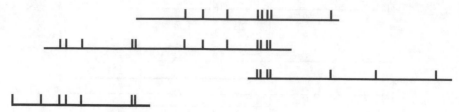

10-9. The two cosmids probably contain DNA sequences that are present as repeat DNAs on many different chromosomes including the one from which the inserts in all the other cosmids came from. Some section of the probe DNA contains this repeated sequence and therefore cosmids from other parts of the genome hybridized to it.

10-10. BAC clones have 200 kb inserts x 15,000 BAC clones = 3,000,000 kb = 3 Gb human genome. However this does not allow for any overlap at the ends of the BAC clones. Such overlaps are necessary to properly align the contigs.

10-11. One method is to search for transcribed regions in the cloned segment using the DNA as a probe versus RNAs from the tissues (Northern blot). A second method is to use computational similarity analysis to search the DNA sequence for open reading frames, intron/exon boundaries and codon usage appropriate for the species. A third method is to compare complete genome sequences from a related mammalian species such as mouse. Non-functional regions of the genomes have significantly diverged while functional regions are relatively conserved. Conserved regions of >25 bp are automatic candidates for coding sequences.

10-12. The whole genome shotgun sequencing strategy involves randomly shearing the entire genomic DNA to make a plasmid library with ~2 kb inserts, a plasmid library with ~10 kb inserts and a BAC library with ~200 kb inserts. Sequence the 2 kb and the 10 kb inserts for a 6-fold and 3 fold coverage, then sequence the ends of the BAC clones for 1-fold coverage. The limitations: some genomic sequences cannot be cloned (e.g. heterochromatin) and some sequences rearrange or delete when cloned (e.g. some tandemly arrayed repeats).

10-13. The 4.1 kb mRNA was processed from a primary transcript that extends through all the DNA between A and G bands. The primary transcript could include part of the DNA in bands A and G or the primary transcript could extend further to the left and/or right, off the DNA map shown. Bands A, D, F and G must contain exon sequences. There are a minimum of 2 introns - one that includes bands

B + C and a second that includes band E. The 3.4 kb mRNA was produced by a primary transcript that extended from somewhere in fragment B to somewhere in fragment E. This gene has a minimum of two exons and one intron in the region covered by fragment D and probably extending into fragments C and E. The 1.8 kb transcript was produced by a primary transcript that extends from D into G. Because the mRNA is only 1.8 kb but the total fragment length in this region is 3.5, there must be at least 1.7 kb of intron in this region. This could be 1 intron in fragment E. If not this then there must be a series of small introns because even the small DNA fragments like G have exon sequences.

10-14.

a. The total size of the pcDNA1 plasmid is 6.93 kb. The insert is therefore 6.93 - 4.2 (vector size) = **2.73 kb**.

b. The pGEN1 plasmid is 7.73 kb. The genomic insert is therefore 7.73 - 4.2 = **3.53kb**.

c. Remember that one fragment in each of the single digestion lanes must include the vector DNA! In cDNA1 there are 3 *Bam*HI sites (one in the vector) and 2 *Pst*I sites (one in the vector). The largest *Bam*HI fragment (4.4 kb) is the only one large enough to include the vector. This *Bam*HI fragment goes from the *Bam*HI site in the vector through the 3.9 kb vector thus placing a *Bam*HI site 0.5 kb in the insert (0.7 kb past the *Pst*I site in the vector). Going back to the *Bam*HI site in the vector, you know that there are no other sites for 0.3 kb because it is still vector sequence. Therefore this 0.3 kb of vector must be in the 2.42 kb *Bam*HI fragment. The 0.11 kb *Bam*HI fragment is between the 2 *Bam*HI sites in the insert. The 4.6 kb PstI fragment must extend from the *Pst*I site in the vector through the vector and 0.6 kb into the insert past the *Bam*HI vector site. This shows where the 0.9 kb *Bam*HI-*Pst*I fragment comes from. You have now unambiguously placed all the restriction sites except for the *Hind*III site in the insert. *Hind*III cuts the 2.33 kb *Pst*I fragment into 1.91 and 0.42 kb fragments. Thus the *Hind*III site is 0.42 kb from either the *Pst*I site in the vector or 0.42 kb from the *Pst*I site in the insert. If it is in the former location the *Pst*I site lies within the 0.5 kb *Bam*HI fragment and so would cleave this fragment in the *Bam*HI + *Hind*III double digest. This does not happen, so the *Hind*III site is 0.42 kb from the *Pst*I site in the vector. For pGEN1: Notice that there is an additional *Hind*III site in this clone and the *Bam*HI and *Pst*I fragments have increased by 0.8 kb. The additional DNA must be in the *Bam*HI 3.22 and *Pst*I 3.13 kb fragments. In the *Hind*III + *Bam*HI digest, the 1.1 kb fragment increases to 1.7 kb and a new 0.2 kb fragment appears. See the map below for the location of the additional HindIII site and the maps of these clones.

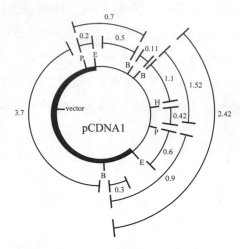

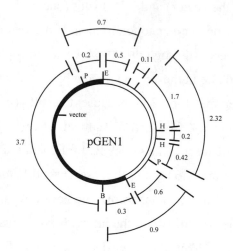

d.   There is a minimum of **one 0.8 kb intron** because there is an additional 0.8 kb of DNA in the genomic clone.

e.   The DNA sequence could have been searched to **locate splice donor and acceptor sites**.

10-15. cDNA clones are made by synthesizing a DNA copy of a processed mRNA. The two different cDNA clones may be due to **alternate splicing** of the same primary RNA (Figure 10.22). The same primary transcript splices out different combinations of introns and exons in the middle of the primary transcript. It is also possible that the transcripts are due to different genes. A **gene duplication** originally gave rise to two identical copies of a gene. The center portion of one of the copies was subsequently replaced by a mobile domain in the process known as **domain accretion** (**see Figure 10.6**). In either case the center portions of the mRNAs have different protein domains.

10-16. Because cDNAs are made from mRNAs, they **lack regulatory region information and intron DNA**.

10-17. If pseudogenes arose from cDNAs they would **lack the introns** found in the genomic copy of the functional gene. Pseudogenes would also have **a polyAT tract** at one end of the gene, corresponding the polyA tail of the original mRNA.

10-18. (i) The repeat sequences found at **centromeres and telomeres** are thought to be important for the proper function of these chomosome structures. (ii) **Transposon-derived repeats have given rise to at least 47 different human genes**. (iii) **Certain genes have repeated sequences** which lead to mutation when these regions are amplified (e.g. Huntington's Disease). (iv) **Transposon-derived repeats have reshaped the genome by aiding the formation of chromosomal rearrangements** like inversions and translocations.

10-19.

a. Once the new gene is sequenced, **the predicted amino acid sequence of the protein product can be examined for protein domains or functional units** (see Figure 10.14). Tell-tale features of protein sequences are motifs that have been shown to have specific functions in other proteins such as transcription factors. **Using comparisons between sequenced genomes of different species you can find orthologous genes.** You can also **look for paralogous genes** (other members of a gene family) within a species. If a function is known for one member of the gene family it is possible to get an idea of the function of another member.

b. **Northern blot analysis to discover in which tissues the gene is transcribed** provide useful clues about function. **Knocking out a gene and looking for a phenotype** caused by the knock-out is another type of analysis that might lead to information on the gene.

10-20.

a. The zinc-finger motif is a DNA sequence that encodes the amino acids capable of forming the zinc finger in a protein. Such domains allow these proteins to bind to specific DNA sequences, suggestions **suggesting that the protein is a transcription factor**.

b. Similarity with a previously identified gene suggests that **the two genes arose by duplication** of a a common ancestor and could be part of a gene family.

10-21. The two mRNAs present in all tissues are different somehow from the third class found only in the brain. The proteins made by the first two mRNAs could be antigenically different enough that the antibody for the brain protein does not recognize them. **The first two mRNAs could be alternately spliced messages or they could be the transcripts of paralogous, diverged genes. Another possibility is that the two other messages are not translated in any tissues.**

10-22. See Table 10.1. All of these organisms are amenable to genetic analysis and they are already extensively studied and mapped. They are presented in order of increasingly complex genomes.

a.   The **genome of yeast is very small** because there is very little repetitive DNA compared with other eukaryotes and therefore the cloning of genes is easier. Any gene involved in basic cell functions would be a good gene to clone in yeast. **Yeast currently has a catalog tryptic peptides related to the genes which code for the as well as a catalog of >7,000 known protein-protein interactions and >500 protein-DNA interactions.**

b.   The **nematode is a fairly complex multi-cellular organism** with a genome of ~20,000 genes with a well understood series of developmental steps. *C. elegans* is small relative to humans but worms have several functions that are similar to, yet more primitive than humans, such as a nervous system. Some of the basic functions of nerve cells can be studied this model system.

c.   The **mouse is the closest model organism to humans** and genes that are homologous to disease genes in humans can be cloned and analyzed. Mouse and humans diverged from a common mammalian ancestor about 85 million years ago. The mouse genome still shows striking similarities in genes and their orthologous order.

10-23.

a.   The gene size is often large due to multiple introns which are sometimes large in size. **You must use cDNAs** as the basis of gene therapy - even for a 100 kb gene the mRNA may only be 1% of that size.

b.   The expression in specific tissues requires **the correct regulatory DNA** so that the therapeutic gene is expressed in the correct tissue. It may also be possible to **target the therapy directly to the correct tissue.**

10-24. ICAT (isotope coded affinity tags) allows quantification of changes in protein concentrations in different cellular states. Of the 30 proteins whose levels changed under the conditions of +galactose and -galactose only half of them showed a corresponding change in mRNA levels. How can the levels of the remaining 15 proteins change if their mRNA levels are constant in the 2 different conditions? It is possible that **the regulation of concentration is at the post-transcriptional level.** For example **perhaps the mRNA is altered so the translation level changes,**

or perhaps the protein is modified by a chemical reaction. **It is also possible that this group of proteins are involved in protein-protein interactions with one or more of the proteins whose level does fluctuate.** In this model the interacting partner's level decreases so there is less to interact with and the non-interacting subunits are destabilized and degraded.

10-25. The region may have many **genome-wide repeat sequences** (see Figure 10.18); or the region may be **duplicated at other sites** in the genome. Other **regions of the genome are unclonable** (e.g. centromeric and telomeric sequences) and therefore it is not possible to get the information to design PCR primers. **Long stretches of repeated DNA** cannot be PCR amplified - the primers must be unique sequences (e.g. AT_n can span hundreds of kb of DNA).

10-26. Not all human genes have been accurately identified: small genes are hard to identify (e.g. small functional RNAs of 25 bp or less); genes expressed only at the RNA level are difficult to identify (tRNAs, rRNAs, snoRNAs and snRNAs); there could be hundred or thousands of genes encoding small peptides that are hard to find. **Genes that are rarely expressed or that have unusual codon usage patterns are difficult to find.** Another factor is that **the human proteome is much more complex** than that of the nematode, and this comes about due to shuffling of functional modules - the human genome evolved many new arrangements that alter the domain architecture. Combinatorial amplification at the DNA level as seen with the antibody gene segments and at the RNA level with alternate splicing also increase the level of complexity of the proteome. **The human genome has more paralogs and chemical modifications of proteins.** Humans have >400 different chemical reactions that affect protein structure, so a human cell can have 20,000 different types of mRNAs and 200,000 different proteins.

Chapter 11   The Direct Detection of Genotype Distinguishes Individual Genomes

Synopsis:

This chapter deals with a familiar genetic theme: variation. The variation described in this chapter is not the easily visible phenotypes discussed in early chapters, but molecular variation – changes in DNA sequence that can be detected by several different molecular techniques. When DNA is examined, the amount of variation (changes in sequence of bases) between individuals and among individuals in a population is much greater than you may have realized. Differences in DNA sequence seen in a population are called polymorphisms. These can be single bases changes or changes in the number of copies of a small repeated sequence (mini or micro satellites). If the single base variation changes a restriction enzyme recognition sequence, the polymorphism can be identified by restriction enzyme digestion of genomic DNA and hybridization to visualize only that region of DNA. Increases or decreases in repeated DNA can be recognized by changes in the size of restriction fragments or of PCR amplified regions. Changes in base sequence that do not affect a restriction enzyme recognition site or the size of a restriction enzyme fragment can be recognized using allele specific oligonucleotides and hybridization.

Much of the DNA variation seen between individuals is silent. That is, it has no effect on phenotype. A molecular genotype detailing the form (alleles) of DNA markers present in an individual can be used to distinguish between individuals and to map disease genes. The concept of linkage that you learned in chapter 4 is very critical again here. Genes and molecular markers that segregate (are transmitted) together more often than 50% of the time are linked. They are physically close on a chromosome and will only be separated if recombination occurs between them. Polymorphic DNA markers can be used to follow the transmission of a disease locus if the marker is linked to the disease gene.

Positional cloning begins with a phenotype and works to identify the gene(s) responsible for the phenotype. This approach is similar to the classical approach you studied in earlier chapters except in earlier analyses a mutation defined a gene as a factor in determining a phenotype, but the actual function of the gene product was harder to understand. In genome analysis, the goal is to get to the sequence of a gene through mapping and use that information to discover how the gene product works. In the second approach, you begin with a gene sequence and try to establish what phenotype the gene causes and therefore the role the gene plays. This bottom-up type of approach has increased in importance as more and more sequence data has been obtained for several organisms.

The characteristics and phenotypes geneticists are studying now are often far more complex than the traits that were studied in early experiments on peas and humans. Genome analysis has provides a wealth of data that allow us to ask more complex questions. Incomplete penetrance, phenocopies, and polygenic inheritance all provide challenges in establishing the relationship between gene(s) and phenotype.

Significant Elements:

After reading the chapter and thinking about the concepts, you should be able to:

♦ Recognize and describe the four different types of DNA polymorphisms (SNPs, microsatellites, minisatellites and deletions/duplications/insertions into non-repeated loci (**see Table 11.1**).

♦ Explain the techniques used to detect the different polymorphisms (**see Figures 11.3, 11.4, 11.5, 11.8, 11.9, 11.11, 11.12 and 11.14**).

♦ Distinguish different forms of a DNA marker and assign heterozygote and homozygote genotypes.

♦ Determine if a particular marker is informative for mapping a disease gene (Is it linked and is it polymorphic in the family you are studying?).

♦ Follow inheritance of a disease allele of a gene using a linked molecular marker.

♦ Determine the likelihood that a particular individual carries a disease allele based on the molecular marker data and the % linkage between the marker and the disease gene.

♦ Describe steps in positional cloning (from phenotype to gene clone, **see Figure 11.7**).

♦ Describe steps for understanding the function of a gene (phenotype it affects) starting with a cloned gene.

♦ Interpret data for locating genes in a DNA sequence.

♦ Know how to analyze genes involved in a complex quantitative trait.

♦ Distinguish between somatic and germ-cell therapy.

♦ Define haplotype (**see Figure 11.26**).

Problem Solving Tips:

♦ Remember that most DNA markers are <u>not</u> in a disease gene (these markers are anonymous markers). Instead the markers are DNA polymorphisms that are near a disease gene, are therefore linked, and are used to trace the disease gene.

♦ RFLP alleles are also referred to as forms of a polymorphism (**see Figure 11.6**).

♦ In any problem, first establish what different forms (alleles) are present in the population being considered

♦   A form (allele) of a DNA marker and the allele of a linked gene are transmitted together in a family unless there is recombination.

♦   A different allele of a marker can be associated with the same disease locus in two different families.

♦   Genomic libraries include all DNA in the genome; cDNA libraries include only expressed (transcribed) genes and lack intron sequences.

♦   Genetic maps are based on recombination frequencies.

♦   The problems in this chapter are more representative of what is actually done in a molecular biology laboratory. Put yourself in the shoes of a researcher; be inquisitive and THINK!

Solutions to Problems:

11-1. a. **5**; b. **3**; c. **8**; d. **6**; e. **2**; f. **7**; g. **1**; h. **4**.

11-2.

a.   The **SNP that is completely linked to the disease locus in this family is part of their haplotype** of this region of the chromosome. The SNP is a DNA sequence change that happened on the same chromosome <1 cM from the mutation in the disease gene. It is never found apart from the disease gene because a recombination event that would separate them is very rare.

b.   **Check the sequence of the SNP in other, unrelated families** - both those with the same disease and those without the disease. If the G allele is the cause of the mutation in the disease gene then the same change could occur in some of the other families. If the G allele is part of the haplotype of the original family but not associated with the disease then there will be no correlation between presence and absence of the G allele and the presence or absence of the disease allele. Also do positional cloning and look for candidates for the disease gene in this area of the chromosome. **Once you have identified the candidate gene you can determine if the G allele is within the disease gene**. For example, does the G allele affect the predicted amino acid sequence of the candidate gene? Does it occur in or near a proposed splice junction?

11-3.

a.   Microsatellites are simple 1-3 bp sequences repeated in tandem 15-100 times. The polymorphisms are different numbers of the simple sequence repeats.

b.   **See Figure 11.3** Changes in the number of repeats are caused by **slippage of DNA polymerase** during replication.

c.   **See Figure 11.4**. Minisatellites are 20-100 bp sequences that are repeated up to thousands of times/locus. Minisatellite polymorphisms occur by **unequal crossing-over if sequences on homologs align out of register during mitosis or meiosis**. The repeat sequence in a minisatellite is too long to cause slippage of DNA polymerase during replication.

11-4. The differences between Southern blot and PCR for detecting restriction site polymorphisms: in the **Southern blot analysis the actual genomic DNA is examined; with PCR you need to know a bit about the flanking sequences so you can design primers**. The similarities are that **both look for the presence or absence of a restriction site**. The **PCR analysis is more efficient** - you can start with a very small amount of genomic DNA (one cell's worth) and get enough material to analyze.

11-5.

a.   **The SNP polymorphism must be within the sequence to which the oligonucleotide hybridizes** so the mismatch destabilizes the annealing of the oligonucleotide and the sample.

b.   **The SNP polymorphism must be at the 3' end of the of the primer** so you can determine if the extension reaction is able to increases the length of the primer by the addition of the single nucleotide in the reaction mix.

c.   In the case of the RFLP probe **the polymorphism must occur in a restriction site** (this is not necessary for parts a and b) and the probe simply has to be homologous to some part of the restriction fragment bounded by the polymorphic restriction site. The probe can actually be kilobases away from the actual polymorphism.

11-6.

a.   **The incubation temperature affects the accuracy of annealing between complimentary sequences**. At 100°C, no hybrids will form because the DNA remains denatured. At 80°C, the conditions are too permissive so that even oligonucleotide probes with a single base mismatch will hybridize. 90°C is a good temperature for differential hybridization using this allele specific oligonucleotide (ASO). Differentiation is made between DNAs having a complete match and those that have a single base difference.

b.   **Individuals 1, 5, 6, 8 are homozygous for the A allele; individuals 2, 7, 9, 10 are AS heterozygotes; individuals 3, 4 are S homozygotes.**

11-7. For ASO analysis, **the potential mismatch should be in then center of the oligonucleotide**. Therefore a 19mer that extends 9 bp in either direction would be a good probe. You need two different oligonucleotides for the analysis - one that contains a sequence corresponding to the mutant and the other corresponding to the wild type sequence. The oligos would be complimentary to:

> 5' CTATAAATGCGCTAGGCGT

and   5' CTATAAATGGGCTAGGCGT

11-8. The affected first child shows which marker allele is linked to the disease allele in each parent. The affected child got Dad's large allele and Mom's small allele.

a. The fetus has Dad's small allele (the normal CFTR allele) and Mom's small allele (the CFTR mutant allele). Thus **the fetus is a carrier, so there is 0% chance the fetus will exhibit the disease**. If Dad's gamete is the result of a recombination event between the microsatellite marker and the CFTR mutation then the fetus could have inherited Dad's small microsatellite allele and the mutant CFTR allele. The probability of this happening is very small as the microsatellite marker is within an intron in the CFTR gene.

b. The child is a carrier, so 1/2 of the gametes produced will be mutant for CFTR. If 3% of the population are carriers for CFTR then the probability of the child marrying a carrier is 0.03. The probability of this couple having an affected child is: 1/2 probability of mutant allele from the carrier form part a x 0.03 probability that spouse is a carrier x 1/2 probability of mutant allele from spouse = **0.0075 probability of affected child**.

11-9. For a marker to be informative in a particular family, individuals must be **heterozygous for a polymorphism**. This **polymorphism must be closely linked to the disease gene**. The two different alleles of the marker allow the inheritance of each homolog and by extension the disease allele to be traced from one generation to the next.

11-10.

a. One strategy is to **transform the mouse with the human gene cloned behind a promoter and an inducible regulatory regio**n. Thus the expression of the human gene can be turned on at will. The effects of over-expression can then be examined in a model organism.

b. Hemophilia is caused by the lack of the factor VIII protein. Therefore, you must knock-out the mouse gene for the factor VIII protein. This knock-out strain will be hemophiliac.

11-11. The pattern seen in the sample from individual 3 is very similar to the pattern of the sperm taken from the crime scene. Thus **it is possible that individual 3 is the perpetrator of the crime**. If this minisatellite probe hybridizes to 10 unlinked lock then there is $.375^{10} = 0.005\% = 1/20,000$ chance that this pattern will be found in another person. If you are looking at 24 separate loci the probability of finding the same pattern in another person is 1 in 17 billion.

11-12. A child should inherit 1 allele of each locus from each parent. Thus if the 10 bands in the daughter's pattern represent 5 heterozygous loci, she should share five in common with one parent and the other five in common with the other parent. The daughter shares 6/11 bands with male 3 and only 1/10 bands with male 4. Therefore, **male 3 is more likely to be the father**. The child may not have exactly the same 5 alleles as one parent - remember that microsatellite alleles change in length due to unequal crossing-over during meiosis, so the allele inherited by the child may be different than either allele in the parent.

11-13. <u>See Figure 11.7</u>. **Eggs are collected** from Angela. The eggs are **fertilized *in vitro*** with George's sperm. The resultant embryos are allowed to mitotically divide to the 8-cell stage. Researchers **take one cell from each embryo. The region of the β-globin gene is amplified from each cell using PCR. The PCR samples are digested with *Mst*II and electrophoresed on a gel**. Two bands on a gel means the β-globin gene has an *Mst*II site = AA genotype; one larger band on the gel means no *Mst*II restriction site = SS genotype; three bands on the gel = AS heterozygote.

11-14. In **humans some of the disadvantages are inability to do controlled crosses, few offspring/family, can't control environmental factors in the development of the disease**. The ultimate proof of a candidate gene is **transforming a mutant individual with the candidate gene and showing that the normal allele of the gene restores the phenotype to normal**. This sort of experiment can not be done in humans! In **mice there are genetically homogeneous inbred lines** which eliminates genetic heterogeneity, **it is possible to cross two inbred lines so the only 2 alternative alleles are at the marker and the disease gene**. This is a good system for dissecting complex traits. It is possible to do the transgenic test of the candidate gene in mice.

11-15. The first step is to identify which of the parents RFLP alleles are linked to the disease gene. Begin by figuring out which allele the affected child inherited from each parent.

a.   In the left hand pedigree **the affected child inherited the 10 kb band from the father which must be linked to his mutant FM allele**. Therefore, even though both parents have a 13 kb allele, the mother's 13 kb RFLP is associated with her mutant allele of FM while the dad's 13 kb allele is associated with his normal FM allele.

b. In the right pedigree the **affected child received the 7 kb allele from his mother** which must contain her FM mutant allele. This child also must have inherited the 10 kb allele from the father which must also have his mutant FM allele.

c. The male child in the left pedigree inherited his mother's 7 kb allele with the normal FM allele and his father's 13 kb allele with the normal FM allele. Therefore this child is homozygous normal. The female child in the right pedigree inherited the mother's 7 kb allele with the mutant FM allele. Because this child is phenotypically normal, she must have inherited her father's 10 kb allele linked to his normal FM allele, so she is a carrier. There is a **0% probability** that this couple will have a diseased child.

d. The male from the left pedigree is homozygous normal and the female from the right pedigree is a carrier, so there is a **1/2 chance that their child will be a carrier**.

11-16. Assign d^+ as the normal allele of the disease gene and d^* as the mutant allele. Use the affected child in generation 2 in each case to determine which allele of the disease gene is linked to which allele of the marker in the affected person in the first generation. In the first pedigree, the individual II-1 marker allele 1 (*m1*) is from the unaffected parent and marker allele 2 (*m2*) is from the affected parent. Therefore, the genotype of II-1 is $m1d^+/m2d^*$. The genotype of II-2 is $m3d^+/m2d^+$. Child A inherited the $m2d^+$ haplotype from II-2 and the *m1* allele from II-1. The *m1* allele is 10 cM from the d^* allele. Individual II-1 can produce the following gametes: $m1d^+$ (0.45 probability), $m2d^*$ (0.45 probability), $m1d^*$ (0.05 probability), $m2d^+$ (0.05 probability). The **probability of disease expression in child A is 0.05 probability of $m1d^*$ / 0.5 probability of inheriting the *m1* allele = 10%**. Child B inherited *m2* from II-1, so there is a probability of 0.45 $m2d^*$ / 0.5 probability of $m2$ = **90% probability of Child B being diseased**. In the second pedigree the genotype of II-1 is $m3d^+/m2 d^*$. **Child C** inherited *m1* from II-1, so 0.05 $m1d^*$ / 0.5 probability of inheriting $m1$ = **10% probability of being diseased**. In the third pedigree III-1 inherited $m1d^*$ from II-1 and $m2d^+$ from II-2. Individual II-1 could be either of two genotypes: genotype (i) is $m1d^*/m2d^+$, 90% probability; or genotype (ii) is $m1d^*/m2d^+$, 10% probability. Child D inherited the *m2* allele from II-1. Therefore, **the probability that child D (III-2) is diseased** is the probability of inheriting $m1d^*$ if II-1 is genotype (i) + the probability of inheriting $m1d^*$ if II-1 is genotype (ii) = 0.9 probability of genotype (i) x 0.1 probability of inheriting $m2d^*$ + 0.1 probability of genotype (ii) x 0.9 probability of inheriting $m2d^*$ = 0.09 + 0.09 = **0.18**.

11-17. The order is (c, d), g, a, d, b, f. It is possible to reverse the order of the steps in ().

11-18.

a. If a DNA fragment hybridizes to a band in the Northern blot then at least part of the DNA fragment was transcribed into the hybridizing mRNA. **A, C and E** contain sequences homologous to genes.

b. Because three differently sized mRNAs are hybridizing, **three different genes have been identified**.

c. **Yes**, it is possible that there are more genes in this region. There could be a gene that is transcribed in very low amounts and would not be detected by Northern blot analysis. Fragments A - E could all be in one large intron of a gene that spans this entire region. Thus these sequences would not be homologous to this mRNA but DNA sequences on either side would hybridize to the same size mRNA.

d. The trancripts recognized by **fragments C and E** are both genes that are expressed in the heart tissue, so they are candidates for the gene causing the disease.

e. The **gene recognized by fragment E** is found only in the heart, so this seems to be the more likely candidate since the disease affects only heart tissue.

f. **If there is a mouse model of this disease you would transform the mice with the cDNA clone of the candidate gene and look for the normal human gene to rescue the mutant phenotype in the mice**. If there is no mouse model of the disease you would compare the DNA sequence of the alleles from unrelated diseased families with the DNA sequence from normal people. Look for obvious mutations in the sequences from the diseased families that would alter the protein - changes to the predicted amino acid sequence of the protein, mutations that would affect intron/exon splicing, etc. Do Northern blot analysis on the heart tissue from affected cadavers - are there differences in the size or the amount of the candidate mRNA?

11-19.

a. The disease is **autosomal dominant**. Dominant because affected children always have an affected parent. It is autosomal because affected male II-2 passes the trait on to both daughters and sons. If it were X-linked he could only pass it on to daughters.

b. No, all individuals that pass the trait on to the next generation show the mutant phenotype, so there is 100% penetrance.

11-20.

a. The disease is **autosomal dominant**. Dominant because all affected children have an affected parent and autosomal because affected male III-1 passes the trait on to both daughters and sons.

b. **Yes**, there are 2 possibilities to account for the disease in III-1. Perhaps either **II-3 or II-4** had the mutant allele but didn't express the phenotype. Otherwise, III-1 inherited a mutant allele because there was **a spontaneous mutation of the normal allele in either the gamete from II-1 or II-2**.

11-21.

a. **The disease is X-linked recessive**. Recessive because it skips generations and X-linked because female II-2 is probably a carrier and she passes the disease on to 1/2 of her sons and none of her daughters. This pattern is also seen with II-7.

b. **Yes**, in this pedigree the carriers don't show the mutant phenotype but they can pass the disease on to their children. Thus, **II-2 and II-7 must be carriers**.

11-22.

a. **The disease is autosomal dominant**. The disease appears to skip generations, which would suggest the disease is recessive. However, if it is recessive then both II-1 and II-4 would have to be carriers. You are told the disease is <u>very rare</u>, so this is extremely unlikely. Therefore, the disease is dominant with incomplete penetrance. It is autosomal affected male I-1 passes the disease to male II-5.

b. **Yes, II-2 and III-1** must have the mutant allele but do not express the disease. II-2 has an affected child and III-1 is an identical twin of III-2 so they must both share the same alleles for all genes!

11-23. In order to find genes in a large cloned and sequenced region you can do the following:
- **use conserved sequence comparisons with evolutionarily divergent species**, for example compare mouse sequences to human sequences. Stretches of highly conserved sequence >25 bp are candidates for genes or regulatory regions. Do a **computational search for open reading frames (ORFs) and exon/intron boundaries. Look for homology to ESTs** (expressed sequence tags) which are cDNA fragments from the organism.

11-24.

a. The **Pinocchio syndrome is autosomal dominant**. Autosomal because there is male-to-male inheritance and dominant because an affected child always has an affected parent.

b. **The Pinocchio locus is linked to the SNP1 locus**. In generation II it is seen that the syndrome (P) is inherited with the $m2$ allele of the SNP locus. Thus the affected haplotype is $m1P$ and the unaffected haplotype is $m1P^+$. In generation III 7/8 children inherit either one of these two haplotypes which suggests very strong linkage. Child III-8 inherits a recombinant haplotype from his affected father, $m2P$, so the genotype of III-8 is $m2P/m1P^+$. This individual passes one or the other of these haplotypes to all of his children except IV-7 who receives a recombinant gamete of genotype $m2P$. Of the sixteen children in generations III and IV, 14/16 inherit a parental haplotype. Thus the SNP1 locus and the locus for Pinocchio syndrome are closely linked, with a recombination frequency between them of 2/16 = 0.125 = 12.5 cM.

c. **It is <u>not</u> likely that the coding region containing SNP1 is the Pinocchio gene**. There are 12.5 cM between these 2 markers. In humans, 1 cM = ~1,000 kb, thus there is roughly 12,500 kb between SNP1 and Pinnochio.

11-25. <u>See Table 11.2</u>.

a. The fact that many people develop heart disease later in life suggests that environmental factors induce the disease over time. Therefore, **choose families where the onset of the disease is early - families where people develop the disease in their 20's and 30's**. These families are likely to have a mutation in a gene that is important in the development of the disease. Look only at diseased individuals and divide complete sets of families on other criteria such as age of onset.

b. **You must have a mouse model of heart disease. You then transform these mice with your candidate gene and examine the transformed mice for less severe symptoms and longer life-spans**.

Chapter 12   The Eukaryotic Chromosome: An Organelle for Packaging and Managing DNA

Synopsis:

This chapter describes the structure of eukaryotic chromosomes and how that structure affects function. The very long, linear DNA molecules are compacted with proteins in the chromosomes to fit into the nucleus. Several structures are essential for duplication, segregation, and stability. Replication origins are necessary for copying DNA during S phase; centromeres are necessary for attachment to spindle fibers and proper segregation during cell division; telomeres are necessary at the ends of the linear DNAs to maintain the integrity of the DNA molecule. Chromatin structure (packaging of DNA in the chromosomes) can have consequences for gene activity. Areas of normally packaged chromosome can become decompacted for expression to occur. Some regions of chromosomes or entire chromosomes are packaged in a different way that decreases gene activity as in, for example, heterochromatin or Barr bodies (**see Figure 12.13**).

Significant Elements:

After reading the chapter and thinking about the concepts, you should be able to:

♦ Describe the essential elements of eukaryotic chromosomes.

♦ Predict the stability of artificially constructed chromosomes based on the components they contain

♦ Analyze data on changes in chromatin compaction.

♦ Understand the role of chromosomal origins og replication.

♦ Explain why centromeres are necessary for proper segregation during mitosis and meiosis (**see Figure 12.11**).

♦ Discuss the role of telomeres (**see Figure 12.10**).

♦ Understand how chromatin packaging influences gene activity.

♦ Explain PEV (position effect variegation) in *Drosophila* (**see Figure 12.14**).

Problem Solving Tips:

♦ Put yourself in the position of being the researcher. When designing experiments consider the aim of the experiment, the concepts that apply to the problem, and think through experimental methods you know to find a relevant methodology.

Solutions to Problems:

12-1. a. **4**; b. **10**; c. **5**; d. **8**; e. **9**; f. **2**; g. **3**; h. **6**; i.**1**; j. **7**.

12-2. Use Pulse-Field Gel Electrophoresis (PFGE) to separate all 16 yeast chromosomes. Then blot the chromosomes in a whole chromosome Southern and probe with your clone. You will be able to match the hybridization signal with the position of a chromosome on the original gel.

12-3. See Table 12.2. In interphase the chromosomes are compacted 40-fold more than naked DNA and during metaphase the chromosomes are compacted 10,000 fold more than naked DNA.

12-4. The core histones (H2A, H2B, H3 and H4) are the core of the most rudimentary DNA packaging unit, the nucleosome. The core is an octamer made up of 2 of each core histone. Roughly 160 bp of DNA wraps twice around the core, leading to a 7 fold compaction over naked DNA. About 40 bp forms the linker that connects one nucleosome to the next. Histone H1 lies outside the core, apparently associating with the DNA where it enters and leaves the core. Removal of H1 causes some DNA to unwrap from each nucleosome, but the core 140 bp of DNA stays intact. H1 is involved in the next level of compaction, formation of a 300 Å fiber.

12-5. Using the cloned gene, mutate the codon coding for the acetylated lysine so that it codes for a similar, non-acetylated amino acid. Then, transform this mutated allele into yeast cells on an autonomous plasmid so that both the mutant and wild type genes are present. If the acetylation is important for function then these transformed cells will grow more slowly.

12-6. Non-histone proteins, which make up ~1/2 the mass of proteins associated with DNA, are a very heterogeneous group of proteins. There are hundreds or even thousands of different kinds of proteins. **Some of these proteins play a purely structural role, (e.g. scaffold proteins, see Figure 12.2a)** while **others are active in replication (DNA polymerase)** and **the processing of recombination (proteins in the synaptonemal compex)**. Still **others are necessary for chromosome segregation (the motor proteins of the kinetochores, see Figure 12.2b)**. The largest class of non-histone proteins are t**hose that foster or regulate transcription and RNA processing. In mammals**. There are 5,000-10,000 of these tissue specific transcription factors that are found in different tissues at different times in the life cycle. The distribution of the non-histone proteins along the chromosome is uneven. They are found in different amounts and in different proportions in different tissues.

12-7.

a. In order to replicate the longest chromosome (66Mb) from one bidirectional origin of replication, 33 Mb would have to be copied along each replication fork during the 8 minute cycle (480 sec) = 33,000,000 bp replicated/480 sec = 68,750 bp/sec = ~69 kb/sec. Therefore, if a single origin of replication was used and replication took the entire 8 minutes of the cycle, **the rate of polymerization would be 0.069 Mb/sec or 69 kb per second**.

b. If bidirectional origins of replication occur every 7 kb, then only 3.5 kb would have to be replicated during the 8 min cell division cycle. **The polymerization rate would be 3.5 kb/480 sec = 7.3 bp/sec, a _much_ more reasonable rate**.

12-8.

a. In order to examine the end of one specific chromosome, **your DNA probe must contain unique DNA found next to the repeated 5' TTAGGG (telomere) sequence**s.

b. The sharpness of the band(s) seen after probing a genomic Southern with most DNA probes is due to the fact that the flanking restriction sites digest all the copies of the DNA into the same set of fragments. **The blurriness of the band seen when probing sequences found at the very ends of the chromosomes indicates that the hybridizing fragments from the end of the chromosome in a population of cells are not homogeneous in length**. In other words, the fragment at the end of the chromosome is not the same size in all cells. **The number of repeat sequences at the telomere, and therefore the telomere length, varies from cell to cell, especially in actively dividing cells.**

12-9. The new sequences that are added on to the end of this chromosome must be specific for the species that is adding them. **Because the YAC was transformed into yeast, telomerase in the yeast cell added on the sequence specific for yeas**t.

12-10.

a. Knowing the amino acid sequences of the proteins associated the human CAF-1 (chromosome assembly factor) complex means that you can **'reverse translate' the protein sequences and predict the degenerate nucleotide sequences of the proteins. The entire yeast genome has been sequenced, so you can search for orthologous yeast genes**. You expect the amino acid sequences, and therefore the DNA sequences, of these proteins to be very highly conserved. **Alternatively, you can clone the yeast genes by using the predicted DNA sequences of the human CAF-1 genes to make degenerate oligonucleotide probes and probe a yeast genomic DNA library.**

b.   Why identify the yeast genes? **You can experimentally manipulate the yeast much more easily! It is possible to make mutations of the CAF-1 proteins and carefully examine s the affects on chromatin assembly in this model organism.** It may be possible to extrapolate your findings to human chromosome assembly.

12-11.

a.   **A plasmid containing only the URA^+ gene must integrate into the chromosome to be replicated and maintained because is has no origin of replication. Once integrated it will be stably maintained.**

b.   **A URA^+, ARS plasmid can be maintained as a plasmid or it can integrate into the chromosome. If it remains as a plasmid, it will not be very stable and would be lost from many of the daughter cells during subsequent rounds of mitotic division. If this plasmid integrated, it would be very stable.**

c.   **The URA^+, ARS, CEN plasmid could only be maintained as a separate plasmid in the cell.** If it did integrate into the chromosome, there would be two centromeres on that chromosome and during mitosis the chromosome would break. **The plasmid would be very stable from one generation to the next because the centromere sequence directs its segregation.**

12-12.

a.   **Use the yeast protein to make antibodies and then use these antibodies to probe the human cDNA expression library.** Alternatively, you could use the cloned yeast gene as a probe to hybridize to clones in a human cDNA library. To identify related genes in distantly related species, the stringency of the hybridization conditions is often lessened so you do not demand that every base be identical.

b.   Use the human protein to make an antibody. This antibody will bind specifically to this protein in fixed cells. **If you label or tag the antibody (usually with fluorescence), you can determine the location of the protein in the cell - nucleus vs cytoplasm, for example.**

c.   There are several possibilities. **The loss of function might disrupt chromosome segregation during mitosis in a haploid cell in a way that is recognized as a signal to stop the cell cycle until the error is corrected.** (Such signals are described more fully in Chapter 18). Or, **if the protein is part of the kinetochore structure, there would be no kinetochore formed at high**

temperature and chromosomes would not migrate to the daughter cell. The result would be a polyploid mother cell. **If the protein acts in chromosome segregation, a temperature-sensitive mutation in the gene encoding the protein at high temperature could lead to loss of chromosomes, or aneuploidy**. This would be lethal in a haploid cell. However, if homologous chromosomes had genetic markers that allowed them to be distinguished in a diploid cell, such aneuploid cells could easily be observed.

12-13. The subcloned fragments that contain the centromeric DNA are those that show a high percentage of Trp$^+$ colonies after 20 generations without selection for the plasmid. These subclones include the 5.5 kb *Bam*HI, the 2.0 kb *Bam*HI-*Hind*III, and the 0.6 kb *Sau*3A. **Because the smallest of these has high mitotic stability and its ends are within the boundaries of the other fragments, the centromere sequence must be contained within the 0.6 kb *Sau*3A fragment**.

12-14. YAC clones can rearrange the insert DNA. BAC clones are not as likely to do this. The clones you have isolated from the BAC and YAC libraries have very different *Hind*III digestion patterns. In order to determine which of these restriction patterns most closely resembles pattern found in the human genome, **digest the BAC, the YAC, and the genomic DNA with several restriction enzymes and compare the restriction patterns of each when they are hybridized with a probe containing the BAC or YAC DNA**.

12-15. Controlled unfolding of the chromosome seems to be important for proper expression of genes. Boundary sequences prevent the opening up or decomposition of chromatin from spreading too far. **Removal of a boundary could cause inappropriate (ectopic) expression of genes in regions adjacent to the boundary**.

12-16. DH (DNase hypersensitive) sites are found at the 5' ends of genes that are actively transcribed (**see Figure 12.12a**). **The DNA in cell type II is very accessible to DNAase. This is characteristic of the open chromatin configuration that is found near a highly expressed gene**.

12-17. Heterochromatin is the regions of darkly staining DNA that is much more condensed than the euchromatin (the rest of the chromosome). Constitutive heterochromatin are the areas of heterochromatin that remain condensed and heterochromatic most of the time in all cells (**see Figure 12.13 a and b**). Facultative heterochromatin is that region of the chromosomes (or even entire chromosomes) that are heterochromatic in some cells and euchromatic in other cells.

a. **In *Drosophila* the centromeric regions of the chromosomes and the Y chromosome are examples of facultative heterochromatin. Constitutive heterochromatin is seen in cases of Position Effect Variegation (PEV).** The example discussed in the text is PEV of the white gene. The mosaic of white and red patches seen in the eyes of these animals suggests that the decision about the heterchromatic spreading is the result of a random process which varies from cell to cell during development. Heterochromatization can spread over >1 Mb of previously euchromatic DNA.

b. **In humans, the centromeric DNA and the great majority of the Y chromosome are also constitutive heterochromatin. The formation of Barr bodies due to the random inactivation of one of the X chromosomes in each cell early in female human fetal development is an example of facultative heterochromatin.**

12-18.

a. *Su(var)* mutations decrease the amount of PEV. **In the presence of a *Su(var)* mutant allele there will be fewer white patches in the eye and more red patches when the eyes are compared to a homozygous *Su(var)*$^+$ fly. The situation would be reversed with more white patches and fewer red (wild type) patches if the fly were heterozygous for the *E(var)* mutation. (See Figure 12.14 a)**

b. The *Su(var)* and *E(var)* mutations both have phenotypes that lead you to think the proteins encoded by the genes are involved in chromatin condensation. Assuming the mutations are loss of function (null) alleles, then **the *Su(var)* $^+$ genes encode proteins that establish and assist spreading of heterochromatin.** Thus, loss of some of the gene product results in engulfment of neighboring genes by heterochromatin. **The *E(var)* $^+$ genes seem to encode proteins that restrict the spreading of heterochromatin**, since loss of one copy of the gene allows heterochromatin to spread into neighboring genes more often. The results also suggest that position effect variegation is very sensitive to amounts of either type of protein because a reduction of 50% of either type of protein causes the mutant phenotype.

12-19. a. 1; b. **0**; c. **1**; d. **1**; e. **3**; f. **0**.

12-20. These twin sisters could still be monozygotic twins. They must both be carriers of the X-linked Duchenne muscular dystrophy (*Dmd*). **In the affected twin, the X^{Dmd+} homolog was inactivated in the cells that are affected by muscular dystrophy are different. In the unaffected twin, the other X chromosome (X^{Dmd}) was inactivated in those same cells.**

12-21. Girls of genotype of X^{CB}X^{cb} could have some patches of cells in the eye in which the X chromosome carrying the *CB* allele was inactivated and therefore those patches would be defective in color vision. Usually, enough cells have the *cb* allele inactivated and the *CB* allele active that there is sufficient color vision and therefore no phenotypic effect of the *cb* mutant cells.

12-22. There are many characteristics of polytene chromosomes that make them extremely useful for cytogenetics. The polytene chromosomes are very large, with ~1,000 copies of each homolog after endoreduplication. That, coupled with the pairing of the homologous chromosomes gives more than 2,000 copies of each chromosome paired in register. **These chromosomes are easily visible in the light microscope. The extremely detailed banding pattern is reproducible from one fly to the next, providing a map of the chromosomes based on the landmark bands (see Figure 12.15).** The limits of various chromosomal rearrngements (duplications, deletions, inversions and translocations) **can be very precisely mapped** in the polytene chromosomes. **It is also possible to do very accurate *in situ* hybridizations to the polytene chromosomes, thus localizing the probe to a region of 30-150 kb of DNA.** In contrast, *in situ* hybridizations in humans only localize the probe to a 5-10 Mb region of DNA.

Chapter 13   Chromosomal Rearrangements and Changes in Chromosome Number Reshape Eukaryotic Genomes

Synopsis:

Rearrangements of sections of chromosomes by duplication, insertion, deletion, inversion, or translocation can affect distances between genes and the function of genes in which they occur. Very large chromosomal rearrangements can be seen microscopically as changes in banding patterns. Many rearrangements are detectable by changes in linkage or effects on meiotic products. Crossing-over during meiosis within an inverted region in an inversion heterozygote leads to imbalanced gametes, so it appears from resulting viable gametes that recombination is suppressed within that region. Transposable elements are segments of DNA that can move from one position to another in the genome. Different types of elements move using a transposase enzyme or by reverse transcription of RNA into a DNA copy.

Changes in chromosome number can be the result of loss or gain of one chromosome (aneuploidy, **see Table 13.1**) or changes in the numbers of sets of chromosomes (e.g., polyploidy). Aneuploid cells are generally inviable in humans, with the exception of those that involve sex chromosomes (where there are still phenotypic consequences of extra or lost chromosomes).

Significant Elements:

After reading the chapter and thinking about the concepts, you should be able to:

- Use deletion data to map a gene.
- Trace products of a crossover within an inverted region of an inversion heterozygote when the inversion is either paracentric or pericentric.
- Predict gametes produced by a translocation heterozygote.
- Design experiments to determine if there are inversions, deletions or translocations in a strain.
- Describe how retrotansposons can cause rearrangements.

Problem Solving Tips:

- Deletions, inversions, and translocations change the linkage of genes that surround or are within the rearrangement.

♦ Draw out chromosomes of parents and progeny from complicated multigenerational crosses so you are very clear about the chromosome composition going into and out of meiosis. This should help you predict the genotypes and phenotypes of progeny.

♦ A set of chromosomes refers to the collection of chromosomes as found in a gamete. For example, in humans, a set of chromosomes is the 22 autosomes + one sex chromosome.

You must understand the effects of each type of rearrangement at several different levels, including pairing of the homologs, effects on recombination and segregation of the chromosomes at meiosis. Furthermore, these effects often differ between an individual that is heterozygous for the rearrangement and one that is homozygous for the rearrangement.

Duplications:

♦ The duplicated regions in duplication homozygotes can occasionally mispair during meiosis. When mispairing is followed by recombination one product has more copies of the repeat and the other has less (**see Figure 13.6**).

Deletions:

♦ If a deletion of a gene on one homolog (deletion heterozygote) uncovers a mutation in the gene on the other homolog the mutant phenotype is seen in the heterozygote (pseudodominance).

♦ You can order genes on a chromosome based on whether they are included in the region of the deletion or not (**see Figure 13.6**).

♦ There is no recombination between genes within a deletion.

♦ Deletions of DNA can be analyzed using restriction analysis. In a diploid organism, a deletion on one chromosome will mean that a restriction fragment that comes from within the deleted region of the genome will be at half the concentration of that found in a normal cell that has the DNA on both chromosomes.

♦ Deletion homozygotes are almost always inviable.

Inversions:

♦ In paracentric inversions the centromere is not included in the inverted region. Pericentric inversions include the centromere within the inversion.

♦ The chromosomes of inversion heterozygotes form inversion loops (**see Figure 13.11**).

♦ Recombination events within the inversion loop give imbalanced gametes which do not give rise to progeny (see Figure 13.13). Recombination events that occur outside the inversion loop give rise to normal, balanced, recombinant progeny. This lack of recombinant progeny from events within the inversion is why inversions are known as suppressors of recombination (**see Problems 13.7 and 13.8**).

♦ All recombination events in an individual that is homozygous for an inversion give rise to balanced gametes, although there will be some changes in recombination frequencies between genes where one is inside the inversion and the other is outside.

Translocations:

♦ During meiosis, the chromosomes of individuals that are heterozygous for a reciprocal translocation form a structure called a cruciform (see Problem 13.6 and **Figure 13.8**). One meiotic segregation (alternate segregation) gives rise to two balanced gametes – the two unaffected homologs in one gamete and the two translocated homologs in the other gamete. The other segregation pattern (adjacent 1) gives two imbalanced gametes that give rise to inviable zygotes (they are semisterile). Thus, genes that are one the chromosomes involved in the translocation do not show independent assortment – they act as though they are linked (pseudolinkage).

♦ Individuals that are homozygous for a reciprocal translocation do not show pseudolinkage nor semisterility.

Solutions to Problems:

13-1. a. **4**; b. **8**; c. **6**; d. **5**; e. **7**; f. **3**; g. **2**; h. **1**.

13-2.

a. **Each of the strains shows pseudodominance for some of the mutant alleles**; that is, each strain is mutant for one or more of the marker genes. Furthermore, after the diploids undergo meiosis, **two of the spores die**. All of these are indications that the **X-rays induced deletion mutations**.

b. Two spores in each ascus die because they receive the deleted homolog. **The deletions remove some essential genes from the chromosomes and this is lethal in a haploid**.

c. There is only one X-ray induced mutation per strain, so **all genes that show pseudodominance are on the same chromosome**. Using this logic, all four genes: $w, x, y,$ and $z,$ are on the same chromosome.

d. **The order is $w\,y\,z\,x = x\,z\,y\,w$**. Genes w and y are deleted in strain 1, uncovering the w^- and y^- alleles, so w and y are adjacent. Genes $x, y,$ and z are deleted in strain 2 so they must be adjacent. Combined from the information from strain 1, this means the order must be $w\,y\,[x\,z]$, with the brackets indicating that you don't know the relative order of x

and *z*.   Strain 3 is deleted for *w, y* and *z*, therefore the gene that follows *y* must be *z*. Note that two answers are given because you cannot determine the left-to-right orientation of this group of four genes.

13-3. In polytene chromosomes there are characteristic banding patterns for each chromosome. If there is a duplication of a region of DNA, **the bands in the duplication loop should be repeated elsewhere on the paired homologs**. That is, there should be three copies of the banding pattern in each genome: one in the duplication loop, one elsewhere on the chromosome carrying the duplication, and one on the wild-type homolog. The latter two copies should be paired with each other. If this is a tandem duplication, the loop should be adjacent to two paired copies. If the mutation is a deletion, **the looped out region in a deletion heterozygote contains the only copy of those bands found on the wild-type homolog**.

13-4. The deletion data allows you to narrow down the region in which the genes lie. All of these deletions remove portions of polytene chromosome region 65 (which turns out to be on chromosome 3, a *Drosophila* autosome). Deletion A shows pseudodominace for javelin and henna, so all or part of both genes must lie within the deleted region - between A2-3 and D2-3. Deletion B, pseudodominant for henna, indicates that *henna* lies between C2-3 and E4-F1. Combining the results for Deletions A and B, *henna* must be between C2-3 and D2-3. Because Deletion B is javelin$^+$, *javelin* must be located between A2-3 and C2-3 (the part of Deletion A that is not removed in Deletion B). Deletions C and D tell you that the *henna* gene cannot lie to the right of bands D2-3 on the figure in the text, delimiting *henna* to the interval between C2-3 and D2-3.

Inversions do not remove genes, they just relocate them. Therefore, if the inversion is made in a wild type chromosome, the inverted homolog will have the wild type alleles for all of the genes. Inversion B gives the expected result, and does not help locate either of the two genes. Inversion A, however, has a mutant javelin phenotype, indicating that there is a mutant allele of *javelin* on the inverted homolog. Thus, the javelin gene was broken by the inversion, so *javelin* is located in band 65A6. (This is consistent with the region containing *javelin* determined from the deletions above.) Very few *Drosophila* genes extend beyond one band, so we can assume that A6 is the location of *javelin*. In summary, **the *javelin* gene is in band 65A6 and the *henna* gene is between 65C2-3 and 65D2-3.**

13-5.

a. Fragments that are deleted on one homolog will be lighter in intensity than those that are not deleted (and are therefore present in two copies per cell). You can tell which fragments are contiguous by analyzing each deletion for the missing bands. Use this information about the deleted fragments in each deletion to order the fragments (as you did to order the genes in problem 13-2). New bands that appear in only one of the deletion strains represent new 'joining' fragments generated by the juxtaposition of the remaining bits of the two restriction fragments around the deletion breakpoints. For each of the strains, the following deleted fragments will be considered:

Strain	complete fragments deleted
Deletion 1	6.3, 5.6, 4.2
Deletion 2	6.3, 4.2, 3.0
Deletion 3	5.6, 0.9
Deletion 4	6.3, 3.0

Deletions 1 and 4 both delete the 6.3 kb fragment. The other deleted bands are not in common between deletions 1 and 4, so they must represent fragments that lie to either side of the 6.3 kb fragment. From strain 4, we know that the 3.0 kb fragment lies to one side of the 6.3 fragment, but since it is not lost in Deletion 1, the 5.6 and 4.2 kb fragments must lie to the other side (although we don't know the order yet). Deletion 2 also has a deletion of the 6.3 kb fragment as well as the 3.0 and 4.2 kb fragments. This tells us that the 4.2 kb fragment must be immediately adjacent to the 6.3 kb fragment and the order is 3.0, 6.3, 4.2, 5.6. Deletion 3, in which the 5.6 and 0.9 kb fragments are deleted, indicates that the 0.9 is adjacent to the 5.6 kb fragment but since it was not deleted in deletion 1 it must be on the side opposite the 4.2 kb fragment.

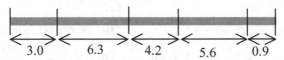

b. The approximate location of the genes can be determined based on which genes are pseudodominant in each deletion strain. Correlating the restriction map (part a) and the phenotypes of the strains, *rolled eyes* and *straw bristles* look like they are found somewhere within the 6.3, 5.6 and 4.2 kb fragments. But since deletion 3 also has the straw bristles phenotype, that gene can be placed at least partly within the 5.6 kb fragment. Deletion 2 has the rolled eyes phenotype as well, so the gene must be in one or both of the fragments that are common to deletions 1 and 2 (6.3 and 4.2 kb). However, since deletion 4 is missing fragment 6.3 but is not mutant for *rolled eyes*, the gene must lie within the 4.2 kb fragment. *Apterous wings* is mutant in deletions 2 and 4 which have in common deletion of the 6.3 and 3.0 bands, but since the mutant phenotype is not seen in deletion 1, the gene lies in 3.0 kb fragment. *Thick legs* is mutant in deletion 3 (5.6 and 0.9 kb fragments), but is not mutant in deletion 1 (5.6 kb fragment), so the gene lies in the 0.9 kb fragment.

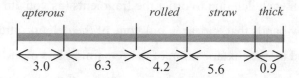

The map above provides only an approximation of gene position. For example, consider the location of the *apterous* gene. It is possible that *apterous* could actually lie in the left half of the 6.3 kb fragment (the part of the fragment removed by Deletion 4 but not by Deletion 1). In addition, because Deletions 2 and 4 (the two deletions that uncover *apterous*) both remove DNA to the left of the map, its also possible that *apterous* lies to the left of the sequences on the map. More accurate mapping would require examination of more deletions and more restriction enzyme sites.

13-6. The genotype of the female is:

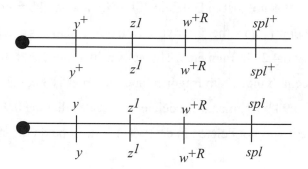

a.   The remainder of the male progeny (76,671) are **the parental types, so they will be** $y^+ z^1 w^{+R}$ spl^+ **/ Y (zeste) and** $y z^1 w^{+R} spl$ **/ Y (yellow zeste split)**.

b.   Classes A and B are a reciprocal pair of products. **They are the result of crossing over anywhere between the *y* and *spl* genes** resulting in the reciprocal classes: $y^+ z^1 w^{+R} spl$ / Y (zeste split) and $y z^1 w^{+R} spl^+$ / Y (yellow zeste).

c.   Remember that w^{+R} allele is really a tandem duplication of the w^+ gene and that the zeste eye color depends on having a mutant z^1 allele in a genome that also contains two or more copies of w^+. **Classes C and D, are the result of mispairing and unequal crossing over between the two copies of the w^+ gene**. The misalignment can occur in two different ways, see the figure below. Misalignment I gives: $y^+ z^1$ [3 copies w^+] spl (zeste split = same phenotype as class B) as one product, and $y z^1$ [1 copy w^+] spl^+ (yellow wild type eye = class C) as the reciprocal product. In misalignment II, the recombinant products are: $y^+ z^1$ [1 copy of w^+] spl (wild type eyes split = class D) and $y z^1$ [3 copies of w^+] spl^+ (yellow zeste = same phenotype as class A). Each of these misalignments thus produces one class of recombinant products that results in a wild-type eye color.

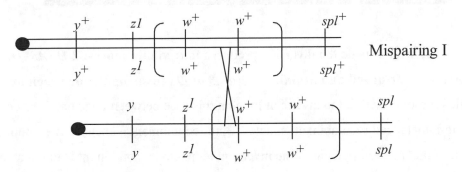

Mispairing I

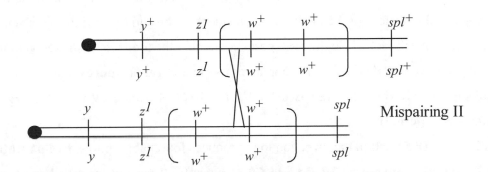

Mispairing II

d. The **genetic distance between y and spl** = # recombinants between y and spl / total progeny = 2430 (class A) + 2394 (class B) + 23 (class C) + 22 (class D) / 81,540 = **5.9 mu**. This recombination frequency includes all the recombinants, because classes A and B contain the reciprocal events to classes C and D.

13-7. The diploid cell contains a pericentric inversion on one homolog. The pairing of the homologous chromosomes during metaphase I of meiosis is shown below. Use this drawing to trace the consequences of crossovers in different regions.

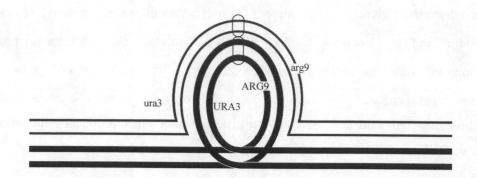

a. A single crossover outside the inversion produces **four viable spores: 2 *URA3 ARG9***
    (prototrophic for uracil and arginine) : 2 *ura3 arg9* (auxotrophic for uracil and arginine).

b. A single crossover within the inversion loop, in this case between *URA3* and the centromere,
    results in imbalanced recombinant gametes. Both recombinant gametes have a duplication and a
    deletion of the material <u>outside</u> of the inversion loop. One recombinant is duplicated for the
    region outside the loop on the left and deleted for the information outside the loop on the right.
    The other recombinant is the reciprocal - deleted for the information on the DNA to the left of
    the loop and duplicated for the information outside the loop and to the right. This genetic
    imbalance is usually lethal, so the two spores containing the products of the recombination will
    die. The single cross over inside the loop gives rise to **2 parental spores and 2 recombinant,**
    lethal spores as in the following ascus: 1 *URA3 ARG9* : 1 *ura3 arg9* : 1 *URA3 arg9* (lethal) :
    1 *ura3 ARG9* (lethal).

c. This 2 strand DCO within the inversion loop produces four viable spores, **two parental spores**
    and 2 recombinant spores: 2 *URA3 ARG9* (1 parental, 1 recombinant) : 2 *ura3 arg9* (1
    parental, 1 recombinant).

13-8. In this problem, the diploid cell contains a paracentric inversion on one homolog. The pairing
of the chromosomes in the inversion heterozygote are shown below. Use this drawing to trace the
consequences of crossovers in different regions.

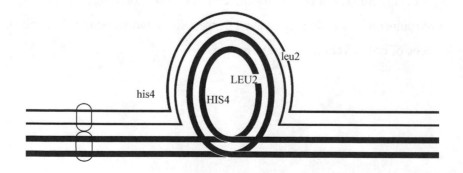

a. A single crossover within the inversion (between *HIS4* and *LEU2*) leads to 2 parental spores (one of each type) and 2 recombinant spores. The recombinants are duplicated and deleted for the regions outside the inversion loop (see problem 13-7). In this case, one of the recombinant spores will be duplicated for the region containing the centromere and deleted for the DNA on the other side of the inversion loop (dicentric). The reciprocal recombinant spore will be deleted for the region containing the centromere and duplicated for the region to the right of the inversion loop (acentric). Neither type of chromosome segregates properly so the spores that receive these chromosomes will definitely die. **The acus will contain: 1 *HIS4 LEU2* (prototrophic for histidine and leucine) : 1 *his4 leu2* (auxotrophic for histidine and leucine) : 1 dicentric (lethal) : 1 acentric (lethal).**

b. This is a 2 strand double cross over within the inversion loop double. All four spores are viable: **2 *HIS4 LEU2* (1 parental and 1 recombinant) : 2 *his4 leu2* (1 parental and 1 recombinant).**

c. A single cross over between the centromere and the inversion loop will give 2 parental spores and 2 balanced recombinant spores: **2 *HIS4 LEU2* (1 parental and 1 recombinant) : 2 *his4 leu2* (1 parental and 1 recombinant).**

13-9. A tetratype ascus means there has been a single crossover between the 2 genes (**see Figure 5.13**). Because both *HIS* and *LEU* are in an inversion loop a single crossover would give inviable spores. In this case the recombination event must be more complicated than a single cross over. **One possibility is a 2 strand double cross over. This could occur in 2 ways: one recombination between LEU and HIS and the second recombination between HIS and the end of the inversion loop; or one crossover between LEU and HIS and the second one between LEU and the end of the inversion loop. In both cases, the second recombination event must occur within the inversion loop.** Such a 2 strand double cross over will give the following tetratype ascus: 1 *HIS4 LEU2* : 1 *his4 leu2* : 1 *HIS4 leu2* : 1 *his4 LEU2*.

13-10. Each of the original strains is true breeding and shows the same recombination frequency of 21 mu between genes *a* and *b*. However, in the F_1 heterozygote genes *a* and *b* are only 1.5 mu apart. The only rearrangements that affect recombination frequency in the heterozygote are deletions and inversions. There are two reasons why this cross cannot involve a deletion: both parental strains are homozygous (true breeding) and homozygous deletions are lethal; and in both parents genes *a* and *b* are 21 mu apart but in a deletion homozygote the genes would be less than 21 mu apart. This reduction of recombination between the 2 genes in the F_1 is therefore caused by an inversion. One parental strain has normal chromosomes and the other parental strain is homozygous for an inversion.

There are 2 possibilities for the inverted region: (i) it includes almost all of the region between genes *a* and *b*, but does not include the genes themselves, or (ii) the inversion includes both genes and the DNA in between them.

There are 2 other possibilities for the location of the inversion with respect to the genes which can be ruled out. (iii) the inversion includes gene *a* and almost all of the DNA between the genes; and (iv) the inversion includes gene *b* and almost all of the DNA between the genes. These possibilities can be ruled out because in both cases the recombination frequency between *a* and *b* in the inversion homozygote would be much less than 21 mu, as the inversion would bring the genes closer together.

In any case, the F_1 progeny is an inversion heterozygote. The inversion loop occupies either (i) almost all of the region between genes *a* and *b*, although the genes are not included in the loop, or (ii) the loop includes both genes and the DNA in between them. Any single crossovers within the inversion loop (the huge majority of crossovers between the genes) will result in inviable gametes. The few recombination events that occur between gene *a* and the inversion loop or between the inversion loop and gene *b* will result in balanced, viable, recombinant gametes in scenario (i), as will some of the double cross over events within the inversion loop in both scenarios (i) and (ii).

13-11.

a. **2, 4**. Inversion loops are seen during MI of meiosis only if the cells are heterozygous for an inversion.

b. **2, 4**. Single crossovers within the inversion loop in inversion heterozygotes generate genetically imbalanced chromosomes. The genetic imbalance involves deletions and duplications of regions outside the inversion loop. If the inversion is paracentric, then the recombinant products are also dicentric and acentric. Crossovers in inversion homozygotes do not cause genetic imbalance.

c. **2**. An acentric (and the reciprocal dicentric) fragment is produced from a single crossover within a paracentric inversion in an inversion heterozygote.

d. **1,3**. In an inversion homozygote, crossovers within the inversion yield 4 viable, balanced, spores all of which have the inverted gene order.

13-12. The data shows unexpectedly reduced recombination frequencies between certain pairs of genes in Bravo/X-ray and Bravo/Zorro heterozygotes. This reduction in recombination will be seen in both deletion heterozygotes and in inversion heterozygotes. You are told that the 3 strains have variant forms of the same chromosome, and that the number of bands in the polytene chromosomes are the same in all 3 strains. Thus, none of the strains is a deletion homozygote. The chromosomal rearrangements here must be inversions.

a. Recombination frequencies between genes in the Bravo/X-ray heterozygote are normal for *a-b* and *b-c* and *g-h* but are reduced for all other gene pairs. Thus, the inversion in the X-ray strain breaks between genes *c-d* and between genes *f-g* and inverts genes *d, e,* and *f.* The *c-d* and *f-g* intervals must still include some non-inverted DNA to allow recombination that produces viable gametes. Similarly, the genetic distance is reduced in the *b-c* and *f-g* intervals for Zorro, where the inversion end points are found and minimal in those intervals completely within the inversion. **The order of the genes in X-ray is: *a b c f e d g h.*** **The order of the genes in Zorro is: *a b f e d c g h*.**

b. **The physical distance in the X-ray homozygotes between *c* and *d* is greater** than that found in the original Bravo homozygotes. The inversion occurred in this portion of the chromosome, so *c* and *d* are now separated by many more genes (all the inverted DNA).

c. **The physical distance between *d* and *e* in the X-ray homozygotes is the same** as that found in the Bravo homozygotes because this interval is completely within the inverted segment. The relationship of *d* to *e* has not changed.

13-13. The semisterile F_1 is a translocation heterozygote and will produce 1/2 fertile : 1/2 semisterile progeny from alternate segregation. Products of adjacent-1 or adjacent-2 segregation are imbalanced and therefore inviable - this is the basis of the semisterility. Because the only viable gametes are the result of alternate segregation, genes that are on the chromosomes involved in the translocation will not show independent assortment. Instead, if the genes are located very close to the translocation breakpoints, they will show only the parental classes; that is, the genes will display pseudolinkage. Genes that are on any other chromosome will assort independently of the translocation (i.e., such genes will assort independently from fertility/semisterility).

a. If the *yg* gene is on a different chromosome than those involved in the translocation, the traits will assort independently. The product rule says you can cross multiply the 2 monohybrid ratios: 1/2 *yg*$^+$ (normal leaf color); 1/2 *yg* (yellow green) and 1/2 fertile : 1/2 semisterile to give: **1/4 fertile *yg*$^+$: 1/4 fertile *yg* :, 1/4 semisterile *yg*$^+$: 1/4 semisterile *yg*.**

b. If the translocation involved chromosome 9, the fate of the fertility and leaf color phenotypes are connected. The original cross was: semisterile *yg*$^+$ x fertile *yg* → F_1 semisterile x fertile *yg* → ?

This means the normal, non-translocated chromosome 9 has the *yg* allele, while the translocated chromosome 9 has the *yg*$^+$ allele. Thus, the chromosomes of the heterozygous F$_1$ at meiosis I would look like:

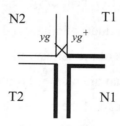

(For simplicity, this figure shows only one chromatid per chromosome.) From the translocation heterozygote, the products of an alternate segregation are balanced and viable. **The progeny (after crossing with a fertile *yg* homozygote) will be: 1/2 fertile *yg* (N1 + N2) : 1/2 semi-sterile *yg*$^+$ (T1 + T2).**

c.   **The rare fertile, *yg*$^+$ and semisterile, *yg* gametes result from recombination between the translocation chromosome and the homologous region on the normal chromosome** with which it is paired. After the recombination event, the N1+ N2 fertile gamete will contain *yg*$^+$ and the T1 + T2 semisterile gamete will contain *yg*.

The frequency of crossing-over, as represented by the rare fertile green and semisterile yellow-green progeny, will give you the genetic distance between the translocation breakpoint and the *yg* gene.

13-14. Any haploid spores with a deletion are dead (white). An octad has 8 spores.

a.   **0 white spores.** The inversion has no effect if recombination does not occur.

b.   **4 white spores.** Only two chromatids are involved in the crossover. The recombination gives 2 unbalanced gametes. The remaining two chromatids survive as haploid products and divide mitotically to form 4 viable spores in the ascus.

c.   **0 white spores.** If a crossover occurs outside the inversion loop all the products are viable.

d.   **8 white spores.** All of the resulting gametes would be genetically imbalanced and would die.

e.   **0 white spores**. Alternate segregation produces balanced gametes.

f.   **0 white spores**. The crossover in the translocated region would simply cause the reciprocal exchange of DNA between homologous portions of the chromosome. All spores live.

13-15.

a.   **1, 3, 5 and 6**. Translocation heterozygotes can produce gametes with any pairwise combination of N1, N2, T1, and T2.

b.   **2 and 4**. Translocation heterozygotes cannot produce gametes with two copies of the same chromatid.

c.   **1 and 3**. These arise from alternate segregation, so they are balanced.

d.   **5 and 6**. These arise from adjacent-1 (5 is a gamete with T1 and N2) and adjacent-2 (6 is a gamete with N1 and T1) segregations.

13-16.

a.   Diagram the cross. Because the two genes are on different autosomes, they should assort independently:

$cn\ cn^+$; $st\ st^+$ $\times$ $cn\ cn$; $st\ st$ → 1/4 $cn\ st$ (white) : 1/4 $cn\ st^+$ (cinnabar) : 1/4 $cn^+\ st$ (scarlet) : 1/4 $cn^+\ st^+$ (wild type).

b.   The genes show pseudolinkage in this male. The $cn\ st$ and $cn^+\ st^+$ allele combinations seem to be linked. **This result suggests that the unusual male has a translocation between chromosome 2 and 3 with the mutant cn and st alleles either on the translocated chromosomes or on the normal chromosomes.** The figures below show two of the four possible genotypes for the unusual male fly. Both diagrams show the mutant alleles on the normal order chromosomes. Instead, the mutant alleles of the genes could both be on the translocated chromosomes. Although both the cn and st genes may actually be on the same chromosome after in the translocation (right hand panel),this is not necessary The pseudolinkage will be seen even if the 2 genes are still on separate chromosomes (left hand panel). Also remember that male *Drosophila* do not recombine.

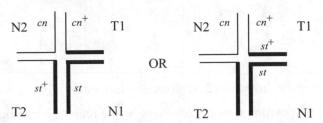

c.   The wild-type F_1 females are translocation heterozygotes containing the cn^+ and st^+ alleles on the translocated chromosomes (she obtained the cn and st alleles from her non-translocated mother). The pairing in meiosis would be the same as shown for the male in part b, except that recombination can occur in the female. **A crossover between cn and the translocation breakpoint or between st and the translocation breakpoint followed by alternate segregation produces gametes with the genotype $cn\ st^+$ (cinnabar) $cn\ st^+$ (scarlet).** These classes allow you to calculate the map distance between the st and cn genes in the translocation. The rf = 10 mu.

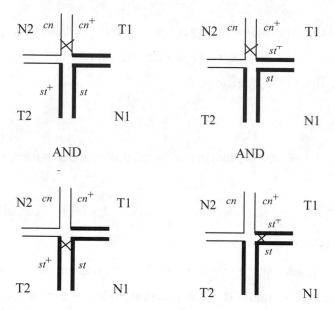

AND

13-17. Remember that sister chromatids are identical to each other. The probe will anneal to homologous DNA on the chromosome and produce a signal at that position.

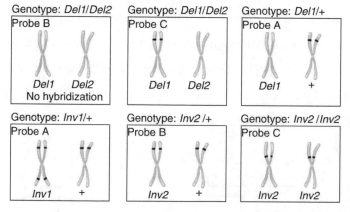

Note that in the paracentric Inversion 2, regions homologous to Probes A and B are not inverted and thus stay at the original position near the telomere, while regions homologous to Probe C in the bottom right figure are inverted and thus move closer to the centromere.

13-18. Individuals that are homozygous for a translocation are fertile. However, **when insects that are translocation homozygotes mate with insects with normal chromosomes, the F₁ progeny will be translocation heterozygotes. The fertility of these F₁s should be reduced by about 50% and half of their progeny will also have reduced fertility**. If the released insects are homozygous for several different translocations, then the fertility of the F₁ individuals should be reduced by a further 50% for each translocation for which they are heterozygous. For example, in an F₁ insect that was heterozygous for 3 different translocations (T1 - T3), only 1/8 of its gametes (1/2 balanced for T1 x 1/2 balanced for T2 x 1/2 balanced for T3 = 1/8 balanced gametes) would be balanced and give rise to progeny.

13-19. Remember that the Y chromosome pairs with the X chromosome during meiosis, so the N1 + N2 chromosomes in the male translocation heterozygote will be the autosome on which *Lyra* is normally found and the X chromosome, as shown on the figure below.

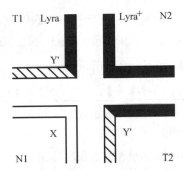

There are only two kinds of genetically balanced gametes produced by the male during alternate segregation. The T1 and T2 segregant has the Y chromosome and the autosome with the mutant *Lyra* allele. This gamete would fertilize an X bearing egg, producing Lyra males. N1 and N2 would yield a gamete with the X chromosome and the autosome with *Lyra⁺*. This gamete would produce wild-type females when fertilized with *Lyra⁺* gametes from the wild-type mother. **The progeny will be 1/2 *Lyra* males : 1/2 *Lyra⁺* females**.

13-20. The homolog with the transposon forms a loop which contains the un-paired transposon sequences. The probe will hybridize to one spot at the base of the loop on the normal, non-inserted homolog. On the homolog with the transposon insertion, the probe will hybridize to the DNA on both sides of the base of the loop.

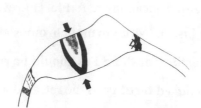

13-21.

a. A translocation heterozygote will make 2 types of gametes as a result of alternate segregation and 2 types of gametes as a result of adjacent-1 segregation in a 1:1:1:1 ratio. The 2 products of alternate segregation are N1 + N2 and T1 + T2, both of which are balanced gametes. The two products of adjacent-1 segregation are N1 + T2 and N2 + T1, both of which are imbalanced gametes. When a translocation heterozygote is crossed to a homozygous normal individual, the imbalanced gametes from the translocation parent never give rise to viable progeny. However, If <u>both</u> parents of a cross are translocation heterozygotes, then it is possible for an imbalanced gamete from one parent to be fertilized by the reciprocally imbalanced gamete from the other parent. For example, a N1 + T2 gamete from one parent can fertilize an N2 + T1 gamete from the other parent, creating a zygote that is a balanced translocation heterozygote!

	N1 + N2	T1 + T2	N1 + T2	N2 + T1
N1 + N2	homozygous normal	translocation heterozygote	imbalanced, lethal	imbalanced, lethal
T1 + T2	translocation heterozygote	translocation homozygote	imbalanced, lethal	imbalanced, lethal
N1 + T2	imbalanced, lethal	imbalanced, lethal	imbalanced, lethal	translocation heterozygote
N2 + T1	imbalanced, lethal	imbalanced, lethal	translocation heterozygote	imbalanced, lethal

Thus, among the viable progeny you would expect a ratio of 2/6 fertile (homozygous normal + translocation homozygote) : 4/6 semisterile (translocation heterozygote) = **1/3 fertile : 2/3 semisterile**.

b. This problem involves the self-fertilization of a particular translocation heterozygote. Instead of producing 2/6 fertile : 4/6 semisterile progeny as in part a, this plant produced a ratio of 1/5 fertile : 4/5 semisterile. These numbers suggest that one out of the two fertile and viable classes in part a did not survive in this cross. Thus, one possible explanation for these results is that **the translocation homozygotes die because the translocation breakpoint interrupts an essential gene.**

13-22. If 2 transposons are near each other on the chromosome and have normal gene-containing chromosomal DNA between them, they could transpose the entire large segment of DNA when transposase acts on the ends the transposons. (See Figure 13.10).

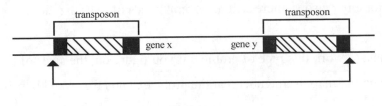

transposase acts on far ends and
transposes the whole section of
the chromosome

13-23. *Ds* is a defective transposable element that does not encode transposase. The *Ac* element is a complete, autonomous copy of the same transposon that encodes transposase. *Ds* can thus transpose (move around the genome) only in the presence of *Ac*. Chromosomal breakage at *Ds* insertion sites could be due to errors in the transposition mechanism catalyzed by the *Ac* transposase. A new insertion of *Ds* into a gene might yield a mutant allele that is unstable in the presence of *Ac* transposase (that is, the transposase could subsequently move *Ds* out of the location where it caused the mutation). Because *Ac* is complete and autonomous, it contains ends that can be recognized by the transposase. Thus, in different strains *Ac* has transposed itself into different chromosomal locations.

13-24. The original *ct* mutant allele was caused by the insertion of the gypsy transposable element. The stable *ct*⁺ revertants are likely to be precise excisions in which the *gypsy* element has moved out of the gene, restoring the normal *ct*⁺ sequence. The unstable *ct*⁺ revertants are likely to be cases in which the transposition process altered the *gypsy* element in the *ct* gene so that the gene could function normally. However, the bit of the transposon that remains may try to transpose when in the presence of the transposase. These attempts might cause deletions or rearrangements of the *ct* gene, thus creating new *ct* mutant alleles. The stronger *ct* mutant alleles could result from imprecise excision of the *gypsy* element, leading to the deletion or alteration of sequences within the *ct* gene that would strongly affect its functions (see Figure 13.26). Alternatively, these stronger *ct* alleles could result from the movement of the *gypsy* transposon into other parts of the gene that would compromise gene function more seriously.

13-25. Create a probe composed of the DNA sequence preceding the 200 A residues. Hybridize the probe to a genomic Southern blot. If other copies of a retroposon are present, you would expect to see several hybridizing bands. You could also use the probe to do *in situ* hybridization to human chromosomes. If the probe is homologous to a retrotransposon (or if the sequence is repeated in the genome for any reason), there will be several bands of hybridization.

13-26. It is easiest to work this type of problem if you figure out the sizes of the genomic fragments and place these on the map. Remember that the probe extends from 2.6 kb to 14.5 kb.

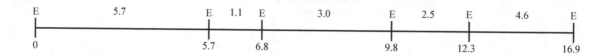

E   5.7   E   1.1   E   3.0   E   2.5   E   4.6   E
0   5.7   6.8   9.8   12.3   16.9

a.   **5**. The base change exactly at coordinate 6.8 will alter the restriction recognition site at that position and therefore the 1.1 and 3.0 kb fragments will disappear to be replaced by one new 4.1 kb fragment.

b.   **3**. A point mutation at position 6.9 has no effect on these restriction fragments since there is no EcoRI site at that position.

c.   **2**. This deletion removes 0.3 kb of DNA within the 2.5 kb fragment, resulting in the loss of the 2.5 kb fragment and the appearance of a new 2.2 kb fragment.

d.   **6**. This deletion removes 0.3 kb of DNA including a restriction site. The 1.1 and 3.0 kb restriction fragments would disapper and be replaced by one new 3.8 kb fragment.

e.   **8**. The insertion of a transposable element at coordinate 6.2 will change the size of the 1.1 kb fragment. The 1.1 kb fragment will disappear and be replaced by and a new larger fragment. If the transposable element has a restriction site in it, the 1.1 kb fragment will be replaced by 2 new fragments, one a minimum of 0.5 kb and the second a minimum of 0.6 kb in size. Because the new fragments actually seen are 4.9 and 2.3 kb long, the results suggest that the transposable element is at least $(4.9 + 2.3 - 1.1) = 6.1$ kb long; it could be even larger if the transposable element has more than one site for this restriction enzyme.

f.   **10**. An inversion with breakpoints at 2.2 and 9.9 will alter the two fragments in which the breakpoints are located but will not affect the 1.1 and 3.0 kb fragments that occur in between. Thus the 5.7 kb and 2.5 kb fragments will disappear and be replaced by a 5.9 kb fragment and a 2.3 kb fragment. The total size of this region of the genome will not change.

g.   **7**. The reciprocal translocation will cause the 2.5 kb fragment to disappear. Two new fragments will be seen, each containing part of the 2.5 kb fragment. One of the new fragments will be a minimum of 0.3 kb and the other will be a minimum of 2.2 kb.

h.   **1**. A reciprocal translocation with a breakpoint at 2.4 will cause the 5.7 kb fragment to disappear. This will be replaced by 1 new fragment with a minimum size of 3.3 kb. The sequences between coordinates 0 and 2.4 will also be connected to DNA from the other chromosome; though this will make a new restriction fragment, it will not be visible in this Southern blot because the probe does not include any of this DNA.

i.   **4**. The duplication will increase the size of the 2.0 kb fragment by 3.0 kb, creating a new 5.0 kb fragment. The restriction map makes it clear that these duplicated sequences do not contain a site recognized by the restriction enzyme.

j.   **9**. The 2.5 kb fragment is maintained but the 4.6 increases to 6.6 kb. Because there is a restriction site within the region that is tandemly duplicated, a new fragment of 2.0 kb is now present.

13-27.

a.   The ***x* number in *Avena* is 7**. This represents the number of different chromosomes that make up one complete set.

b.   **Sand oats are diploid ($2x = 14$); Slender wild oats are tetraploid ($4x = 28$); Cultivated wild oats are hexaploid ($6x = 42$)**.

c.   The number of the chromosomes in the gametes must be half of the number of chromosomes in the somatic cells of that species. **Sand oats: 7; slender wild oats: 14; cultivated wild oats: 21**.

d.   **The *n* number for each species is the number of chromosomes in the gametes and therefore is the same as the answer in c**.

13-28.

a.   **15** ($= 2n + 1$)

b.   **13** ($= 2n - 1$)

c.   **21** ($= 3n = 3x$ because the original species was diploid)

d.   **28** ($= 4n = 4x$)

13-29. Remember that nondisjunction in MI causes one copy of each homolog to be found in the gamete, while nondisjunction in MII causes both copies of a single homolog to be found in the gamete. Both nondisjunction in MI and nondisjunction in MII also produce gametes without the chromosome (nullo). See Synopsis for Chapter 4 and Problem 4-26. **Possibility A** received 2 copies of the smaller band from Fred; this is **due to non-disjunction during meiosis II in the father**. **Possibility B is caused by non-disjunction in meiosis I in the mother. Possibility C arose by non-disjunction in meiosis I in the father. Possibility D arose by non-disjunction in the mother in meiosis II**.

13-30.

a. **Perhaps the most obvious mechanism for uniparental disomy is the fusion of a nullo gamete from one parent (the result of nondisjunction in either MI or MII in that parent.** <u>see Problem 4-26</u>**) with an MII nondisjunction gamete from the other parent** (the mutant homolog must be the one that undergoes MII nondisjunction). **However, several other scenarios are also possible**, of which a few are described here. A second mechanism begins with the fusion of a nullo gamete from one parent (due to MI or MII nondisjunction) with a normal gamete carrying the mutant allele from the other parent. This generates a monosomic embryo, which then undergoes mitotic nondisjunction of the monosomic chromosome very early in development to create an individual homozygous for that chromosome. A third possible mechanism begins with the formation of a normal heterozygous embryo. Very early in development this embryo undergoes mitotic nondisjunction, leading to loss of the normal allele and retention of one homolog with the mutant allele. This loss of the normal homolog is then followed by a second mitotic nondisjunction to generate a homozygous mutant genotype. A fourth mechanism is fusion of a normal wild type gamete from one parent with a gamete carrying 2 copies of the affected homolog (the result of nondisjunction in MII). This would generate a trisomic embryo. Early in development a mitotic nondisjunction or chromosome loss event would cause the loss of the normal allele, leaving the 2 mutant alleles.

There is no easy way to distinguish between these possibilities, because they all lead to the same result of uniparental disomy. In very rare cases, it might be possible to discriminate between these scenarios if the individual were a mosaic and you could find some cells with aneuploid chromosome complements predicted by one of the proposed mechanisms.

b. **Girls with unaffected fathers that are affected by rare X-linked diseases could be produced by any of the mechanisms described in part a**. The mother must be a heterozygous carrier, and the father's chromosome is the one that must be lost. **The transmission of rare X-linked disorders from father to son could only be explained by the first mechanism described in part a.** For the son to be a male and also affected, the son <u>must</u> inherit <u>both</u> his father's X and Y chromosomes. The maternal X chromosome must also be lost; this could occur by a meiotic nondisjunction event in the mother, or by a mitotic non-disjunction/chromosome loss event early in the development of an XXY zygote.

c. Another way in which a child could display a recessive trait if only one of the parents was a carrier involves mitotic recombination early in the development of a heterozygous embryo. The recombination event must occur between the mutant gene and its centromere **(see Figure 5.26)**. The zygote then forms from the recombination product that is homozygous for the mutant allele. To detect the occurrence of mitotic recombination, **you must examine several DNA markers along the chromosome arm containing the gene involved in the syndrome**. In particular, you want to compare markers close to the centromere with those near the telomere. **If the presence of the disease were due to mitotic recombination, then the person would be heterozygous for markers near the centromere but homozygous for those near the telomere. If uniparental disomy caused by any of the mechanisms described in part a was instead involved, the individual would be homozygous for all of the DNA markers** (assuming that the individual is not a mosaic).

13-31. Meiotic nondisjunction should give roughly equal numbers of autosomal monosomies and autosomal trisomies. In fact, the total number of monosomies would expected to be greater, because chromosome loss produces only monosomies. The actual results are the opposite of these expectations. The much higher frequency of trisomies seen in the karyotypes of spontaneous abortions suggests that human embryos tolerate the genetic imbalance for 3 copies of a gene much better than one copy. Also, one copy of a chromosome will be lethal if that copy carries any lethal mutations. Monosomies usually arrest zygotic development so early that a pregnancy is not recognized, and thus they are not seen in karyotypic analysis of spontaneous abortions.

13-32. Both types of Turner's mosaics could arise from chromosome loss or from mitotic nondisjunction early in zygotic development. Chromosome loss would involve the loss of one of the X chromosomes early in development in an XX embryo (producing a mosaic with both 46, XX and 45, XO cells) or the loss of the Y chromosome in the XY embryo (yielding a mosaic with 46, XY and 45, XO cells). Mitotic non-disjunction in a normal XX embryo should produce an XXX daughter cell in addition to an XO, while mitotic non-disjunction in an XY embryo yields an XO and an XYY daughter cell. If the XXX or XYY daughter cells did not expand into large clones of cells during development, karyotype analysis would not be able to detect their presence. Note that for mitotic non-disjunction to have given rise to the described mosaic individuals, the nondisjunction event must have occurred after the first mitotic division so that there would be some XX or XY cells.

13-33. You have 3 marked 4th chromosomes: $ci^+ ey$, $ci ey^+$ and $ci ey$. *Drosophila* can survive with 2 or 3 copies of the 4th chromosome, but not with 1 copy or 4 copies. You are looking for mutations which are defective in meiosis and cause an elevated level of nondisjunction.

a. **Mate potential meiotic mutants that are $ci^+ ey$ / $ci ey^+$ with $ey ci$ / $ey ci$ homozygotes. The normal segregants should be $ci^+ ey$ / $ey ci$ (ey) and $ci ey^+$ / $ey ci$ (ci). Nondisjunction in MI will be seen as the rare $ci^+ ey$ / $ci ey^+$ / $ey ci$ (wild type) progeny.** Nullo-4 gametes without any copy of chromosome 4 would produce zygotes with only 1 copy of this chromosome that would not survive.

b. **The cross in part a will detect nondisjunction in MI, but it will not distinguish MII nondisjunction.**

c. Diagram the test cross: $ci^+ ey$ / $ci ey^+$ / $ey ci$ x $ey ci$ / $ey ci$ → ?

   Remember that in a trisomic individual, two of the three copies of the chromosome pair normally at metaphase I of meiosis, while the third copy assorts randomly to one pole or the other. There are 3 different ways to pair the 4th chromosomes in the trisomic individual. The first option is: 1/3 ($ci^+ ey$ segregating from $ci ey^+$ with $ey ci$ assorting independently) = 1/6 probability of (1/2 $ci^+ ey$ / $ey ci$: 1/2 $ci ey^+$) and 1/6 probability of (1/2 $ci^+ ey$: 1/2 $ci ey^+$ / $ey ci$). The second option is: 1/3 ($ci^+ ey$ segregating from $ey ci$ with $ci ey^+$ assorting independently) = 1/6 probability of (1/2 $ci^+ ey$ / $ci ey^+$: 1/2 $ey ci$) and 1/6 probability of (1/2 $ci^+ ey$: 1/2 $ey ci$ / $ci ey^+$). The third option is: 1/3 ($ci ey^+$ segregating from $ey ci$ with $ci^+ ey$ assorting independently) = 1/6 probability of (1/2 $ci ey^+$ / $ci^+ ey$: 1/2 $ey ci$) and 1/6 probability of (1/2 $ci ey^+$: 1/2 $ey ci$ / $ci^+ ey$). Each option gives the following progeny phenotypes (which is the same as the gamete genotypes, as this is a test cross) in the following frequencies: option 1 = 2/12 + ey : 2/12 ci + ; option 2 = 1/12 wild type : 1/12 ci ey : 1/12 ci + : 1/12 + ey; option 3 = same as option 2. The 3 options can be summed to give the final result: **1/3 ci + : 1/3 + ey : 1/6 wild type : 1/6 ci ey**.

d. These compound 4th chromosomes (att4) can be used in crosses to assay potential mutants. For instance, cross a potential mutant that is $ci^+ ey$ / $ci ey^+$ (as in part a) to a fly with attached 4th chromosomes that are not marked (that is, both are $ci^+ ey^+$); that is, **potential meiotic mutants of genotype $ci^+ ey$ / $ci ey^+$ x att4 $ci^+ ey^+$ / $ci^+ ey^+$ →** ? In this cross all of the normal progeny would have 3 copies of the 4th chromosome and would be phenotypically wild type.

Nondisjunction in MII would be seen as unusual + ey progeny or ci + progeny. This occurs because half the gametes in the att4-containing parent would have no copy of the 4th chromosome (nullo-4) since the att4 chromosome does have a partner. The products of nondisjunction in MII in the other parent would thus be *ci⁺ ey / ci⁺ ey* (+ ey) or *ci ey⁺ / ci ey⁺* (ey +). **In this case, nondisjunction in MI is not distinguishable** because the resulting progeny would be *ci⁺ ey / ci ey⁺* (wild type like the normal progeny). **Another possible cross would be: potential meiotic mutants that are *ci⁺ ey / ci ey⁺* x att4 *ci ey / ci ey* →** . In this cross the normal progeny will be + ey and ci +. **Nondisjunction in MI will give unusual wild type progeny** (*ci⁺ ey / ci ey⁺* gametes from the potential mutant; nullo-4 gametes from the att4 parent). Nondisjunction in MI or MII would yield nullo-4 gametes from the potential mutant, so the progeny (which could only be formed with att4 *ci ey / ci ey*) would be phenotypically ci ey. **With this second cross you can screen for nondisjunction both in MI and MII.**

13-34. The genotype of the diploid *Neurospora* cell is shown below. For a review of recombination in fungi and tetrad analysis see Chapter 5 Synopsis and Problem Solving Tips. For a review of meiotic segregation and non-disjunction see Chapter 4. The figure below shows prophase of meiosis I that occurs during this mating. In the answers to parts a-f, asci are written as though they have 4 (rather than 8 spores), and all spores are written in order relative to the equator of the ascus:

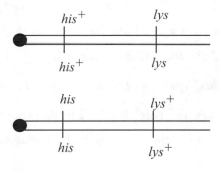

a. A single crossover between the centromere and *his* will give a PD ascus that shows MII segregation for both *his* and *lys*: **1 *his⁺ lys* : 1 *his lys⁺* : 1 *his⁺ lys* : 1 *his lys⁺*.**

b. A single crossover between *his* and *lys* will give a tetratype ascus showing MI segregation for *his* and MII segregation for *lys*: **1 *his⁺ lys* : 1 *his⁺ lys⁺* : 1 *his lys* : 1 *his lys⁺*.**

c. Non-disjunction during MI causes one daughter cell to have both homologs (4 chromatids) and the other daughter cell to have none (nullo). At metaphase II of meiosis, the homologous chromosomes align independently of each other on the metaphase plate, and 1 chromatid from each segregates into the gametes. This leads to 2 gametes that are diploid for the chromosome that underwent non-disjunction. (The nullo cell will have two nullo daughters.) The gametes after MII segregation are: **2 *his*$^+$ *lys* / *his* *lys*$^+$ (his$^+$ lys$^+$) : 2 nullo (aborted, white)**.

d. Non-disjunction during MII affects one of the 2 daughter cells formed after MI segregation. In the affected daughter cell, both chromatids go to one gamete and the other gamete is nullo. Two different results are possible, depending on which daughter cell undergoes MII non-disjunction: **2 *his*$^+$ *lys* : 1 *his* *lys*$^+$ / *his* *lys*$^+$ (his lys$^+$) : 1 nullo (aborted) OR 1 *his*$^+$ *lys* / *his*$^+$ *lys* (his$^+$ lys) : 1 nullo (aborted) : 2 *his* *lys*$^+$**.

e. The single crossover between the centromere and *his* involving chromatids 2 and 3 gives the following meiotic structure:

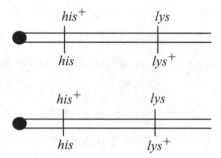

Non-disjunction in MI gives one daughter cell with both homologous centromeres and a second nullo daughter cell. At metaphase of MII the homologous centromeres line up on the metaphase plate independently of each other. MII segregation then causes one of each sister chromatid to segregate into the 2 resultant gametes. After the crossover the bivalents are no longer homozygous. As a result, there are 2 different segregation patterns that can happen, leading to the following types of asci: **1 *his*$^+$ *lys* / *his*$^+$ *lys* (his$^+$ lys) : 1 *his* *lys*$^+$ / *his* *lys*$^+$ (his lys$^+$) : 2 nullo (aborted) OR 2 *his*$^+$ *lys* / *his* *lys*$^+$ (his$^+$ lys$^+$) : 2 nullo (aborted)**.

f. The single crossover between *his* and *lys* gives the following meiotic structure at metaphase I of meiosis:

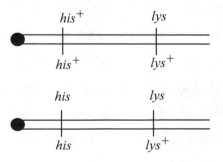

Non-disjunction at MI causes both centromeres to segregate to one daughter cell while the other daughter cell is nullo. Again, the homologous chromosomes line up independently of each other at metaphase II of meiosis, leading to 2 different possible segregation patterns: **1 his^+ lys / his lys (his^+ lys) : 1 his^+ lys^+ / his lys^+ (his^+ lys^+) : 2 nullo (aborted) OR 1 his^+ lys / his lys^+ (his^+ lys^+) : 1 his^+ lys^+ / his lys (his^+ lys^+) : 2 nullo (aborted)**.

13-35. <u>See Figure 13.30c</u>. Haploid plant cells in culture can be **treated with colchicine to block mitosis (segregation of chromosomes) during cell division and create a daughter cell having the diploid content of chromosomes. Once the diploid resistant cell is obtained, it is grown into an embryoid. Proper hormonal treatments of the embryoid will yield a diploid plant.**

13-36. Only plants that are F^A- F^B- will be resistant to all 3 races of pathogen. Because the chromosomes of same ancestral origin still pair, i.e. F^a pairs with F^A and F^b pairs with F^B, the cross can be represented as a dihybrid cross between heterozygotes. This treats the resistance genes from the two ancestral species as independently assorting genes. You are doing a cross between parents that are heterozygous for 2 different genes. Thus, **9/16 of the progeny will have the F^A- F^B- genotypes and these plants will be resistant to all three pathogens**.

13-37. Karyotype analysis is done at metaphase of mitosis. The chromosomes are stained such that each different kind of chromosome has a characteristic banding pattern. **The banding patterns of the homologs in the autopolyploids should be the same (for example, in an autotetraploid you would see four copies of each type of banding pattern), but the chromosomes of the different species that formed the allopolyploids would probably have different banding patterns (so in an allotetraploid you would see two copies of each type of banding pattern)**. Note that karyotype analysis allows you to determine the x number (the number of chromosomes in the basic set), because this will equal the number of different patterns you see. In turn, this allows you to determine the ploidy (the number of copies of the basic set), which is equal to the number of copies of each pattern.

Chapter 14   The Prokaryotic Chromosome: Genetic Analysis in Bacteria

Synopsis:

This chapter describes characteristics of genetic analysis in bacteria, with a focus on gene transfer in *E. coli*. Two key features have made *E. coli* a powerful model organism for understanding basic cell processes. First, the ability to grow massive numbers of bacteria on defined media allows easy and quick isolation of mutants. Second, there are many naturally occurring ways to transfer DNA from one cell to another (**see Figure 14.9**). The DNA transfer mechanisms - transformation, conjugation, and transduction (general and specialized) and the ways in which geneticists use these are described. In addition to their value for research, DNA transfer mechanisms are important for survival and evolution of bacterial species.

Significant Elements:

After reading the chapter and thinking about the concepts, you should be able to:

♦ Distinguish between selection (only one specific genotype can grow) and screening (more than one genotype can grow, but additional analysis is needed to establish the genotype of each cell).

♦ Describe how Hfr and F' cells are formed and the uses of each for mapping genes and complementation analyses (**see Figures 14.13 and 14.18**).

♦ Set up Hfr crosses to map genes (describe the genotypes of donor and recipient and the selective media used, **see Figure 14.14**).

♦ Analyze time of entry data to map genes and origins of transfer for different Hfr strains (**see Figure 14.15**).

♦ Map genes using cotransduction or cotransformation frequencies.

♦ Describe the differences between transformation, transduction, and conjugation.

♦ Analyze three point crosses (**see Figure 14.17**).

Problem Solving Tips:

♦ The F plasmid can integrate at different locations around the *E. coli* chromosome to generate Hfr strains that have different origins of transfer.

♦ Genes closest to the origin of transfer are more likely to be transferred into a recipient than those further from the origin.

♦ The time at which genes are transferred into a recipient in an Hfr cross is a reflection of the distance from the origin of transfer.

♦ Hfr crosses are usually used to get a low resolution map; P1 transductions are useful for finer genetic mapping.

♦ Bacteriophage lambda (λ) can integrate into the chromosome (lysogeny) to generate a lysogen (**see Figure 14.22**).

♦ F plasmids and bacteriophage lambda can excise imprecisely, picking up adjacent chromosomal genes.

Solutions to Problems:

14-1. a. **4**; b. **5**; c. **2**; d. **7**; e. **6**; f. **3**; g. **1**.

14-2. The initial tube of bacteria has 2×10^8 cells/ml. The first step of the dilution series is a 10^{-2} dilution (0.1 ml of the initial tube into 9.9 ml of diluent = 0.1 ml/10.0 ml = 1/100). The second step of the dilution is again 10^{-2}, so at this point the total dilution is 10^{-4}. The third step is a 10^{-1} dilution (1 ml/10.0 ml = 1/10) for a total dilution of 10^{-5}. You then put 0.1 ml (10^{-1} ml) of this dilution on the petri plate. **Therefore, you are putting $10^{-5} \times 10^{-1}$ ml $\times$ (2×10^8 cells/ml) = 2×10^2 cells on the first petri plate, which should grow into 200 colonies**. The fourth step of the series is another 10^{-1} dilution, for a total dilution of 10^{-6}. You again plate 0.1 ml of this dilution, so you expect $10^{-6} \times 10^{-1}$ ml $\times$ (2×10^8 cells/ml) = 2×10^1 cells = **20 colonies on this second plate.**

14-3. In this problem, "minimal media" contain no carbon sources, so they require supplementation with the indicated sugars. In most cases, if there is no carbon source specified (e.g. rich media + X-Gal) then the sugar is glucose. Otherwise, all sugars in the media must be specified. Note that X-Gal is a substrate for the ß-galactosidase enzyme, but it cannot serve as a carbon source.

a. **(iv)** Lac$^+$ cells are able to use lactose as the sole carbon source for growth, while Lac$^-$ cells would not be able to grow if the only sugar in the media were lactose. Medium iv is thus selective, because Lac$^+$ cells can grow on it but Lac$^-$ cells cannot.

b. **(iii)** A screen is different than a selection. For a genetic screen, you need to be able to examine the phenotype of each individual cell or colony. To screen for Lac$^+$ cells, you therefore need a medium on which both Lac$^+$ and Lac$^-$ cells can grow but on which they have different visible phenotypes. In a selection, only certain genotype(s) of cells can grow (min + lac selects for Lac$^+$

cells and against Lac⁻ cells). The rich medium (iii) allows both Lac^+ and Lac^- to grow, so it is not selective. However, the X-Gal in these plates distinguishes between the two phenotypes (Lac^+ cells are blue; Lac^- cells are white).

c.   (ii) To select for Met^+ cells, the medium should lack methionine, demanding that the bacteria must be able to synthesize methionine in order to grow. Medium ii is the only choice that lacks methionine.

14-4.

a.   In order to determine the number of nucleotides necessary to identify a gene, you must calculate how many bases represent a unique sequence in a DNA molecule of the size of the *E. coli* genome (4.6 Mb or 4,600,000 bases). There are 4 bases possible at each position in a sequence, so 4^n represents the number of combinations found in a sequence n bases long. For example, 4^2 is the number of unique sequence combinations that could be made with 2 positions. Therefore, if you were looking for a unique 2 nucleotide sequence, you might expect to find it, on average, every 1/16 nucleotides. A sequence of 11 nucleotides would appear $1/4^{11}$ or once in every 4×10^6 bases (4 Mb); a sequence of 12 nucleotides would appear uniquely $1/4^{12}$ or one in 16.8 Mb. Therefore **you need a sequence of about 12 nucleotides in order to define a unique position in the *E. coli* genome**. This problem can also be solved using the equation $4^n = 5 \times 10^6$ (5 Mb). To solve for n, rearrange the equation: $n\log 4 = \log 5 \times 10^6$; $n = 11.1$. Thus you need more than 11 nucleotides to find a unique nucleotide sequence.

b.   This problem assumes that you have obtained a sequence of contiguous amino acids within a protein. The 12 nucleotides shown in part a to define a unique position in the genome would encode 4 amino acids. However, because of the genetic code's degeneracy, you actually only know the identity of about 8 of these nucleotides. (For amino acids specified by 6 codons, you would potentially know one less nucleotide per codon; for tryptophan and methionine, which are specified by only a single codon, you would know all three nucleotides in the codon.) **If you had a sequence of six amino acids, you would probably know at least 12 unique nucleotides**. Because many genes evolved through a pattern of duplication followed by divergence, some protein domains of 6 amino acids might appear in more than one protein. As a result, knowing a few more than 6 amino acids would make the case that you have identified the correct gene even stronger.

14-5. Isolate genomic DNA separately from *E. coli B* and *E. coli K*, digest these DNAs with the *Eco*RI restriction enzyme, electrophorese the DNAs on a gel, and transfer the DNAs to a nitrocellulose filter for a genomic Southern hybridization using an IS1 DNA as a probe. The number of bands that appear correspond to the number of IS1 elements in the genome because *Eco*RI does not cut within the element. Instead, the enzyme will cut at the nearest recognition sites that flank each IS1 element. If a band is twice as intense as other bands, this would indicate there are two different fragments of the same size that contain an IS1, and such a band represents two different IS1 elements.

14-6. Do a mating between the mutant cell with 3-4 copies of F and a wild-type F⁻ recipient. If the mutation were in the F plasmid, you expect the recipient strain into which the plasmid is transferred (the exconjugant) to have the higher copy number. If the mutation is in a chromosomal gene, the higher copy number phenotype would not be transferred into the recipient. There is however a potential complication with this experiment. When you do an F^+ x F^- mating, some of the F^+ cells will have converted to Hfr cells that can transfer bacterial DNA. An alternative experiment could avoid this complication. **You could isolate the F plasmid DNA from the mutant cell, and then transform this plasmid into new recipient cells. By examining the number of copies of the F factor in the transformed cells, you could tell whether the trait was carried by the plasmid.**

14-7. Transfer the plasmid (by transformation) into a non-toxin producing recipient strain. If the gene is encoded on the plasmid, the transformant cells will produce the toxin.

14-8. The *purE* and *pepN* genes will be cotransformed at a lower frequency if the *H. influenzae b* pathogenic strain was used as a host donor strain. There is a lower likelihood that the two genes will be on the same piece of DNA because they are separated by 8 more genes-worth of DNA (~ 8 kb of DNA) than in the *H. influenzae Rd* non-pathogenic strain.

14-9. Selecting for Pyr⁺ exconjugants selects for an early marker transferred from the donor into the recipient. The frequency with which genes beyond *pyrE* are transferred decreases with distance from this early marker. This presumably occurs because of the fragility of the contact between the cells. The further a gene is from *pyrE*, the more likely it is that the connection between the donor and recipient will be broken before that gene can be transferred. This is very similar to what is depicted in **Figure 14.15b**. **The order of genes is: *pyrE xyl mal arg met tyr his*.**

14-10.

a. To avoid confusion, we will use two conventions in describing the composition of bacterial media in this and subsequent problems. (1) If no carbon source is specified, the medium has glucose as a carbon source. (2) If a carbon source is specified, then all sugars included in the medium will be listed. **The Hfr strain will grow on min + cys. The F⁻ strain will grow on min + trp + his + tyr + thr.** The F⁻ strain does not need streptomycin to survive, so this does not need to be included in the media. (But it would not be a bad idea to include this antibiotic to minimize contamination.) This strain is also *man⁻* which means the cells cannot employ mannose as a carbon source, but they will grow on the glucose provided in the above medium.

b. The selective media used in both experiments selects for exconjugants that have received Trp⁺. **In experiment 1, the matings were not disrupted as they were in experiment 2, so some mating pairs continued to transfer DNA even after the cells were diluted and plated. In experiment 2, Trp⁺ must have been transferred before the cells are interrupted in order to get exconjugants.** Thus, the Trp⁺ gene is first transferred after 8 min.

c. **Man⁺ exconjugants are selected by plating on min + man + cys + trp + his +tyr +thr + strep.** (According to the second convention above, this medium cannot contain glucose or any sugar other than mannose (man) as a carbon source.) Note that the strep antibiotic is needed to kill the Hfr donor cells.

d.

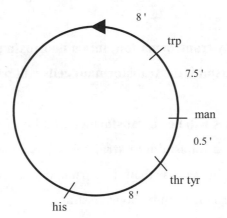

e. cross 3: Hfr *man⁺ thr⁻ tyr⁺* x F⁻ *man⁻ thr⁺ tyr⁻*

    cross 4: Hfr *man⁺ thr⁺ tyr⁻* x F⁻ *man⁻ thr⁻ tyr⁺*

<u>See Figure 14.17 parts b and c</u>. You know from part d that *man⁺* enters the F⁻ cell first, followed by the *tyr⁺* and *thr⁺* genes. When considering a circular F⁻ chromosome and a linear piece of bacterial DNA from the Hfr, viable exconjugants result from even numbers of crossovers. In this

experiment, wild type exconjugants were selected, and many more were seen in cross 3 than cross 4, therefore a double crossover is sufficient to give man^+ thr^+ tyr^+ exconjugants in cross 3, while a quadruple crossover is needed to give wild-type exconjugants in cross 4. Because you know the *man* gene is at one end, there are 2 possible orders for these 3 genes. The figure below shows how Cross 3 would look given these two possible orders. One order requires 4 crossovers to get a wild type exconjugant, while the other possible order only requires a double crossover. As you know that only a double crossover is needed to produce wild-type recombinants in cross 3, you can conclude that **the order is *man tyr thr*.** The results from cross 4 confirm this order, as you can see if you draw out the possibilities for cross 4.

cross 3:

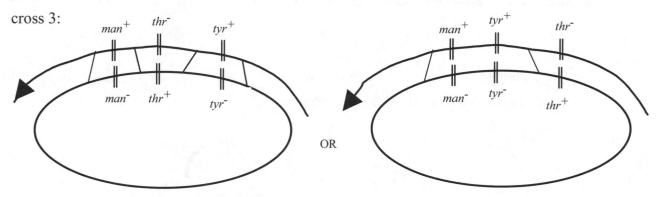

14-11. See Figure 14.17 parts b and c and problem 14-10 above. The exconjugants scored in the data table are Arg^+ Lac^+. The recombination events must place wild-type sequences of the *arg* gene in the F⁻ chromosome. Depending on the order of mutations, it takes either a double or a quadruple crossover to reconstitute the arg^+ gene. If the wild type exconjugants occur frequently (0.5%), then there is a double crossover. If the wild type exconjugants occur rarely (0.06% or lower), then they are the result of a quadruple crossover. Note that all the frequencies are low because in order to be Arg^+, one of the crossovers must occur in the very small region between the two *arg* mutants. In cross A, the Arg^+ phenotype occurs at a relatively high frequency. The order of the mutations must be *lac arg102 arg101*, as a double crossover with the mutations arranged in this order could regenerate an arg^+ gene. This pattern will be true for each pair of crosses shown. That is, in the cases where the Arg^+ exconjugants occur frequently, the wild type allele from the Hfr is closer to the *lac* gene. **The order is *lac arg102 arg101 arg103*.**

14-12. See Figure 14.17 parts b and c and problems 14-10 and 14-11. This is a three factor cross in which the selected marker, *ilv,* is the farthest from the origin of transfer. This ensures that all 3 genes were transferred into all exconjugants. The largest class of exconjugants is the one in which all of the genes recombined into the donor chromosome, which suggests that the genes are close together. The smallest class is the one in which quadruple crossovers occurred. In this case, Ilv$^+$ Bgl$^-$ Mtl$^+$ is the smallest class, so the *bgl* gene must be between *ilv* and *mtl*. The Ilv$^+$ Mtl$^-$ Bgl$^-$ class results from a crossover between *ilv* and *mtl* and another crossover on the other side of *ilv*. The Ilv$^+$ Mtl$^-$ Bgl$^+$ class comes from a crossover between *mtl* and *bgl* and another crossover on the other side of *ilv*. **The order is *ilv bgl mtl*.** Note that you can also obtain an estimate of the distances separating the three genes by looking at the frequencies of the latter two classes. The distance between *ilv* and *bgl* is about 3.3 times the distance between *bgl* and *mtl* (60:18).

14-13.

a. The F plasmid integrated into the chromosome near the *mal* genes, creating an Hfr, and then the F factor excised incorrectly, picking up adjacent *mal* genes.

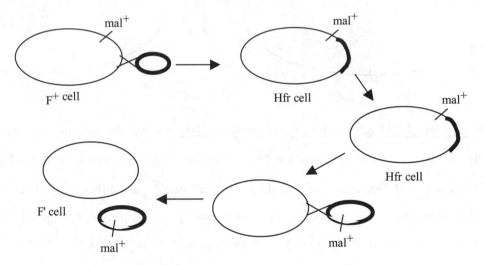

b. The F' *mal* must have recombined with the chromosome to generate an Hfr strain.

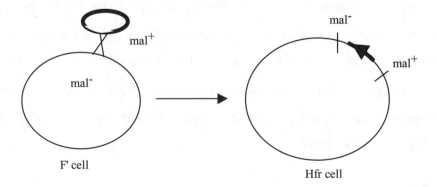

There are two details in the above picture that are worth considering. First, as long as the *mal⁻* mutation was not a deletion removing the whole *mal* gene, it is most likely that the recombination event would occur between the two copies of the *mal* gene, as this is the region of greatest homology between the episome and chromosome. However, the integration event could have happened elsewhere, particularly if the circle with *mal⁺* also had an IS or Tn element. Second, regardless of where the integration event occurs, this Hfr cell will have two copies of the *mal* gene: one wild type, and one mutant.

14-14.

a. **The order of the genes is *pab ilv met arg nic (trp pyr cys) his lys.***

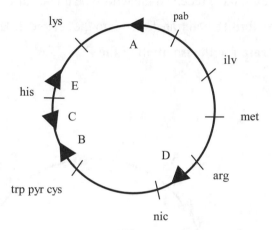

b. See <u>Problem 14-12</u>. After mating with HfrB selecting for Nic⁺ meant the *trp⁺ pyr⁺ cys⁺* alleles had all been transferred to the F⁻ cell. Trp⁺ was selected and most of the exconjugants (790/1,000) also recombined in the *pyr⁺ cys⁺* alleles from the donor. This shows that the 3 genes are closely linked. Usually the smallest class of recombinants requires four crossovers instead of two. After the crossovers have occurred, the two outside genes will have the genotype of the Hfr parent and the gene in the middle will have the genotype of the F⁻ parent. In this problem, however, the smallest class of exconjugants is represented by Trp⁺ Pyr⁻ Cys⁻, so only the <u>middle</u> gene has recombined in the allele from the prototrophic Hfr parent. This happens when the selected gene is the gene in the middle, so **the order is *cys trp pyr*.** That is, the quadruple recombinant would have been Trp⁻ Pyr⁺ Cys⁺, but you can't see this class because you selected for Trp⁺. The 3 genes are very closely linked, so the double crossover giving Pyr⁻ Trp⁺ Cys⁻ is rare. The two other classes represent double crossovers between *cys* and *trp* or *pyr* and *trp*. The recombination frequency between the genes = the number of recombinants between the two genes / by the total number of exconjugants. **Rf between *cys* and *trp* = (145+5)/1,000 = 0.15 = 15%. Rf between *pyr* and *trp* = (60+5)/1,000 = 6.5%.**

c.   F' plasmids are formed by inappropriate excision of an F plasmid from the chromosome. To isolate an F' *trp, pyr, cys* episome, the Hfr strain from which the F' is derived should have the F factor integrated next to these genes. HfrB and HfrC have the closest integrated F factors, so these would be at first glance the best candidates. However, there is a technical issue that means only one of these possibilities would work. Remember that you must be able to isolate the F' mating from the much more numerous Hfr matings that can also transfer the same three genes. It would thus be advantageous to choose an Hfr that transfers the *trp, pyr* and *cys* genes only late in the conjugation. If you interrupted the mating before these genes were transferred, the only way an exconjugant could have received the wild-type alleles of the *trp, pyr,* and *cys* genes is if an F' plasmid had been formed and transferred into the recipient. **HfrC would therefore both be the best candidate strain for the isolation of the F'.**

14-15.

a.

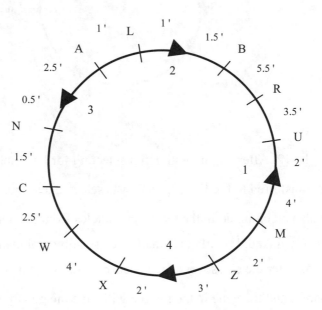

b.   <u>**See Figure 14.14**</u> and Problem 14-14c. You need to start with an Hfr strain that has the F plasmid integrated near the genes you would like to isolate on the F' plasmid. The Hfr strain must transfer thegenes you are interested in late in the mating. **In this case you would use Hfr3 to isolate an F' carrying the N$^+$ gene. Then if you screen for a derivative of that Hfr strain that transfers the N$^+$ gene early, the genes should be on an F' rather than being transferred by the Hfr.** An interesting alternative is that you can screen for the transfer of *N*$^+$ into a recipient that is a *recA⁻* mutant. To get an N^+ exconjugant cell from an Hfr mating requires *recA⁻*mediated recombination (the fragment bearing N^+ would be lost from the cell if it did not recombine into the chromosome), but no recombination is required for transfer and maintenance of an F' plasmid in the recipient cell.

14-16. To get an exconjugant cell from an Hfr mating requires *recA*-mediated recombination in the F⁻ recipient cell (see problem 14.15 above). Therefore, **this assay would detect *recA⁻* mutants in the F⁻ cell based on the inability to form stable exconjugants**.

14-17. See Figure 14.19. Transduction is the transfer of bacterial genes mediated by phage. The DNA is protected inside the protein head of the phage. Transformation is gene transfer using naked DNA, which is not enclosed in any protective structure. The DNA being transferred by transformation will therefore be susceptible to degradation by DNase. **If the transfer of the *ampʳ* allele still occurs after DNase treatment, transduction must be occurring**.

14-18.

a.   Arbitrarily place the first Hfr insertion site (HfrA) on the bacterial chromosome and then order the genes that are transferred by that Hfr. When you place HfrB on the same map notice that the first gene transferred by this Hfr is lys. HfrB could be placed on either side of this gene. However, the second gene transferred determines on which side of the first gene the F factor is inserted and the directionality of transfer. The time of transfer for four of the markers *(gly, phe, tyr, ura)* is indistinguishable, so we cannot put them in an order on the map, but *lys, nic* and the cluster of four genes can be placed.

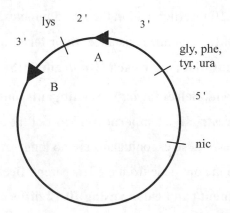

b.   **See Figure 14.20**. *Phe* was cotransduced with *ura* more frequently than with *tyr*. Also note that all of the Tyr⁺ transductants are also Ura⁺. These two facts both indicate that the order is *phe-*

ura-tyr. None of the cotransduction classes is very rare, so none of these result from quadruple crossovers. The relationship of these three genes to other markers and the order of *gly* gene is still unknown, and this is represented by placing *gly* in parenthesis and the flanking marker *lys* and *nic* in brackets.

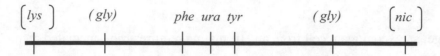

c.   **To map the *gly* gene with respect to other markers, select for Gly$^+$ transductants and score the other markers to determine which genes are cotransduced with Gly$^+$ at the highest frequency**. The Gly$^+$ transductants would be selected on min + lys + phe + tyr + ura, and then individual transductants would be tested for growth on media lacking one or more of the unselected amino acids.

14-19. In generalized transduction, the bacterial chromosomal DNA is packaged randomly into transducing phage particles that contain no phage DNA, only bacterial DNA. Thus it is possible for any region of the bacterial chromosome to be transferred. In specialized transduction, the phage in the lysate contains bacterial DNA covalently attached to phage DNA. The bacterial DNA that is transduced from any single lysogenic strain is always from the same region of the chromosome, which is adjacent to the insertion site of the integrated prophage.

14-20.

a.   The Hfr strains have a Tn10 insertion within the first 5 minutes of transfer, so the exconjugants can always be selected for tetracycline resistance. **In order to quickly map the location of the temperature sensitive mutation, mate each Hfr strain to the temperature sensitive mutant F$^-$ strain for a short period. Select for *tet*^{R} exconjugants and then plate the exconjugants and screen them at the restrictive temperature.** Most of the cells from one of the matings will form colonies on the plates. These exconjugants are no longer mutant because they have received the wild type allele of the mutant gene from the Hfr parent. Because the entire *E. coli* genome is 100 minutes, **this experiment can be done using 10-12 different Hfr strains if the F factor insertion sites are evenly spaced around the bacterial chromosome and the matings are allowed to go for about 10 minutes each.**

The matings could be done as patch replica platings. Grow up a plate spread with recipient cells so it produces a confluent lawn of cells. Grow up a plate containing patches of the different Hfr strains. Replica plate each of these onto the same velvet and print that onto the selection plate (tet + strep). (Tet selects for cells with Tn10; strep is used to select against the donor bacteria, assuming that they are tet sensitive while the recipients are resistant.) Check the selected exconjugants for the mutant phenotype by replica plating from the selection plate onto two rich plates, one incubated at 30° (permissive temperature for the mutation) and the other at 42° (restrictive temperature). Patches of exconjugants that grow at the restrictive temperature reflect parental Hfr strains in which Tn10 (at a known location) was located close to the gene of interest.

b.   The effective average plasmid size is 20 kb. If you allow are 2×1 kb for overlap on the two ends, each plasmid contains 18 kb of DNA not contained on other plasmids. The *E. coli* genome is 4.6 Mb in size, so 4600 kb/18kb = **255 plasmids** and therefore 255 transformations. Because of the difficulties inherent in constructing a set of plasmids with exactly these characteristics, in practice you would want to use more than this number of plasmids to be sure you could locate the mutant gene.

14-21.

a.   **(1) Pass an extract from *E. coli* cells which have been induced for β-galactosidase expression through an APTG-agarose resin.** β-galactosidase will bind to the resin along with any proteins that in turn bind to β-galactosidase. **You remove these proteins from the column using a detergent. (2) Digest the mixture of proteins with trypsin to generate a large number of smaller peptides. (3) Subject this mixture of peptides to mass spectrometry to get the molecular weight of the peptide fragments.** The molecular weights of each of the peptide fragments will be unique because of the range of molecular weights for individual amino acids. **(4) Take this profile of peptide molecular weights and compare it to a list of all the expected molecular weights for the tryptic fragments of all proteins in *E. coli*.** This list has already been determined using the genomic DNA sequence information. The comparison will tell you which proteins are present in the mixture, and therefore which genes encode proteins binding to β-galactosidase.

b.   **Make a fusion between the gene for your protein of interest and the *lacZ* gene. Clone the gene fusion into an expression plasmid that will produce the fusion protein in bacterial cells. Then repeat the steps you performed in part a.** When an extract from these cells is passed over an APTG-agarose column, you will isolate proteins that bind to your protein of interest as well as those that bind β-galactosidase. Mass spectrometry of the fragments resulting from tryptic digests will identify the proteins. Ignore the proteins that were also seen in the experiment discussed in part a, as this will be those that bind to β-galactosidase, rather than to the protein of interest.

Chapter 15   The Chromosomes of Organelles Outside the Nucleus Exhibit Non-Mendelian Patterns of Inheritance

Synopsis:

Mitochondria and chloroplasts are organelles that contain their own genomes. The genomes of both share similarities with prokaryotic genomes- in the types of genes, organization of genes, as well as chromosome structure. The presence of several organelles per cell in most organisms leads to many different combinations of organelle types in the same cell. Heteroplasmy refers to the presence of more than one type of genome per cell – homoplasmy is the state in which there is uniformity of the genome copies. Inheritance of organelle genomes is not dependent on the same cell machinery as nuclear chromosomes use for mitosis and meiosis. In most organisms, organelles are inherited from one generation to the next from the mother. This leads to distinct patterns of inheritance as seen in pedigrees.

Significant Elements:

After reading the chapter and thinking about the concepts, you should be able to:

♦ differentiate between heteroplasmy and homoplasmy experimentally

♦ recognize organelle inheritance in human pedigrees

♦ suggest explanations for unusual phenomenon involving organelle genomes that may arise from heteroplasmy and different proportions of affected cells

♦ Describe the endosymbiont theory and discuss the supporting evidence.

Problem solving tips:

After reading the chapter and thinking about the concepts, you should be able to:

♦ Explain the characteristics of mitochondrial gene inheritance in humans, including maternal inheritance and variable levels of expression in affected individuals.

♦ Understand the implications of the fact that there can be several organelles in a cell and several copies of the genome in each organelle.

Solutions to Problems:

15-1. a. **6**; b. **8**; c. **7**; d. **2**; e. **1**; f. **3**; g. **5**; h. **9**; i. **4**.

15-2. a. **both**; b. **mitochondria**; c. **chloroplast**; d. **chloroplast**; e. **mitochondria**.

15-3. a. **both**; b. **b**; c. **neither**; d. **both**.

15-4. If the red alga nuclear DNA sample is significantly contaminated with chloroplast DNA, a positive hybridization signal with the gene you are interested in might be a false positive. Your gene may not be nuclearly encoded; instead the hybridization to the nuclear DNA sample could be due to contaminating chloroplast DNA in the nuclear DNA preparation. In order to be sure that there is not significant cross-contamination of the nuclear and chloroplast DNAs, probe blots of both types of DNA with a gene known to be nuclearly encoded in plants and red algae. Also probe with a gene known to be encoded in the chloroplast in both organisms.

15-5.

a.   <u>See Figure 8.3</u>. The 'universal' code is always true for nuclearly (n) encoded genes.

		Trp	His	Ile	Met
mRNA	5'	UGG	CAU/C	AUU/C/A	AUG
nDNA mRNA-like	5'	TGG	CAT/C	ATT/C/A	ATG
nDNA template	3'	ACC	GTA/G	TAA/G/T	TAC

b.   <u>See Table 15.3</u>. The mitochondrial (mt) genomes of some organisms have differences from the universal code.

		Trp	His	Ile	Met
mRNA	5'	UGA/G	CAU/C	AUC/U	AUG/A
mtDNA mRNA-like	5'	TGA/G	CAT/C	ATC/T	ATG/A
mtDNA template	3'	ACT/C	GTA/G	TAG/A	TAC/T

15-6.

a.   **The large subunit of Rubisco is encoded in the chloroplast genomes of all three species. The small subunit is nuclearly encoded in green alga and chloroplast encoded in the red and brown alga**.

b.   **In red and brown algae the size of the transcripts** recognized by the large and small subunit probes is the same, so they appear to be **cotranscribed**. This is consistent with the fact that both genes are chloroplast encoded. The transcripts for the green algal Rubisco subunits are different sizes and therefore represent different transcripts. **In the green alga the genes are not cotranscribed**, which is impossible for genes located in different genomes.

15-7. a, c, and e are characteristics of chloroplasts and mitochondria that are like those found in bacteria. Choice b is not related to bacteria because bacteria don't have mitochondria. Introns (choice d) are not a universal feature of bacterial genomes.

15-8. The best way to determine if the entire sequence of the gene is present and intact is to **clone the fragment of nuclear DNA that contains the homology to the probe and determine its DNA sequence**.

15-9. Repeated backcrossing of the female gametes from the male sterile plants to the male fertile line causes the nuclear genome of the hybrid to become more similar to the original male fertile nuclear genotype. The fact that sterility still occurred with each generation indicates that sterility was not a nuclear gene effect. Remember that in each generation the female parent is passing on the cytoplasmic elements (mitochondria and chloroplasts).

15-10. In order to determine if individual organelles are heteroplasmic, you **use method a**. Differentially tag probes (different fluorescent tags, for example) that are specific to each of the two genomes. Then do a double hybridization to cells *in situ* and see if both probes hybridize to the same organelles in individual cells. PCR amplification of DNA from a population of cells will <u>not</u> indicate the genotypes of <u>individual</u> organelles. For example, there could be cells containing all a^+ mitochondria and other cells in the population containing all a^- mitochondria. When DNA from such a population is used for PCR amplification, both mitochondrial genomes would be represented, suggesting heteroplasmy although individual cells are homoplasmic.

15-11. a. **3**; b. **1**; c. **2**.

15-12. There are many mechanisms that cause the mitochondrial DNA from one parent to be absent from the zygotes. **The small size of the sperm can mean that organelles are excluded; cells can degrade organelles or organellar DNA from the male parent; early zygotic mitoses distribute the male organelles to cells that will not become part of the embryo; the details of the fertilization process may prevent the paternal cell from contributing any organelles (only the**

sperm nucleus is allowed into the egg); and in some species that zygote destroys the paternal organelle after fertilization.

15-13. Offspring will resemble the mother because the sperm does not contribute significantly to the cytoplasm of the zygote.

15-14. Heteroplasmic cells have a mixture of the 2 genotypes in their organelles. Homoplasmic cells have only one type of genome. Thus, homoplasmic cells can either be totally normal or totally mutant. **If the mutation is very debilitating to the cell, either because of the loss of energy metabolism in the case of mitochondria or of photosynthetic capability in the case of chloroplasts, a cell that is homoplasmic for the mutant genome will die.** Therefore, you will only find the mutant plastid genome in heteroplasmic cells.

15-15. In *Chlamydomonas* mt^+/mt^- diploid cells the mt^+ chloroplast genome is degraded while the mt^- mitochondrial genome is degraded in the diploid.

a. **This diploid will be resistant to erythromycin and will have the genotype ery^r.** The ery^s mitochondria will be destroyed.

b. In the case of a nuclear gene, **the diploid cell will be heterozygous, ery^r/ery^s and the phenotype will reflect the dominant allele.**

c. Sporulate the diploid so the diploid undergoes meiosis. If the gene is nuclear, there will be 2 ery^s : 2 ery^r spore. If the gene is mitochondrial, there will be 4 ery^r : 0 ery^s.

15-16. <u>See Figure 15.13</u> and problem 15-15.. **Mate the C^r MATα strain × a C^s MATa strain and sporulate the diploid. If the gene is nuclear, the chloramphenicol phenotype would segregate 2 C^r : 2 C^s. If the gene is mitochondrial, there will be 4 C^r :0 C^s segregation.**

15-17. a. **2**; b. **1**; c. **4**; d. **3**.

15-18. One characteristic of a mitochondrial mutation is **maternal inheritance**. Most of the offspring of an affected female are affected. None of the offspring of affected males are affected. Another indication of mitochondrial inheritance is **differing levels of expression of the mutant phenotype in different progeny** due to differing amounts of heteroplasmy in either the egg or the cells of the embryo.

15-19.

a.   **The mother (I-1) may have had very low levels of mutant mitochondrial chromosomes**, while the proportion of mutant genomes in the daughter was much higher. Alternatively, there **may have been a spontaneous mutation either in the mitochondrial genome of the egg that gave rise to individual II-2 or in the early zygote of individual II-2**.

b.   You could look at the mitochondrial DNA from somatic cells from various tissues in the mother. If the mutation occurred in her germline and was inherited by II-2, the mother's somatic cells would not show any defective DNA. Remember that it is impossible to make a firm conclusion based on negative results. Therefore, if no mutant mitochondrial genomes are found in I-1 it is possible that you haven't looked at enough different tissues. Taking the tissue samples is very unpleasant for the donor!

15-20.

a. **3^d pedigree**; if a mutation occurred in the somatic cell of an embryo this individual will be affected by will not be able to pass the mutation on to her progeny through her germline.

b. **1st pedigree**; in this case I-1 will be unaffected but she will pass the disease on to all her offspring.

c. **2nd pedigree**; in this case the individual will express the disease and she will pass it on to her offspring, as the mutation will occur in both her somatic cells and her germline.

15-21. The variation in affected tissues is due to differences in where and when during development the mutation occurred. In different individuals, the mutation may occur in cells that give rise to different sets of tissues. If the mutation occurs early in development then more tissues will be affected. **Variation in the severity of the disease can be due to the proportion of mutant genomes (the degree of heteroplasmy) in the cells of different tissues.**

15-22. Gel electrophoresis is better suited for an overview of the differences between mitochondrial genomes. Deletions can be very large and might not be amplified by PCR. In addition, sequences to which primers bind might be deleted in some mutations so no information will be obtained about the sizes of these deletions.

Chapter 16   Gene Regulation in Prokaryotes

Synopsis:

This chapter describes gene regulation in bacteria including the genetic analysis that led up to the postulation of the operon theory - the paradigm of gene regulation. A basic principle derived from experiments on the *lac* operon is that proteins bind to DNA to regulate transcription (**see Feature Figure 16.10**). How mutations in the components of the regulatory system proved Jacob and Monod's theory is an instructive lesson in the power of the genetic approach for understanding basic cellular processes. With the development of molecular biological techniques and increased study of protein structure, the operon theory of gene regulation as been refined. We now understand more about DNA binding proteins and their interactions with regulatory regions. Experiments on the lactose operon lead to the development of fusion technology in which the *lacZ* gene is placed next to a regulatory region of another gene. Expression of that gene could then be monitored by measuring expression of β-galactosidase. Another type of fusion was developed in which the regulatory region of the *lac* operon was fused to a gene whose expression was then controlled using the induction of the *lac* genes.

In addition to the negative and positive control described for the *lac* operon (**see Figure 16.20, note that CAP = CRP**), global transcriptional regulation based on changes in RNA polymerase and its subunits is described as is attenuation- a mechanism for fine-tuning transcription (**see Figure 16.20**).

Significant Elements:

After reading the chapter and thinking about the concepts, you should be able to:

♦ Predict overall cellular expression of proteins when there are site or gene mutations on a chromosome and other mutations on the F plasmid (merodiploid analysis).

♦ Distinguish between positive and negative regulators based on the behavior of mutants.

♦ Propose models for regulation of a set of genes based on mutant and molecular analyses.

♦ Describe the process of attenuation and how mutations in the component parts might affect expression of the *trp* genes or other amino acid operons that are regulated by attenuation.

♦ Design molecular experiments to test predictions of models based on mutational analysis.

♦ Describe use of and distinguish between *lacZ* fusions (geneX-*lacZ*) and lac regulatory fusions (*lac*-geneX).

Problem Solving Tips:

♦ Negative regulation blocks transcription.

♦ Positive regulation increases transcription.

♦ Regulatory sites in the DNA (P, O) only affect DNA and structural genes adjacent to them.

♦ <u>Regulatory proteins diffuse</u> in the cytoplasm and therefore can act on any copy of their binding site in a cell (**see Figure 16.7**).

♦ Sites and proteins that bind to sites form the control system of gene regulation.

♦ Regulatory proteins that bind to several molecules (e.g., DNA, another protein, inducers) have distinct regions or domains in the protein for these interactions.

♦ Learn the nomenclature for the different sorts of mutations - P^-, O^c, I^-, I^s, etc.

♦ Operons are transcribed as a single, polycistronic mRNA. Each structural gene has its own start codon (5' AUG) and its own stop codon, so the structural genes are each translated as an individual polypeptide chain. However, there is only one ribosome binding site on the mRNA, and that is just 5' to the start codon of the first open reading frame. Thus, the ribosome binds to the binding site and translates the first structural gene. When the ribosome reaches the stop codon at the end of that gene, it releases the completed polypeptide, but it does not disassociate from the mRNA. Instead, by a poorly understood process, the ribosome 'moves' to the next start codon and translates the second structural gene. What happens if a nonsense mutation is introduced into the sequence for the *lacZ* gene? All such mutations give a *lacZ⁻* phenotype. There are two different possibilities for the downstream genes, *lacY* and *lacA*. Some nonsense mutations in *lacZ* cause the ribosome to dissociate from the mRNA. In these cases, the downstream structural genes are not translated, even though their DNA (and mRNA) sequences are totally normal! Nonsense mutations of this sort are called polar nonsense mutations. Other nonsense mutations cause the production of a defective *lacZ* gene product (*lacZ⁻*), but they do not terminate translation for the other structural genes. These are called non-polar nonsense mutations.

♦ When asked to determine regulation in a partial diploid (merozygote or merodiploid or a cell with an F' plasmid) examine the promoters first. If one of the promoters is non-functional, then you can ignore all of the cis-regulated structural genes. Next, consider the O/repressor interactions - are they wild type or mutant? Then examine the structural genes for their ability to give a functional enzyme.

Solutions to Problems:

16-1. a. **4**; b. **8**; c. **5**; d. **2**; e. **7**; f. **1**; g. **3**; h. **6**.

16-2. The lack of *rho* function must be lethal for the cell, so **the *rho* gene function is essential**. Conditional mutations are the only sort of mutation that can be isolated for essential genes.

16-3. When the λ phage infects a cell with an integrated λ phage (lysogen), the *c*I repressor protein is already present in the cytoplasm, so it can bind to operators on the incoming phage DNA. Thus the expression of genes on the incoming phage chromosome is blocked, preventing the synthesis of proteins needed to make new progeny phages. **The non-lysogenic recipient cell has no *c*I repressor protein in the cytoplasm, so the incoming infecting phage goes into the lytic cycle, producing progeny phage.**

16-4. Mutations in the promoter region can only act in cis to the structural genes immediately adjacent to this regulatory sequence. This promoter mutation will not affect the expression of a second, normal operon.

16-5. The *lacZ* gene codes for the β -galactosidase enzyme. The *lacY* gene codes for a permease. For an overview of the different types of mutations in the *lac* operon, see 'How to Begin Solving Problems' at the beginning of this chapter.

a. ***lacZ* is constitutive; *lacY* is constitutive.** Wild-type copies of both genes are downstream of a constitutive operator that cannot bind to the repressor protein, so both genes will be expressed even in the absence of inducer.

b. ***lacZ* is constitutive; *lacY* is inducible.** Here, wild type *lacZ* is downstream of a constitutive operator, so it will be made even in the absence of inducer. The only functional copy of *lacY* is downstream of a functional operator, so it will be expressed only in the presence of inducer.

c. ***lacZ* is inducible; *lacY* is inducible.** This cell has active repressor made from one wild-type copy of the repressor gene. This will repress the expression of both wild-type genes (*lacZ* and *lacY*) unless inducer is present.

d. **no expression of *lacZ* ; *lacY* is constitutive.** Neither gene can be expressed from the copy of the operon without a promoter. The other copy of the operon lacks a functional *lacZ* gene, while *lacY* is downstream of a constitutive operator that cannot bind to the repressor protein.

e. **no expression of *lacZ* ; no expression of *lacY*.** In the presence of a "superrepressor" protein that cannot be removed from the wild-type operators even when inducer is added, neither gene can be expressed.

16-6. These revertants of the constitutive expression of the lac operon could be changes in the base sequence of the operator site that compensate for the mutation in *lacI*. For example, if an amino acid necessary for recognition of operator DNA is changed in the *lacI* mutant, a compensating mutation changing a recognition base in the operator could now cause the mutant LacI protein to recognize and bind to the operator.

16-7. See 'How to Begin Solving Problems' at the beginning of this chapter. As just one example, graph a below shows that all three proteins will be expressed only after the lactose (the inducer) is added to the medium. This cell has one wild-type copy of *lacI* making functional repressor protein, and all wild-type copies of the structural genes *lacZ, lacY,* and *lacA* are downstream of wild-type promoters and operators. The phenotype of this cell will thus be the same as a wild-type *E. coli* cell.

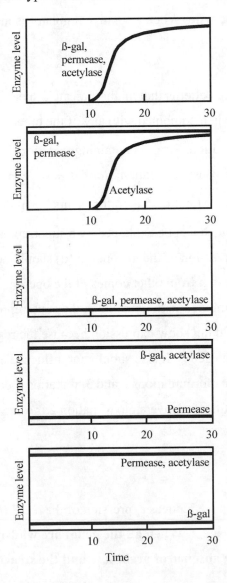

16-8.

a. The DNA binding protein is a negative regulator because when it is mutant the result is constitutive expression. Therefore, **the wild type regulatory protein blocks transcription = negative regulation**. This is of course similar to the *lacI* protein in the *lac* operon of *E. coli*.

b. The assumption is that *reg2* encodes the repressor for this operon, while *reg1* is the operator to which the repressor binds. **Strain (i) will have inducible *emu1* and constitutive *emu2* expression, while strain (ii) will have inducible *emu1* and *emu2* expression.**

c. The assumption is the opposite of that in part b. That is, in this case *reg1* encodes the repressor and *reg2* is the operator site. **Strain (i) will have inducible *emu1* and *emu2* expression while strain (ii) will have inducible *emu1* and constitutive *emu2* expression**.

16-9. Statement (b) will be true. When a positive regulator is inactivated there is no expression from the operon.

16-10. First look for striking patterns that might indicate a specific mutation. For example, mutant 6 shows expression of *lacZ* when combined with any of the other mutations. Of the possible choices, this could only be an o^c mutation. Strains containing mutation 5 and any of the mutations other than mutation 6 also have a very consistent pattern. *lacZ* is never expressed, except when combined with mutation 6 (the o^c mutation). Mutation 5 therefore shuts down the other copy of the operon in addition to its own copy, which can be explained by a superrepressor (I^s) mutation. Looking at the remaining mutations, the inversion of the *lac* operon (d) should not have an effect on expression and should not influence expression from other copies of the operon. Mutation 4 leads to inducible *lac* expression except when combined with mutation 5, the superrepressor. The inversion that does not include *lac I, p,* and *o* should not show expression because the regulatory region is now in the opposite orientation from the genes. It also should not influence expression from the other mutant copy. Looking at the patterns for mutations 2 and 3, mutation 3 does not show expression except when combined with 4, so fits with inversion. Mutation 2 is a *lacZ* mutation. Therefore, **a. 1; b. 6; c. 2; d. 4; e. 5; f. 3.**

16-11.

a. The evidence that arabinose induces expression of the *araBAD* genes is seen in the genotypes and media conditions 1 and 2. **When all the genes are wild-type, there is no expression of the structural genes in the absence of arabinose and the structural genes are expressed in the presence of arabinose.**

b. The conclusion that *araC* encodes a positive regulator of the operon is based on the data from the strains and conditions in 3and 4. **The mutant version of *araC* blocks synthesis of the three structural gene products in the presence of arabinose**.

16-12.

a. In the presence of glucose, **the repressor protein (LacI) is bound** to the operator.

b. in the presence of glucose + lactose there will be **no proteins bound to the regulatory region** of the lac operon.

c. in the presence of just lactose, **the CAP-cAMP complex will bind to the promoter region** of the lac operon.

16-13. This problem can be approached by starting with the expression data and assessing what types of mutations could produce that particular phenotype or by starting with each mutant and matching it with an expression pattern. Using the latter approach, the superrepressor mutant (a) will show no expression under any conditions and therefore is either mutant 3 or 4. The operator deletion (b) will give constitutively high levels of expression with glycerol or lactose but will give low expression with lactose + glucose because it will be catabolite repressed. Therefore the operator deletion is mutant 5 or 6. The amber suppressor tRNA would have no effect on its own and is mutant 7. The defective CAP-cAMP (= CRP-cAMP) binding site (d) produces the same low levels of expression with lactose or lactose + glucose because the CAP-cAMP complex cannot bind to the promoter to increase expression. Thus mutant 1 or 2 contains a mutation in the CAP-cAMP binding site. The nonsense mutation in β-galactosidase (e) will give no expression, so is mutant 3 or 4. The nonsense mutation in the repressor gene (f) leads to constitutive expression of β-galactosidase, as in mutants5 and 6. A defective *crp* gene (g) means the CAP-cAMP complex cannot form so it cannot bind to the promoter region as seen in mutants 1 and 2. To summarize, mutations 1 and 2 are the mutant CAP-cAMP binding site and the defective *crp* gene; mutations 3 and 4 are a superrepressor and a nonsense mutation in the β-galactosidase gene; mutations 5 and 6 are an operator deletion and the nonsense mutation in the repressor gene; and mutation 7 is an amber suppressor tRNA.

The next stage of the analysis involves understanding the results of the double mutant and merodiploid genotypes. An amber nonsense mutation of the *lacZ* gene would be suppressed by mutation 7 (amber suppressor tRNA) in the same cell. Therefore, mutant 3 contains the nonsense mutation in β-galactosidase (*lacZ*) gene and mutant 4 must be a superrepressor mutation. Likewise, a nonsense mutation in the repressor gene would be suppressed by mutation 7 in the same cell, so mutant 5 is the mutant repressor gene and mutant 6 is the operator deletion. The defective *crp* gene can be distinguished from the defective CAP-cAMP binding site by the merodiploid genotypes presented. In both merodiploids the bacterial chromosome has a mutant gene for β-galactosidase, but all other parts of the lac operon expression system are wild type. The F' element has either a mutant CAP-cAMP binding site or a defective *crp* gene. In the latter case the presence of the trans-acting wild type crp gene on the bacterial chromosome will give normal regulation of the lac operon. However if the mutation on the F' element is in the cis-acting CAP-cAMP binding site the phenotype of the merodiploid will still be mutant. Thus, mutant 1 is the *crp* mutation and mutant 2 is the altered binding site. Therefore, **a. 4; b. 6; c. 7; d. 2; e. 3; f. 5; g. 1.**

16-14. In order to determine the effect of the various genotypes on the expression of the operon(s), focus on the mutant portions of the operons.

a. If the promoter is deleted **there is no expression of either *trpE* or *trpC*** because transcription cannot occur.

b. The repressor is defective so transcription of the structural genes is constitutive. However, the attenuator is normal and this results in partially constitutive expression of the structural genes. There will be a **lower level of expression of *trpE* and *trpC* when tryptophan is present and high level of expression when there is no tryptophan**.

c. The repressor cannot bind tryptophan and thus cannot bind to the operator so transcription of the structural genes is constitutive. However, the attenuator is normal and this results in partially constitutive expression of the structural genes. Thus there will be a **lower level of expression of *trpE* and *trpC* when tryptophan is present and high level of expression when there is no tryptophan**.

d. The repressor cannot bind to the operator, so expression of *trpC* and *trpE* is constitutive. The attenuator is also mutant, so the operon will always be transcribed at the maximal level. Thus, ***trpC* and *trpE* will show completely constitutive expression**.

e. There is **inducible expression of *trpC*** (remember that r^+ is a <u>trans</u> regulatory protein), and **partially constitutive expression of *trpE*** because the cis acting operator is defective but the attenuator is still functional.

f. There is **inducible expression of *trpC*** (remember that r^+ is a <u>trans</u> regulatory protein), but **no expression of *trpE*** because the cis acting promoter is defective

g. Expression of ***trpE* is fully constitutive** because the cis acting operator is defective and the cis acting attenuator site is defective. The ***trpC* expression is partially constitutive** because the $trpC^+$ gene is cis to a defective operator and a functional attenuator.

16-15. If the three genes make up a single operon they will be cotranscribed as one polycistronic mRNA. Thus, **when a Northern blot is probed with DNA from each of these genes you will always see one band of the same size. This mRNA will be large enough to include the RNA of all three genes. If the genes are not part of an operon, each one will be transcribed separately so each probe will hybridize to a differently sized mRNA in the Northern blot.**

16-16.

a. The operon **seems to be a biosynthetic operon** because in a wild type operon the addition of Z decreases the expression of the A, C and D genes. **The operon is repressible.**

b. **A, C, and D are structural genes needed for the biosynthesis of compound Z.** Compound B is constitutively produced, so it is not part of the transcriptional unit that includes the structural genes. A mutation in B leads to a less pronounced turn-off of the structural genes A, C and D than in wild-type. This implies that **B could be a repressor that shuts off synthesis of the structural genes when compound Z is present.** Obviously, the total inhibition of transcription requires a second component - an attenuator. This explains why there is still some transcription even when B is mutant. The nonsense mutation in C must be a polar nonsense mutation, as it affects not only the presence of the C protein, but also the A and D proteins, so C must be the first structural gene in the operon. A nonsense mutation in gene A blocks production of A and D while a nonsense mutation in D only affects the synthesis of D. **The order of structural genes is therefore C A D.** Deletion of site G blocks all expression of the three structural genes, so **site G is probably the promoter of the operon.** The removal of either E or F lead to reduced repression which occurs if there is a repressor binding to an operator and an attenuator site in the operon, each being responsible for a 10-fold repression when compound Z is available. Sequence E codes for a small peptide, and since small peptides are part of attenuator sites in amino acid biosynthetic operons, we can assume that **deletion of E and adjacent DNA would delete an attenuator site. Site F then is the operator site.** The operator and promoter are first, then the attenuation region including the leader peptide, then the structural genes. B is not cotranscribed with the other genes, so we do not know its location and it is shown on the map below in brackets {}.

16-17. Analyze this mRNA sequence. There is one open reading frame beginning with the first nucleotide. The predicted amino acid sequence is:

N Met Thr Arg Val Gln Phe Lys His His His His His His Pro Asp C

There are 7 histidines in a row out of a total of 16 amino acids! Thus, if the cell is starving for histidine, the ribosomes will pause at the His codons (CAC or CAU) in the sequence because the tRNA^{His} molecules will not be completely charged with histidine. This ribosome pausing allows RNA polymerase to complete transcription of the operon almost 100% of the time, giving maximal production of the polycistronic mRNA and maximal expression of the structural gene proteins which synthesize histidine. This is very similar to attenuation in the trypotophan operon of *E. coli* as shown in **Figure 16.20.**

16-18. Any mutations that prevent the appearance of a functional LamB protein would be resistant to lambda. These will include: ***lamB* point mutations or deletions; *malT* mutations that prevent positive regulation (superrepressors); mutations in the site to which *malT* binds; *malK lamB* promoter mutations; and polar nonsense mutations in *malK* which will also block expression of LamB.** Two other kinds of mutations would lead to lower levels of LamB expression: mutations in the gene encoding the CAP protein involved in catabolite repression, and mutations the in CAP-cAMP binding site of the *malK lamB* operon. It is unclear whether the amount by which LamB is decreased in such mutants would prevent lambda phage infection or would instead just make it less efficient.

16-19.

a. **The operator begins at the left end of the sequence give**n, after the endpoint of deletion 1 and before the endpoint of deletion 5. The right endpoint of the operator cannot be determined by this data.

b. **The deletion may have removed bases within the promoter that are necessary for the initiation of transcription**.

16-20.

16-21. Mutations in O_2 or O_3 alone have only small effects on synthesis levels and would therefore be difficult to detect in screens for mutations that affect the regulation of the lac operon.

16-22.

a. **If you only screen for Lac$^+$ revertants, some of the revertants will be compensating mutations in just the lac operon CAP-cAMP binding site. Demanding that the suppressor mutations affect both Mal$^+$ and Lac$^+$, will give more general revertants, which affect <u>all</u> CAP-cAMP binding or activity.**

b. **The α subunit of RNA polymerase interacts directly with the CAP protein.**

16-23. The protein-coding region of your gene must be in the same reading frame as the *lacZ* gene.

16-24. The loss of LexA function leads to the new expression of many genes. Therefore, **the wild type LexA protein negatively regulates the expression of these genes.** One mechanism by which this could be accomplished is if the LexA protein were a repressor that binds to the operators of these genes to shut them off; biochemical studies have in fact shown this to be true. Because these target genes are turned on by DNA damage, they are likely to encode proteins that are important in the ability of cells to repair DNA damage. These considerations suggest that the LexA protein is important in preventing the expression of these DNA repair genes unless they are needed. Interestingly, it turns out that DNA damage activates a protease activity of the protein encoded by the *recA* gene.that cleaves and thus destroys the LexA repressor protein. This explains how DNA damage leads to the induction of the DNA repair genes.

Chapter 17   Gene Regulation in Eukaryotes

Synopsis:

This chapter describes different ways in which gene expression is regulated in eukaryotes. Regulation of transcription is a major means of controlling expression. Regulatory proteins (activators and repressors) bind to sites in the DNA near the gene to turn up or down expression. The low level of basal transcription of a gene requires a set of transcription factors binding at the promoter. Gene expression that occurs in only certain cells or is specific for a stage in development or occurs in response to external factors is mediated by additional proteins binding near the coding region. Characteristic motifs that are important for DNA binding or interaction with RNA polymerase or with other activator proteins can be recognized in regulatory proteins that bind to DNA.

Other controls of gene expression include alternate splicing events (producing different mRNAs from the same gene), mRNA stability, control of translation, different packaging of region of DNA in chromatin, and protein modifications.

Significant Elements:

After reading the chapter and thinking about the concepts you should be able to:

♦ Distinguish between positive and negative regulators based on the effects of mutations.

♦ Design experiments to determine the sequence(s) of a regulatory region of a gene that are important for basal transcription and which sequences are important for tissue specific expression.

♦ Interpret expression of a gene using a reporter gene fused to a regulatory region.

♦ Identify interacting proteins in regulation of a gene.

♦ Interpret expression of a paternally or maternally imprinted gene (**see Feature Figure 17.15**).

♦ Describe the importance of chromatin structure for eukaryotic gene regulation (**see Figures 17.13 and 17.14**).

♦ Distinguish the similarity and difference between sex determination in humans and *Drosophila* (**see Figure 17.21**).

Problem Solving Tips:

♦ When a regulatory region of gene "X" is cloned next to the *lacZ* gene and reintroduced into eukaryotic cells, the expression of β-galactosidase will reflect expression patterns of gene "X".

♦ Deletions of sites to which activators bind will lead to lower expression; deletions of sites to which repressors bind lead to increased expression (**see Figures 17.4 and 17.5**).

♦ Several regulatory proteins can interact to regulate one gene.

♦ Mutations in regions of proteins that interact with another protein or a site will only affect that one specific function.

♦ DNase hypersensitive (DH) sites indicate that the DNA is in a less compacted form and is more available for transcription.

♦ For maternally imprinted genes, the copy of the gene inherited from the mother will not be expressed in either her sons or daughters. For paternally imprinted genes, the copy of the gene inherited from the father will not be expressed in either his sons or daughters.

Solutions to Problems:

17-1. a. **8**; b. **5**; c. **1**; d. **7**; e. **3**; f. **9**; g. **6**; h. **4**; i. **2**.

17-2. See Table 17.1.
a. **eukarotyes**; b. **both prokaryotes and eukaryotes**; c. **eukaryotes**; d. **prokaryotes**; e. **both prokaryotes and eukaryotes**.

17-3. Introns are spliced out of a primary transcript (see Figure 8.16), a ribonuclease cleaves the primary transcript near the 3' end to form a new 3' end to which a poly-A tail is added (see Figure 8.14), and a methyl-CAP is added to the 5' end of the transcript (see Figure 8.13).

17-4. a. **iii**; b. **i**; c. **ii**.

17-5. See Figure 17.2. a. **pol III**; b. **pol II**; c. **pol I**.

17-6. Maximal gene expression occurs when there is both a promoter and an enhancer upstream of the gene. There is <u>no</u> expression of a gene without a promoter (P) sequence. In the presence of a promoter sequence, there will be a low (basal) level of β-galactosidase reporter gene expression. An enhancer sequence alone shows no expression, while promoter + enhancer will give high levels of expression. In order to discuss the fragments shown in the restriction map, number them from left to right. Fragment #1 is M to M, fragment #2 is M to H, fragment #3 is H to M, fragment #4 is M to M and fragment # 5 is M to H. **The promoter must be located in M-H fragment #5** since each of the clones showing basal levels of transcription contain this fragment and clones without this fragment produce no β-galactosidase. **The enhancer is present in H-M fragment #3** since that sequence is found in the two clones (sixth and eighth) showing high expression. Clone #2 also contains this fragment but does not show any expression because there is no promoter upstream of the reporter gene.

17-7.

a. **DNA binding, see Figure 17.5b**;

b. **DNA binding**;

c. protein-protein **dimer formation, see Figure 17.7**;

d. **transcriptional activation** - protein-protein interactions between activating factors and basal proteins bound to the DNA at the transcriptional activation region, **see Figure 17.5a**;

e. **DNA binding.**

17-8.

a.   The deletion to -85 has full activity indicating that the binding site is to the right of -85. The binding region must begin after the endpoint of the first deletion (-85) as the deletion of the -85 region (to -75) abolishes all activity above the basal level. Within the 19 bp region that is protected in the DNase-I footprinting experiment, there is a short sequence that is almost repeated (between nucleotides -83 and -65). It is possible that this is the monomer binding site. This site must be repeated to get active dimer binding, as seen in the **region including base pairs -83 to -65**.

b.   The likely consensus for the monomer binding would be: **CTG(C/T)G**.

17-9. Isolate mRNA from cells expressing the 3 genes. Do a Northern blot analysis with this mRNA using probes from each of the three genes. If the 3 genes are cotranscribed as one polycistronic mRNA, you will see the same large band when the Northern blot is probed with DNA from any of the 3 *gal* genes. However, **you will see a differently sized mRNA band for each gene when the blot is probed with each gene separately**. None of these bands will be large enough to contain the coding information for all three genes.

17-10. The constitutive expression of the galactose genes in yeast will be seen if there is no way to prevent the binding of the GAL4 protein to the DNA. Therefore, **a *GAL80* mutation in which the**

protein is not made or is made but cannot bind to the GAL4 protein will prevent repression and lead to constitutive synthesis. A *GAL4* mutation that inhibits binding to the GAL80 protein will also be constitutive. A mutation of the DNA at the binding site for the GAL4 protein will also give constitutive synthesis.

17-11. If the Id protein binds the DNA enhancer sequence then antibodies against the Id protein should bind to the chromatin in *in situ* hybridizations. Also, there should be a DNA binding domain (for e.g. a zinc finger or helix-turn-helix or helix-loop-helix sequence) in the *Id* gene. If the Id protein acts by quenching, it interacts with the MyoD protein, so there is no need for a DNA binding sequence in the *Id* gene. However the *Id* gene should have a protein-protein binding sequence, like a leucine zipper. See Figure 17.8. In this case, the Id protein would bind to the MyoD protein but not to the chromatin.

17-12. Isolate mRNA from sporulating cells and make cDNA copies. The cDNAs then could be cloned into a vector to generate a cDNA library representing genes expressed during sporulation. The cDNAs can also be used to probe a genomic library to clone the genomic versions of the expressed genes. These methods will isolate all genes that are expressed in sporulating cells, both the sporulation-specific ones and the genes that are generally expressed (the house-keeping genes).

An alternate method that allows isolation of the genes that are specifically expressed during sporulation as well as those whose level of expression increases or decreases during sporulation involves **microarrays**, discussed in Chapter 10 (**see Figure 10.24**). The DNA sequence of the entire genome of yeast is known and all potential genes have been identified by computer alogorithm. Oligonucleotides from the 3' ends of all the putative genes can be arranged in a microarray. These arrays can be **probed with two types of differentially labeled mRNA** - that isolated from sporulating cells (labeled with a red fluorescent dye, for instance) and mRNA isolated from normal, vegetatively growing diploid cells (labeled with a green fluorescent dye). When a mixture of the 2 groups of mRNA are mixed and probed to the arrays, you will be able to distinguish genes that are expressed only in the vegetatively growing cells (red spots) from those that are only expressed in sporulating cells (green spots) from those expressed in both types of cells (**see Figure 10.25**). Those genes expressed in both states but whose level of expression changes will be seen as intermediate shades. It is possible to measure the levels of the two dyes for each spot on the microarray to determine patterns of gene expression.

17-13. The conclusion that chromatin structure has a major effect on the level of gene expression is based on many observations. The earliest involved the **differing levels of gene expression depending on their association with highly compacted, heterochromatic DNA vs euchromatic DNA** (see Chapter 12). One example is **Position Effect Variegation in *Drosophila***, where the expression of a gene can be abolished when it is placed next to heterochromatin. Another is **Barr body formation in human females**. In this case, almost all of the genes on the inactive, heterochromatic X chromosome are permanently inactivated in the mitotic descendants of that cell. Many genes in euchromatin must be decompacted in order to be transcribed (**see Figure 17.13**). **Decompaction affects the location of the nucleosomes, and gives rise to DNase I hypersensitive sites where nucleosomes have been removed and the DNA is available for binding by RNA polymerase or regulatory proteins**. Such work lead to the discovery of boundary elements which are DNA sequences that help define the limits of the localized decompaction of the DNA. **Transcriptional silencing, on the other hand, involves methylation of the DNA (see Figure 17.14**). A more recent example is imprinting (**see Figure 17.15**), which is a sex-specific inactivation of genes that occurs during meiosis and is passed on to the zygote. In imprinting, the inactivation of genes is associated with increased methylation of the DNA. Studies on imprinted genes lead to the discovery of insulators, which are DNA sequences that seem to determine the limits of such methylation. The evidence is now compelling that chromatin structure plays a major role in gene expression in eukaryotes.

17-14. The gene is more likely to be transcribed in liver cells, based on the DNase I profile, than in muscle cells. **Liver cell DNA has a DNase I hypersensitive (DH) site 4 kb from 1 end of the *Eag*I fragment. This site is probably the promoter region for your gene.** This site is not be exposed in the muscle cells, indicating lower (or no) transcriptional activity in muscle cells.

17-15. Option b. is suggested by the presence of a DNase I hypersensitive site.

17-16.

a.   **1, 2** and **4 (see Figures 17.15d, 17.14 and 17.15)**;

b.   **1** and **4**;

c.   **1** and **4**;

d.   **3 (see Figure 17.14)**.

17-17. (See Figure 17.15). The boy received an imprinted (transcriptionally inactivated by methylation) allele of this gene from his father. The boy should be transcribing and translating the supposedly normal, active allele from his mother. If the boy has a mutant phenotype, it must be due to a **mutant allele of the gene that he received from his mother**.

17-18. Remember that Prader-Willi syndrome is caused by a mutation in a <u>maternally</u> imprinted gene. In order to answer this question, you must be able to figure out the genotype of the original affected individual. Remember that all individuals have only one transcriptionally active allele of this gene, and that is the allele they inherited from their father.

a. An affected male has a mutant allele of the gene that he inherited from his father and an imprinted (transcriptionally silenced) allele of the gene that he inherited from his mother. This imprinted allele is normal, even though the affected male can not express it. When the affected male makes gametes, he removes all imprinting. As this gene is maternally imprinted, he does NOT imprint either allele of this gene. Therefore, half of his sperm will have the non-imprinted normal allele and the other half will have the non-imprinted mutant allele. All of his children will receive an inactivated (presumably wild type) allele of this maternally imprinted gene from their mother and so **half of his sons** and half of his daughters **will be affected**.

b. **True**, see part a.

c. **False.** An affected female has a mutant allele of the gene that she inherited from her father and an imprinted (transcriptionally silenced) allele of the gene that she inherited from her mother. This imprinted allele is normal, even though the affected female can not express it. When the affected female makes gametes, she removes all imprinting. As this gene is maternally imprinted, she DOES imprint BOTH alleles of this gene. Therefore, half of her eggs will have the imprinted normal allele and the other half will have the imprinted mutant allele. All of her children will receive an activated wild type allele of this maternally imprinted gene from their father, so none of her children will be affected.

d. **False**, see part c.

17-19. This is a paternally imprinted gene. Neither parent in generation I expresses the disease, but 2 out of 3 children do. At first glance, either parent could have the mutant allele of the gene. Because neither expresses the disease, they must have received an inactivated mutant allele from their father and a normal, transcriptionally active allele from their mother. However, further consideration shows that it can not be the male parent (I-2) who is the source of the mutant allele. He can only provide transcriptionally inactive alleles. If the children show the disease phenotype, then they must have received an inactive normal allele of the gene from him and an activated, mutant allele of the gene from their mother, I-1. In the genotypes below, an allele in {} brackets is imprinted (transcriptionally inactivated) and *a* is the mutant allele of the gene.

a.   The genotype of I-1 is **A/{a}**; see explanation above.

b.   The genotype of II-1 is **A (from I-1)/{A} (from I-2)**.

c.   The genotype of III-2 is **A (from II-2)/{A} (from II-3)**.

17-20. Draw out the pedigree, including the information given in the problem. As in problem 17-19, {} represents an allele that is transcriptionally inactivated (imprinted) and the alleles of the maternally imprinted *IGF-2R* gene are *60K* and *50K*.

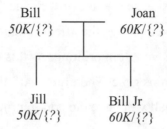

Bill
50K/{?}
Joan
60K/{?}

Jill
50K/{?}
Bill Jr
60K/{?}

a.   Both of Joan's alleles will be imprinted in her gametes, so Jill and Bill Jr each express the allele they received from Bill. Therefore, **Bill Sr's genotype is 50K/{60K}**. With this information, nothing further can be said about **Joan's genotype (*60K/{?}*)**, and there is no way of telling which of Joan's alleles Jill and Bill Jr received.

b.   Bill Sr's complete genotype and the further information is shown in the pedigree below about Pat and Tim are presented in the pedigree.

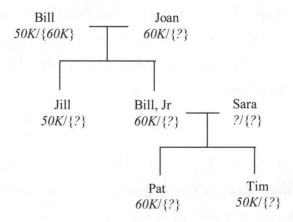

Bill
50K/{60K}
Joan
60K/{?}

Jill
50K/{?}
Bill, Jr
60K/{?}
Sara
?/{?}

Pat
60K/{?}
Tim
50K/{?}

As in part a for Joan, both of Sara's alleles will be imprinted in her gametes. Therefore, Pat and Tim express the transcriptionally active alleles they received from Bill Jr. Because Tim's phenotype is 50K, Bill Jr must be heterozygous *60K/{50K}*, with his imprinted *{50K}* allele having come from Joan. Therefore, **Joan's genotype is** *60K/{50K}* **and Bill Sr's genotype is** *50K/{60K}*.

17-21. <u>See Figure 17.15</u>. This pedigree tracks a paternally imprinted gene.

a. **The alleles of the gene are not expressed in the germ cells of male I-2** (<u>see Figure 17.15b</u>). The gene is paternally imprinted, so the imprinting is removed during gametogenesis. By the end of meiosis both alleles of this gene will be imprinted, so male I-1 will not express the gene in germ cells.

b. **The allele of the gene from male I-2 will not be expressed in the somatic cells of II-2.** The gene inherited from her father was turned off in his germ cells, so the somatic cells arising after fertilization will contain an inactive copy from I-2.

c. **The allele of the gene from male I-2 will be expressed in the germ cells of II-2.** The imprinting from the I-2 is erased during gametogenesis and a female will not inactivate this gene, so both alleles of this gene will be active in the germ cells of II-2.

d. **The allele of the gene from male I-2 will not be expressed in the somatic cells of II-3.** The allele inherited from I-2 is imprinted, so the son's somatic cells will not express that allele.

e. **The allele of the gene from male I-2 will not be expressed in the germ cells of II-3.** The imprinting from I-2 is erased during gametogenesis but the male will re-imprint both alleles of the gene in his germ cells.

f. **The allele of the gene from male I-2 will be expressed in the germ cells of III-1.** The III-1 inherited the non-imprinted allele of the gene from his mother (II-2), so the gene is expressed in the somatic cells of III-1.

g. **The allele of the gene from male I-2 will not be expressed in the germ cells of III-1.** In the his germ cells male III-1 will have imprinted both alleles of this gene.

17-22. The mRNA for this gene is made in all tissues tested, while the protein is made in only muscle cells. **Expression of this gene is translationally regulated**.

17-23.

a. Gene *A* = transcription factor 1; gene *B* = transcription factor 2. If both transcription factors are required for expression, only the *A- B-* genotype will produce blue flowers. Diagram the cross: *Aa Bb* x *Aa Bb* → 9/16 *A- B-* (blue) : 3/16 *A- bb* (white) : 3/16 *aa B-* (white) : 1/16 *aa bb* (white) = **9/16 blue :7/16 white flowered plants**.

b.   If either transcriptional factor is sufficient to get blue color, then three of the four genotypic classes from a dihybrid cross will produce blue flowers: 9/16 *A- B-* (blue) : 3/16 *A bb* (blue) : 3/16 *aa B-*(blue) : 1/16 *aa bb* (white) = **15/16 blue : 1/16 white flowers**.

17-24. The 1.2 kb mRNA in the skin cells and the 1.3 kb mRNA in the nerve cells could be due to **either a different start site for transcription in the two cell types or alternative splicing of the primary transcript** in the two cell types.

17-25. The 5' and 3' untranslated regions (UTRs) could be cloned at the 5' or 3' ends of a reporter gene (for example, *lacZ*) that is transformed back into *Drosophila* early embryos to see if either sequences cause translation of the reporter protein at the appropriate times during development.

17-26. The protein in the fat cells may be post-translationally modified (for example, phosphorylated or de-phosphorylated) so that it is only active in fat cells. Alternatively, the protein may need a cofactor to be activated, and this cofactor is only transcribed in fat cells.

Chapter 18   Cell-Cycle Regulation and the Genetics of Cancer

Synopsis:

This chapter describes characteristics of the eukaryotic cell cycle and the cancer phenotype. Genetic analysis of the regulation of the cell cycle has lead to our understanding of the genetic basis of cancer. Cancer is an uncontrolled growth of cells in which the cell cycle is no longer correctly regulated (**see Feature Figure 18.16**). There are genes whose products are necessary for the cells to proceed in the cell cycle and genes whose products monitor progress of the cell cycle, stopping the cycle if there is something amiss. Signals for beginning cell division are relayed through a series of steps (signal transduction). A series of mutations, each one either inherited or newly arising in a somatic cell, lead to a cell dividing out of control. Inherited predisposition to particular cancers is due to a mutation in one of the genes controlling cell cycle growth.

Significant Elements:

After reading the chapter and thinking about the concepts, you should be able to:

♦ Distinguish between oncogenes and tumor suppressor genes based on the behavior of cells containing a mutation in a gene (**see Figure 18.22**).

♦ Describe the role of growth factors and the pathway of signaling that leads to the start of the cell cycle (**see Figures 18.14 and 18.5**).

♦ Describe the role of cyclins in regulating the cell cycle (**see Figures 18.7, 18.9**).

♦ Explain inherited predisposition to a non-scientist.

Problem Solving Tips:

♦ Think about gene regulation, cell signaling and how they all might fit together in a system for regulating the cell cycle.

♦ Remember that there will be regions of genes coding for interacting proteins where mutations will affect the ability to the proteins to interact.

♦ Use your knowledge of molecular techniques (hybridizations, PCR, transcription analysis, etc) to design experiments.

♦ Multiple occurrences of cancer and/or early onset of cancer are suggestive indicators of inherited predisposition alleles, but further experiments are necessary to identify the mutant alleles especially if that type of cancer is very prevalent in the population.

Solutions to Problems:

18-1. a. **7**; b. **6**; c. **8**; d. **2**; e. **1**; f. **9**; g. **3**; h. **5**; i. **4**.

18-2. <u>See Figure 18.2</u> for an overview.

a.   **M** (mitosis)

b.   **M**

c.   **S** phase

d.   **G₁** phase (<u>see Figure 18.6</u>)

18-3. (<u>See Figure 18.3</u>) **Plate mutagenized haploid yeast cells at 30°C (permissive temperature) and replica plate colonies onto two sets of plates- one incubated at 30°C and the other at 23°C.** Genes necessary for the cell cycle are essential, so cold-sensitive mutants defective in cell cycle genes should die at the non-permissive temperature (23°C). Those mutants that are cold-sensitive can be studied more extensively by growing a liquid culture of the strain at permissive temperature, shifting to the non-permissive temperature and watching the cells microscopically to see if they stop with the same phenotype. If they have the same phenotype when they stop dividing, this indicates the stage in the cycle at which they stop.

18-4. There are three complementation groups and therefore three genes. Mutant 1 does not complement mutants 4 or 5, so these three mutations define one complementation group. Mutant 2 does not complement mutant 8, so these make up a second complementation group. Mutant 3 does not complement mutants 6,7, or 9 so these are in a third complementation group.

18-5.

a.   Although condensation begins during prophase, chromosomes are most visible during mitosis. Mitosis is 1/24 of the cycle, so 1/24 or **4% of the population of cells are in M**.

b.   **The remaining 96% (23/24) of the cells are in interphase** (G1, S and G2 combined).

18-6. There is **cyclical regulation of transcription** that allows certain genes to be expressed only at certain points in the cell cycle. There is also **cyclical regulation of translation** for certain mRNAs and **cyclical control of posttranslational modifications** that lead to proteins that are active at certain points in the cell cycle and inactive at other times.

18-7.

a. **False** - CDKs (<u>c</u>yclin <u>d</u>ependent <u>k</u>inases) function in CDK/cyclin complexes; only when the CDKs are complexed are they are active kinases (**<u>see Figure 18.7</u>**).

b. **True** - their appearance and disappearance activates the correct kinases and regulates the phosphorylation/dephosphorylation (activation/deactivation) of other proteins.

c. **False** - Checkpoint monitoring proteins check for aberrant cell cycle events (**<u>see Figure 18.12</u>**).

18-8. a. **2**; b. **3**; c. **1** (**<u>see Figure 18.12</u>**).

18-9

a. Molecules outside the cell that regulate cell cycle include **hormones and growth factors (<u>see Figure 18.14</u>**).

b. The whole of Chapter 18 is a discussion of the molecules inside the cell that regulate the cell cycle! These include **cyclins, cyclin dependent protein kinases, any molecule in the signal transduction pathway, receptors or molecules that transmits a second signal (<u>see Table 18.1</u>**).

18-10.

a. RAS is inactive when it is bound to GDP (RAS-GDP) (**<u>See Figure 18.5d</u>**). When RAS is bound to GTP it is activated and in turn activates three protein kinases (the MAP kinase cascade). This cascade activates a transcription factor which causes cells to divide. Therefore, **a RAS mutant that stays in the GTP-bound state is permanently activated and will cause the cell to continue dividing**.

b. This second RAS mutant is blocked in the RAS-GDP form at the restrictive temperature. Therefore, **under the restrictive conditions the cells will not divide**.

18-11. c, e, a, b, d (**<u>see Figure 18.15</u>**).

18-12.

a. T antigen binds to p53 and prevents it from acting in cell cycle regulation by binding a transcription factor. By supplying excess p53 from the high level promoter there is enough now in the cell to bind all of the T antigen (quenching, as described in Chapter 17) and still have enough unbound p53 to regulate the cell cycle. Thus, **the effect of the T antigen is minimized**.

b. Mutants 1 and 2 have the phenotype seen in cells with the non-elevated levels of p53 in the presence of T antigen. These mutants can no longer rescue the cells from the effect of T antigen, so they must **decrease the ability of p53 to function in cell cycle control**. The further presence of the T1 antigen blocks the ability of the mutant p53 to function even more.

c. Mutant 3 still rescues the cells from T antigen so this mutation does not seem to alter a the ability of the p53 protein to bind the T antigen, nor does it alter the ability of the p53 to regulate the cell cycle. Therefore, this mutation must be **in a functional domain other than those that bind the T antigen and the transcription factor**.

18-13. For an overview, <u>see Feature Figure 18.16</u>. The four main characteristics of cancer are: **uncontrolled growth; genomic instability; potential for immortality; and the disruption of local tissues and the ability to invade distant tissues**.

18-14.

a. Amplification of a specific DNA sequence can be observed using hybridization to DNA from tumor and normal cells. **Use two different probes- one representing the specific sequence that you are analyzing and the other representing an unamplified control sequence**. In tumor cells, look for an increase in the signal of the amplified region while the level of the control region does not change.

b. To find gross rearrangements look for **alterations in the chromosomal banding patterns in a karyotype analysis**.

18-15. These 'conflicting views' can be easily reconciled. We know that **some of the environmental agents that are implicated in increased cancer risk cause increased level of mutations**, so this is consistent with the fact that mutations in genes are necessary to cause cancer. **The inherited mutations that lead to predisposition to cancer inactivate one allele of a gene (often a tumor suppressor gene) that inhibits cell growth**. This inherited mutation is just one of several mutations that must occur within a cell to lead to cancer. The environmental and inherited factors are therefore both affecting the same thing – the genes that are important for regulating the cell cycle.

18-16. You need to do epidemiological studies of the colon cancer profile looking for a correlation between diet and the incidence of cancer. To assess **the role of diet, studies can be set up within the recent immigrant population vs the United States-based native Indian population**

examining the effect of a Westernized diet vs a diet resembling that of the ethnic group. Also, the effects of the same diets on non-Indian Westerners should be examined.

To assess **the role of genetic differences, you need to keep other factors, for example, diet, as constant as possible. You could look at the incidence in Indians and Americans who have similar diets**. This is often done by studying cancer rates in immigrants. For example, people who moved from India to Toronto could be compared with Toronto natives.

18-17. For the background on positional cloning, **see Figure 11.7**. First, you must localize the trait to a specific region of the genome. Then you must sort through the candidate genes in that region to decide which one (or two) to focus your studies on. Lastly, you should sequence your clone and predict the protein structure and function from the predicted amino acid sequence. Step d can actually be fit in anywhere after step b is done (after you have the clones). You should also look for or create mutations (knock-out mice, for example) in the gene in other species in order to see what mutant phenotype is produced. **The order of the steps is: e, b, a or d, c, f.**

18-18. Steps d, a and c must be done in that order. First you must localize the human homolog of the yeast gene in the human genome and find nearby markers that can be easily followed. Next, see if that region of the genome is linked to a hereditary predisposition to any sort of cancer. If you do find families with a predisposition that is linked to that region of the genome, then you must isolate the human gene and examine that gene in the affected and unaffected members of these families. **The order of the steps is: d; a; b, c.**

18-19. Protooncogenes are genes that code for proteins that stimulate cell division. Therefore, oncogenic mutations in such genes usually increase the level of expression of the gene product. This results in cells that divide too much or at the wrong time. A deletion of a protooncogene will decrease the level of expression, not increase it. Therefore, **choice c. will not be associated with cancer**. All of the rest of the changes will increase the probability that the affected cell, or its mitotic descendants, will become cancerous.

18-20.

a. The fact that everyone develops the disease at an early age is often a clue that an inherited mutation is an important factor in the development of colon cancer in this family. However, it is not an absolute predictor. Likewise, the appearance of so many affected individuals in one family is not a necessarily a predictor of an inherited basis for the prevalence of colon cancer in this family either. Both of these may indicate that there is an important environmental factor that contributes to development of the disease. Examination of the pedigree suggests that **predisposition to colon cancer in this family could be an autosomal dominant trait. If this is true, then individuals II-2, and either I-1 or I-2 must have the mutation, but not express it.**

b. **Individuals I-1, I-2 and II-2 are not among the high coffee consumers. Perhaps the predisposition to colon cancer is a combination of a particular genotype and the environmental factor of consumption of the special coffee.** Thus, the interaction of a certain genotype and an environmental factor results in the phenotype of colon cancer. In this scenario, the genotype alone is not enough to predispose you to cancer, as seen in individuals I-1, I-2 and II-2. On the other hand, note that a substantial number of individuals who drink coffee are not affected (for example III-8, III-9, III-10, III-11, III-12) so coffee alone does not appear to have a major effect on the early onset of colon cancer.

18-21. If there is a dominant predisposing allele segregating, **individual II-2 carries the mutant allele** since male III-2 has colon cancer and thus has the mutant, predisposing allele. Two individuals in the second generation carry the mutation, so one of the generation I parents must have had the mutant allele and passed it on to individuals II-2 and II-3. Thus, **either I-1 or I-2 also carries the mutation**. Note that if only one member of generation 2 inherited the mutant allele (developed early onset colon cancer), it is possible that the mutation arose in development of a germ cell in a generation I individual. In this case, the somatic tissues of that parent would <u>not</u> contain the mutation.

18-22. Technique d will be most useful in detecting a specific point mutation in the p53 gene. Unlike techniques b and c, hybridization with ASOs is a high throughput method that looks directly at the DNA change. Method a will give you a lot of DNA, but you still have to assay for the DNA sequence at the nucleotide of interest by sequencing the DNA that was amplified. If the mutation is one in which a restriction site is altered, you could use restriction enzyme digestion followed by Southern blot.

Chapter 19   Using Genetics to Study Development

Synopsis:

The problems for Chapter 19 are organized in a very different fashion than those for the other chapters in the textbook: here, all of the problems focus on a single comprehensive example concerning the development of mechanosensory bristles on the bodies of *Drosophila* adults. Each problem takes advantage of information supplied by the preceding problems, so it is important to work on these problems in the order in which they are presented! You will find that many of these problems are quite difficult, which is not surprising given the complexity of the system. However, the effort you expend will be rewarded by an increased understanding of how genetics can be used to dissect complicated biological systems. As the science of biology continues into the twenty-first century, this kind of analysis will become of greater and greater importance.

The most pertinent information for the solution of these problems is discussed in **Figures A and B and the associated explanation in the text**; in addition, **Figure 19.28** shows the biochemical events that occur during Notch-mediated cell-to-cell signaling. It will be easiest if you think of the end result of all of the events described as a binary decision: will a cell become a neural sensory organ precursor (SOP) that can give rise to a bristle, or will the cell instead become an epidermal cell that cannot develop into a bristle? All the cells in a "proneural cluster" originally have the potential to become SOP cells because they express the transcription factors encoded by the *achaete-scute* genes. But as development proceeds, only one cell is selected to become an SOP cell, and signals emanating from this cell turn off *achaete-scute* expression in all of the other cells in the same cluster so that they become epidermal cells instead. The process that changes the fate of the cells surrounding the presumptive SOP cell (the preSOP) is called "lateral inhibition".

Significant Elements:

First, make a summary of the biochemical events that allow lateral inhibition to take place. A condensed version of **Figure 19.28** is shown below to help you visualize some of the steps of the process. The major points in the figure are discussed below.

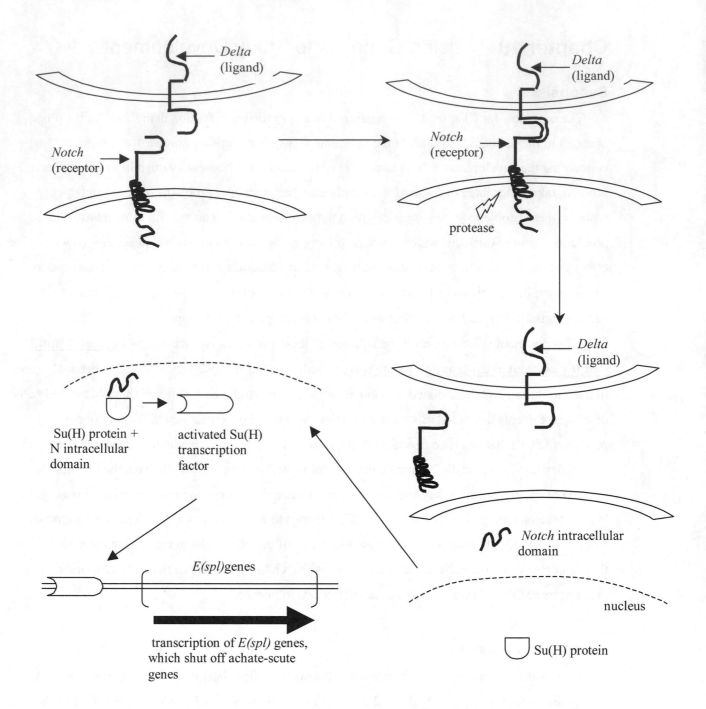

In the proneural cluster, all of the cells express both Delta and Notch proteins on their cell surfaces. Binding of the Delta ligand on one cell (the signaling cell) with the Notch receptor on an adjacent cell (the receiving cell) causes the activation of a membrane bound protease in the receptor cell. The activated protease cleaves the Notch protein at the inner edge of the cellular membrane; this is the first step of a signal transduction pathway in the receiving cell that will ultimately cause the cell to turn off the *achaete-scute* genes and thus adopt an epidermal fate. The interior portion of the Notch protein produced by cleavage moves into the nucleus, where it binds to an inactive transcription factor, Su(H). When the binding occurs, the Su(H) transcription factor is activated, and it binds to the regulatory region of a clustered group of related genes known as *E(spl)*. The E(spl) proteins are also transcription factors that are thought to bind to the regulatory region of the *achaete-scute* genes and prevent their transcription. Thus, the receptor cell is converted to an epithelial cell while the signaling cell continues to express *achaete-scute* and becomes the sensory organ precursor (SOP) cell, making the neuron and bristle structures shown in Figure A that allow flies to detect foreign materials on their bodies.

Although the above discussion summarizes some of the most pertinent information, you will still need to synthesize many of the ideas described elsewhere in this chapter to work your way through this problem set. The major challenges you will face (as were indeed faced by the researchers investigating lateral inhibition) will be in trying to convert the results of genetic experiments into models for what is occurring at the biochemical level. This means that you will have to think very clearly about the relationship between genotype and phenotype: how can changes to either one or both copies of a gene change the amount or activity of the gene product in ways that affect phenotype? This kind of model building can become very "hairy" (pardon the pun) when considering the interactions of more than one gene, as can occur when mutations in one gene suppress or enhance the phenotypes associated with mutations in a second gene. You should always remember that the most fundamental consideration in your model building is potentially the greatest source of confusion. That is, the whole purpose of these exercises is to try to infer the normal function of wild-type proteins in lateral inhibition, but what you are observing is what happens when the cells contain mutations in the genes encoding these proteins.

Solutions to Problems:

19-1.   All cells in the proneural cluster express the *achaete-scute* genes. At this point all of the cells in the proneural cluster are potential SOP (sensory organ precursor) cells. **After lateral inhibition has taken place only the preSOP cell will still be expressing the *achaete-scute* genes. The surrounding cells will express the *E(spl)* transcripts but not the *achaete-scute* genes**. The preSOP cell will also be expressing other genes (not presented in this discussion) indicative of cells that have acquired a neural fate. You can visualize these cell types with 2 different techniques. You can **localize specific transcripts using RNA *in situ* hybridization, as in <u>Figure 19.17</u>**. For instance, if you probe the cells of the epithelium with a probe for the *achaete-scute* genes, the cells which are transcribing that complex of genes can be visualized. This will detect the groups of cells that correspond to the proneural cluster or single cells which are the preSOP cells. The use of an *E(spl)* probe would show the cells that are epithelial in nature. **The second method of visualizing the different cell types employs labeled antibodies which recognize specific proteins**. Again, an antibody against the achaete-scute proteins would show cells in the proneural cluster and the single preSOP cells while antibodies against the E(spl) protein would bind to the epithelial cells. If the antibodies were labeled with compounds that fluoresce in different colors, you could even look at achaete-scute and E(spl) proteins simultaneously, as was done for other proteins in **Figure 19.18b.**

19-2.   In lateral inhibition the preSOP cell turns off the *achaete-scute* gene expression in the other proneural cells. In order to do this, **the *Delta* gene product (the ligand) must be expressed in the preSOP cell. In contrast, the proneural cells that are inhibited (and will thus become epithelial cells) must express the Notch protein (the receptor).**

19-3.   The effects of a signal transduction pathway can differ depending of the local concentrations of ligand and receptor molecules. It is likely that the cell that by chance has the most Delta protein on its surface (and/or the least Notch protein) will gain an advantage over the surrounding cells in a race to become the preSOP cell. It is not clear however that this simple model is correct. The final answer is not known, but it is currently believed that Notch signaling causes mutual inhibition of all of the cells in a cluster, yet one cell in the cluster (the preSOP) may have an elevated neural potential that allows it to overcome Notch signaling. This idea is supported by evidence presented in later problems that too much Notch signaling can prevent all of the cells in a proneural cluster from becoming SOPs. No one really knows what might cause this elevated neural potential in one particular cell, but there may be local differences in the expression of genes like *achaete-scute* that promote a neural phenotype. Alternatively, there may be local differences in the expression of genes like *Su(H)* that govern later steps in Notch signaling. In this latter model, a cell less capable of responding to Notch signaling would be the cell selected to be the presumptive SOP.

19-4.

a.   **The epithelium outside the clone should be normal** because those cells still have a wild-type allele of *Notch*. (Most of the cells in the animal are N^+/N^-; only the cells in the clone are N^-/N^-.) Each proneural cluster outside the clone should thus make a single SOP cell. **Within the clone the cells are homozygous for a null allele of *Notch*. These cells cannot form the Notch intracellular domain. Thus, there is no activation of the *Su(H)* transcription factor and no transcription of the *E(spl)* genes. Because the *achaete-scute* genes cannot be shut off, there** will be more SOP cells and fewer epithelial cells. When the cluster of cells homozygous for the null allele of *Notch* intersects a proneural cluster, the N^+ cells in the cluster will become epithelial while the cells in the clone will become preSOPs.

b. Classically, nonautonomous traits are expressed by mutations in genes encoding ligands, as the absence of a ligand from a signaling cell could affect the phenotype of a receiving cell. On the other hand, autonomous traits are seen when the mutations are in genes encoding receptors, as the absence of the signal transduction pathway in the receiving cell affects only that cell. Thus, *Notch* **mutants display autonomy.**

c. (**See Figure 19.19b.**) If L2 cells are $AGAMOUS^-$, adjacent L1 cells do not differentiate properly even if they are $AGAMOUS^+$. This means that the phenotype caused by mutants in $AGAMOUS$ is a non-autonomous trait. This suggests (but does not prove) that the $AGAMOUS$ gene in L2 cells encodes a ligand that binds to a receptor in L1 cells.

19-5.

a. The mutant clone in this case is homozygous for a null allele of *Delta*. Again, **the epithelium outside and well away from the clone should be normal. Within the clone none of the cells have the ligand/receptor interaction that allows the Notch protein to be cleaved. Thus none of the cells in the clone can repress the expression of the *achaete-scute* genes, so all of the cells within the proneural cluster should become preSOP cells and then SOP cells. If this** $N^+N^+ D^nD^n$ (**n = null mutant) clone bisects a proneural cluster (so that the other half of the cells in the cluster are normal ($N^+N^+ D^+D^+$)), then most of the normal cells will become epithelial cells because they can express *E(spl)* and inhibit the expression of *achaete-scute*.** It is also possible that one of the normal cells in the cluster can become a preSOP cell. In addition, **most of the cells from the clone that are in close contact with the normal cells in this bisected clone should become epidermal cells.** The mutant cells express the Notch receptor, so they can receive the Delta protein signal on the surface of the adjacent normal cells.

b. *Delta* **mutants display nonautonomy.** This result was the first indication that the Delta protein functions as a ligand, a supposition that was proven in later biochemical experiments.

19-6. The expression of the *achaete-scute* genes is necessary to determine the development of an SOP cell. The end result of lateral inhibition is to suppress the SOP/neural fate. If the *E(spl)* gene product is a transcription factor that binds to the regulatory region of the *achaete-scute* complex, the E(spl) protein must inhibit the expression of the *achaete-scute* genes, so the cell expressing *E(spl)* becomes epithelial. Thus, **the E(spl) proteins are repressors of *achaete-scute* transcription**.

19-7. The Hairless (H) protein plays a role in lateral inhibition. In clones of cells that are homozygous for a null allele of the H gene ($H^n H^n$), none of the cells become SOP cells. Furthermore, the H protein binds to the Su(H) protein.

a.   In the absence of the H^+ protein, all of the cells in the proneural cluster are laterally inhibited. Therefore, the H^+ protein works to counteract lateral inhibition. Remember that the binding of the intracellular domain of Notch activates $Su(H)$ so that the $E(spl)$ genes are transcribed. **The binding of the H^+ protein to Su(H) seems to block activation of the Su(H) transcription factor. H protein could prevent Su(H) from binding to DNA or otherwise prevent Su(H) from activating transcription of $E(spl)$.** The intracellular domain of Notch seems to competitively bind to $Su(H)$, so in the absence of any H protein ($H^n H^n$) the Su(H) protein is very sensitive to the level of the Notch intracellular domain.

b.   The H protein must be inside the same cell as the Su(H) protein to have any effect, so **the phenotype caused by mutations in H is autonomous**.

19-8. $H^+ H^n Su(H)^+ Su(H)^+$ cells have a reduced number of bristles relative to wild type. In other words, when there is only half (roughly) as much of the H protein present (1 part H : 2 parts Su[H]), the Su(H) protein can more readily bind the N intracellular domain and thus activate $E(spl)$ gene expression to higher levels. The fact that $H^+ H^n Su(H)^+ Su(H)^n$ flies have nearly normal levels of bristles suggests that the relative dosages of the 2 proteins is important (1 part H : 1 part Su[H]). The Su(H) protein from the unusual allele of $Su(H)$ enhances the effect of a H^n allele, in that cells of the genotype $Su(H)^+ Su(H)^{abn} H^+ H^n$ have fewer bristles than the $H^+ H^n Su(H)^+ Su(H)^+$ cells. **The effect of the $Su(H)^{abn}$ allele is to increase the expression of $E(spl)$. The problem states that the $Su(H)^{abn}$ protein can bind to the H protein but cannot activate $E(spl)$ transcription. These facts suggest that the $Su(H)^{abn}$ protein sequesters the H protein into inactive complexes. This would titrate the amount of H protein, so there are lower levels of H to bind to the normal $Su(H)^+$ protein encoded by the wild-type allele.** As we have seen in Problem 19-7, this would result in greater activity of $Su(H)^+$ and thus more lateral inhibition.

This problem shows why it is important to determine whether a mutant allele is a null allele or not. If you thought $Su(H)^{abn}$ was a null mutation that was causing fewer bristles, then you would conclude that the normal Su(H) protein is involved in producing more bristles, in other words decreasing the level of *E(spl)* expression. Of course this is exactly the wrong conclusion about the role of *Su(H)*$^+$.

19-9. As described in <u>problem 19-8</u>, animals with the genotype $H^+H^n Su(H)^+Su(H)^n$ have a bristle number that is close to wild type.

a.   Animals that are $H^+H^+ Su(H)^+Su(H)^n$ also have the wild type number of bristles. Thus these animals still have lateral inhibition. In this genotype, the ratio of these proteins is 1 Su(H)$^+$: 2 H$^+$ yet there is no effect on the level of expression of *E(spl)* genes. This suggests that **the bristle phenotype is more sensitive to increases in Su(H) activity rather than to decreases of this activity**.

   If heterozygosity for null *Su(H)* mutations alone produces a normal phenotype, why aren't flies that are heterozygous for null *Notch* mutations also wild type in appearance? This suggests that either the Notch protein can activate signal transduction pathways other than that shown in **Figure 19.28**, or that reduction in the amount of Notch protein causes a non-linear reduction in the level of active Su(H). Thus, the Notch null heterozygotes would somehow have less than half the normal amount of active Su(H).

b.   The implication in this question is that surrounding tissue is heterozygous for both genes, so that a clone of tissue that is homozygous null for both mutations can arise. Within the clone, there is no Su(H) transcription factor, so the level of H protein is irrelevant. Thus, the *Su(H)* homozygous null mutant phenotype is epistatic to the *H* gene. **All the cells in proneural clusters within this mutant clone will become preSOP cells because there is no lateral inhibition**.

19-10. In the N^+N^{null} individuals one half of the *N* gene product is not enough to produce wild type phenotype, that is **they show haploinsufficiency.**

19-11. Animals that are heterozygous for a null allele of *Notch* have more bristles because these cells have less Notch receptor, and so produce less of the Notch intracellular domain. Thus there less of the transcriptional activator and less expression of the *E(spl)* genes. The *achaete-scute* genes are not turned off, and more of the cells in the proneural cluster become SOP cells.

a.   **A null allele of *E(spl)*** means the production of lower levels of the inhibitor of *achaete-scute* expression. This means the expression of *achaete-scute* will be enhanced, in turn giving more bristles. This mutation **will thus be a dominant enhancer of N^{null} mutants**.

b. **A null allele of *H*** will increase the level of activate *Su(H)*, leading to fewer bristles. This mutation **will be a dominant suppressor**.

c. **A null allele of *Su(H)*** means a decrease in the expression of the *E(spl)* genes, and thus more bristles - **a dominant enhancer**.

d. **A null allele of *Delta*** means less ligand/receptor binding, decreased levels of the Notch intracellular domain, decreased activation of the *E(spl)* genes and so more bristles. Thus this mutation **is a dominant enhancer**.

19-12.

a. **Proneural refers to the idea that the function of the <u>wild type</u> *achaete-scute* genes is to promote the formation of neurons or cells that are ancestors of neurons like the SOP cells. However, neurogenic is exactly the opposite. The wild type activity of genes like *Notch* suppresses neuronal differentiation. The term 'neurogenic' was used in reference to the phenotype of a <u>mutant</u> allele of *Notch*.** In the absence of *Notch* activity, more cells turn into SOPs and then neurons. This historical ambiguity is the basis for the two different terms. The *Hairless* gene (*H*) is named after the phenotype caused by a loss-of-function mutation. The actual wild type gene product actually promotes the formation of hairs (bristles). This is true for many genes in *Drosophila* - the gene is named based on the phenotype seen when the gene is mutant. For example, the wild type product of the *white* eye gene makes red eyes. A lack of that gene product causes white eyes.

b. (i) ***E(spl)* is neurogenic** because animals with a null mutation of this gene make more neurons than normal. (ii) **The *H* gene counteracts the neurogenic genes**, because null mutants in *H* make no SOP cells. However, **most *Drosophila* geneticists would not consider the *H* gene to be 'proneural'** in the same sense as *achaete-scute*. The wild type product of the *H* gene doesn't directly promote neural differentiation; instead it counteracts lateral inhibition. (iii) ***Su(H)* is neurogenic** because null mutations of this gene cause more neurons to develop. Likewise, (iv) ***Delta* is also neurogenic**.

19-13.

a. **Animals homozygous for mutations in *Notch* or *Delta* do not survive to adulthood, and so can not make eggs.** The homozygous null germline clones are necessary to produce the holonull embryos.

b. There are 3 conclusions you can draw from this data. (i) *Notch* holonull embryos have a more severe phenotype than the *Notch* zygotic nulls. Thus, **maternally supplied Notch protein plays a role in early development**. This role cannot be fulfilled by the zygotically produced Notch protein. (ii) This is not the case for *Delta*, where **lack of the maternally supplied protein does not appear to be very important**. (iii) **The *Notch* holonull has a more severe phenotypic effect than the *Delta* holonull**. If Delta was the only Notch ligand, you would expect the severity of the phenotypes to be the same. **Therefore, there must be Notch ligands other than Delta**. In fact, this is true. There are other ligands, including the Serrate and Scabrous proteins. These proteins are encoded by genes which are related to but not identical to *Delta*. These other ligands may turn on Notch activity in other developmental processes.

19-14. Heterozygosity for the N^{414} allele causes a reduction in the number of bristles; the opposite effect of a null allele of *Notch*. This seems to reflect an increase in the amount of Notch signaling that blocks any of the cells in the proneural cluster from becoming an SOP. The 2 models that could account for the are (i) the mutation leads to an increase in affinity of the Notch receptor for the Delta ligand, or (ii) the mutation causes a constitutive activation of Notch signaling independent of the ligand binding.

a. The number of bristles in the $N^{414}N^+$ $Delta^{null}Delta^+$ double heterozygotes is the same as the $N^{414}N^+$ individuals. **This finding is consistent with model (ii)** but not with model (i) **because the mutant phenotype is unaffected by the amount of the Delta ligand.** If model (i) were correct, you would expect the *Delta* null allele to suppress the phenotype caused by heterozygosity for N^{414} because there is less ligand to turn on lateral inhibition.

b. **You would generally expect the duplication to enhance the phenotype**: the more Notch protein (whether constitutively activated or not) the stronger the Notch signaling in lateral inhibition and thus **fewer bristles**. However, the answer actually depends on the exact mechanism by which the N^{414} allele causes a gain-of-function. One could imagine scenarios that are not too far-fetched in which the N^+ gene product competes with the N^{414} product, so the duplication would weakly suppress the phenotype; in other words, an $N^{414} / Dp(N^+)$ animal could end up with a few more bristles than an $N^{414}/ +$ animal.

19-15. The wild type Hairy protein is likely to repress the transcription of the *achaete-scute* genes in the regions of the animal that develop into the wings and legs. If *hairy* function is lost, the *achaete-scute* genes are expressed in these regions, leading to ectopic bristles.

19-16. There are many ways to screen (look for) flies that are mutant in the ability to groom themselves. One clever screen that was actually performed took the progeny of flies that were exposed to a mutagen and **dusted these flies with a fluorescent powder. After a certain period of time to allow the flies to groom themselves, the flies were put under an ultraviolet light. Flies that cannot groom themselves will glow because they cannot remove the fluorescent powder.** This scheme will identify dominant mutations that lead to grooming defects.

Chapter 20   The Genetic Analysis of Populations and How They Evolve

Synopsis:

This chapter involves the study of how genetic laws impact the genetic makeup of a population. Mendelian principles are the basis for the Hardy-Weinberg law that allows one to calculate allele and genotype frequencies from one generation to the next. The Hardy-Weinberg law can be used only if other forces are not acting on the allele frequency. Those forces include selection, migration, mutation, and population size.

Population geneticists try to determine the extent to which a trait is determined by genetic factors and how much is determined by environmental factors. Knowing if a trait is largely determined by genetic factors introduces the possibility for animal and plant breeders to select and maintain populations with desired traits.

Significant Elements:

After reading the chapter and thinking about the concepts, you should be able to:

♦ Determine allele frequencies in a population given the frequencies of genotypes.
♦ Determine genotype frequencies in a population given the frequencies of alleles.
♦ Determine genotype and/or allele frequencies in the next generation given the genotype or allele frequencies in the present generation.
♦ Determine if a population is in equilibrium.
♦ Determine allele and genotype frequencies after migration has occurred.
♦ Describe how heritability of a trait can be determined.

Problem solving tips:

♦ p, q are representations of allele frequencies.
♦ p^2, 2pq, q^2 are representations of genotype frequencies.
♦ Once a population is at equilibrium, allele and genotype frequencies do not change.
♦ If a genotype is selected against or if populations are combined (by migrations) or if there is significant mutation (usually together with selection), allele frequencies will change.
♦ selection for one genotype and selection against another genotype are balanced at a particular allele frequency (equilibrium frequency).
♦ Genetic drift is most often seen in small populations.

♦   Polygenic traits are controlled solely by alleles of two or more gene; multifactorial traits include polygenic traits and traits that are influenced by both genes and environment.

♦   Genetic and environmental contributions to a phenotype are sorted out by setting up conditions in which the genetic background is consistent (to analyze environmental contributions) or conditions in which the environment is constant (to analyze genetic contributions).

Solutions to Problems:

20-1. a. **3**; b. **5**; c. **8**; d. **7**; e. **6**; f. **1**; g. **9**; h. **2**; i. **4**.

20-2.

a.   To calculate genotype frequencies, divide the number of frogs with each genotype by the total number of frogs.

$G^G G^G$ = 120 / 200 = **0.6**; $G^G G^B$ = 60 / 200 = **0.3**; $G^B G^B$ = 20 / 120 = **0.1**.

b.   The allele frequencies are determined by totaling all alleles within each genotype.

$G^G G^G$ = 120 individuals with 2 G^G alleles = 240 G^G alleles

$G^G G^B$ = 60 individuals with one G^G allele = 60 G^G alleles

$G^G G^B$ = 60 individuals with one G^B allele = 60 G^B alleles

$G^B G^B$ = 20 individuals with two G^B alleles = 40 G^B alleles

There are 300 G^G alleles/400 total alleles so the **frequency (p) of G^G is 0.75**. There are 100 G^B alleles/400 total alleles so the **frequency (q) of G^B is 0.25**.

c.   The expected frequencies can be calculated using the allele frequency and the terms of the Hardy-Weinberg law, $p^2 + 2pq + p^2 = 1$; **0.5625 $G^G G^G$ + 0.375 $G^G G^B$ + 0.0625 $G^B G^B$ = 1**.

20-3.

a.   The allele frequencies are calculated from the proportion of individuals with different genotypes. The problem becomes easier and more intuitive if you assume the population has 100 individuals.

0.5 *M M* individuals = 50 individuals = 100 *M* alleles

0.2 *M N* individuals = 20 individuals = 20 *M* alleles + 20 *N* alleles

0.3 *N N* individuals = 30 individuals = 60 *N* alleles

120 *M* alleles / 200 total alleles = **0.6 allele frequency for *M***

80 N alleles / 200 total alleles = **0.4 allele frequency for** N

The expected genotype frequencies in the next generation are calculated using these allele frequencies are $p^2 + 2pq + p^2 = 1$; **0.36** MM **+ 0.48** MN **+ 0.16** $NN = 1$.

b. **Non-random mating, small population size, migration into the population, mutation, or selection** could cause a departure from the Hardy-Weinberg equilibrium.

20-4.

a. There are 60 flies with normal wings D^+D^+ out of 150 total flies and 90 flies with the heterozygous genotype DD^+.

   60 D^+D^+ flies = 120 D^+ alleles

   90 D^+D flies = 90 D^+ alleles + 90 D alleles

   120 + 90 D^+ alleles = 210 / 300 or allele **frequency of 0.7;**

   90 D alleles / 300 = **allele frequency of 0.3**

b. In the second generation, the following offspring would be produced:

   $p^2 = (0.7)^2 = 0.49$

   $2pq = 2(0.7)(0.3) = 0.42$

   $q^2 = (0.3)^2 = .09$

   The homozygous DD do not live, so the 0.49 and 0.42 remaining makeup a total of 0.91

   0.49 / 0.91 = 0.54 = proportion of the viable offspring with the D^+D^+ genotype

   0.42 / 0.91 = 0.46 = proportion of the viable offspring with the D^+D genotype

   The numbers of individuals with these two genotypes in the viable population would

   be 0.54 × 160 = **86.4** D^+D^+ and 0.46 × 160 = **73.6** D^+D.

20-5. For each of the populations, determine the allele frequency using the genotype frequencies in the population, then calculate the expected genotype frequency for that allele frequency when a population is in equilibrium.

a. **Population a is in equilibrium**.

   0.25 AA = 50 A alleles

   0.50 Aa = 50 A alleles + 50 a alleles

   0.25 aa = 50 a alleles.

   100 A alleles / 200 = 0.5 = p

   100 a alleles / 200 = 0.5 = q

$p^2 = 0.25$; $2pq = 0.5$; $q^2 = 0.25$.

These genotype frequency numbers match those seen in the population, so it is at equilibrium.

b. **Population b is not in equilibrium**

0.1 *AA* = 20 *A* alleles

0.74 *Aa* = 74 *A* alleles + 74 *a* alleles

0.16 *aa* = 32 *a* alleles

94 *A* alleles /200 alleles = 0.47 = p

106 *a* alleles / 200 alleles = 0.53 = q

$p^2 = 0.22$; $2pq = 0.5$; $q^2 = 0.28$ (Numbers do not match that seen in the population.)

c. **Population c is not in equilibrium**

0.64 *AA* =128 *A* alleles

0.27 *Aa* = 27 *A* alleles + 27 a alleles

0.09 *aa* = 18 *a* alleles

128 +27 = 155 / 200 alleles = 0.78 = p

27 + 18 = 45 / 200 alleles = 0.22 = q

$p^2 = 0.61$; $2pq = 0.34$; $q^2 = 0.05$ (Numbers do not match the frequencies seen in the population.)

d. **Population d is not in equilibrium**

0.46 *AA* = 92 *A* alleles

0.50 *Aa* = 50 *A* alleles + 50 *a* alleles

0.04 *aa* = 8 *a* alleles

92 +50 *A* alleles / 200 = 142 / 200 = 0.71 = p

58 *a* alleles / 200 = 0.29 = q

$p^2 = 0.50$; $2pq = 0.41$; $q^2 = 0.08$ (Numbers do not match those seen in the population.)

e. **Population e is in equilibrium**

0.81 *AA* =162 *A* alleles

0.18 *Aa* = 18 *A* alleles + 18 *a* alleles

0.01 *aa* = 2 *a* alleles

162 + 18 *A* alleles / 200 = 180 / 200= 0.9 = p

18 + 2 *a* alleles / 200 = 0.1 = q

$p^2 = 0.81$; $2pq = 0.18$; $q^2 = 0.01$ (Numbers do match the genotype frequencies seen in the population.)

20-6.

a. Consider the Q and R genes separately. In each case, you need to determine the frequency of each allele, then calculate the genotype frequencies expected if the population is in equilibrium for that gene. The number of Q^F alleles in the population is $202 + 202 + 101 + 372 + 372 + 186 + 166 + 166 + 83 = 1850$. The number of Q^G alleles is $101 + 101 + 101 + 186 + 186 + 186 + 83 + 83 + 83 = 1110$. The total number of alleles is 2960.

The frequency of the Q^F allele is $1850/2960 = 0.625$.

The frequency of the Q^G allele is $1110/2960 = 0.375$.

From these numbers you can calculate the expected genotype frequency. $p^2 = 0.39$; $2pq = 0.47$; $q^2 = 0.14$.

   The genotype frequencies in the population are the following.

For $Q^F Q^F$, $202 + 372 + 166 = 740/1480 = 0.5$. For $Q^F Q^G$, $101 + 186 + 83 = 370/1480 = 0.25$.

For $Q^G Q^G$, $101 + 186 + 83$ or $370/1480 = 0.25$.

Since the expected and observed are not similar, the **population is not in equilibrium for the Q gene**.

   The number of R^C alleles is $202 + 202 + 101 + 101 + 101 + 101 + 372 + 186 + 186 = 1552$.

The number of R^D alleles is $372 + 186 + 186 + 166 + 166 + 83 + 83 + 83 + 83 = 1408$.

The frequency of the R^C allele is $1552/2960 = 0.52$.

The frequency of the R^D allele is $1408/2960 = 0.48$.

The expected genotype frequencies for these allele frequencies are $p^2 = 0.27$; $2pq = 0.5$; $q^2 = 0.23$.

The observed genotype frequencies are the following. For $R^C R^C$: $202 + 101 + 101 = 404/1480 = 0.27$; for $R^C R^D$: $372 + 186 + 186 = 744/1480 = 0.5$; for $R^D R^D$: $166 + 83 + 83 = 332/1480 = 0.22$. These numbers are very close to that expected, so **the population is in equilibrium for the R gene**.

b. In the next generation, with random mating, the fraction that will be $Q^F Q^F$ will be that which we calculated as the expected genotype frequency in part a: **0.39 for the $Q^F Q^F$ genotype**.

c. The fraction that will be **$R^C R^C$ in the next generation** will again be the expected frequency calculated based on the allele frequency: **0.27**.

d. The prediction on the genotype of a child of two specific parents in the population is not influenced by allele frequency. This is a standard probability question starting with specific parents $Q^FQ^G\ R^CR^D$ and $Q^FQ^F\ R^CR^D$. The probability is the sum of the individual probabilities for each of the genes. There is a 1/2 chance that the female will contribute the Q^G allele and a 1/1 chance that the male will contribute the Q^F allele. There is a 1/4 chance that the parents, heterozygous for the R^C and R^D allele, will be homozygous. There is a 1/2 chance the child will be male. The overall probability is (1/2)(1/2)(1/4) = **1/16 that the child will be Q^FQ^G R^DR^D male**.

20-7. Each different allele frequency has a different set of genotype frequencies at equilibrium. There is not one equilibrium point that an allele will go to.

20-8. The 3:1 ratio is seen when two heterozygous **individuals** are crossed. This ratio is <u>not</u> relevant for a <u>population</u>. **The ratio in the population will depend on allele frequency**.

20-9.

a. For the sailor population:

324 *MM* sailors = 648 *M* alleles

72 *MN* sailors = 72 *M* alleles + 72 *N* alleles

4 *NN* sailors = 8 *N* alleles

72 + 8 = 80 *N* alleles / 800 total alleles = **0.1**

b. When the population is mixed together (random mating), the allele frequency in the mixed population can be calculated and the genotype frequency after random mating determined. You need to determine the genotype frequency and use that to determine the number of Polynesians with each genotype to be able to determine the number of alleles they contribute to the pool.

Polynesian allele frequency p=.06, q=.94.

p^2 (*MM*) = .0036; 600 individuals $\times$ .0036 = 2 *MM* individuals

2pq (*MN*) = .1128 600 individuals $\times$ .1128 = 68 MN individuals

q^2 (*NN*) = .8836 600 individuals $\times$ .8836 = 530 *NN* individuals

Now determine the allele frequency in the mixed population.

	sailor	Polynesian
MM	648 *M* alleles	4 *M* alleles
MN	72 *M* alleles	68 *M* alleles
	72 *N* alleles	68 *N* alleles

NN   8 *N* alleles   1060 *N* alleles

M alleles = 648 + 4 + 72 + 68 = 792/2000 = 0.396

N alleles = 72 + 68 +8 + 1060= 1208/2000 = 0.604

2pq (*MN*) frequency will be .478. **The number of children with the *MN* genotype will = 0.478 × 1000 or 478**.

c.   genotype   number of children   number of alleles

   MM   50   100 *M*

   MN   850   850*M* + 850*N*

   NN   100   200 *N*

The frequency of the *N* allele is 200 / 2000 or 0.1.

20-10. Evaluate the information you are given.

In the French population, individuals with the Ugh phenotype (could be either *AA* or *Aa* because Ugh is dominant) number 90. But there is only 50% penetrance, so twice as many individuals have those genotypes as the number that express the phenotype. That means that 180 individuals have the *AA* or *Aa* genotype and 320 have the *aa* genotype. Because the allele is in equilibrium, you know that q^2 = *aa* frequency. q^2 therefore is 320 / 500 or 0.64. q = 0.8 and p =1-q or 0.2.

In the Kenyan population, 75 individuals had the Ugh phenotype and therefore were homozygous *AA* or heterozygous *Aa*. 150 individuals must have the *AA* or *Aa* genotype and the remaining 50 individuals had the *aa* genotype 50 / 200 or 0.25 = q^2 q=0.5 and p=0.5.

To determine the number of *A* and *a* alleles in the mixed population, you need to calculate the number of individuals with each genotype, then tally up the alleles.

For the French population,

p=0.2; q=0.8

p^2 *(AA)* = 0.04   0.04 × 500 = 20

2pq *(Aa)* = 2(0.2)(0.8) = 0.32   0.32 × 500 individuals = 160 individuals

q^2 *(aa)* =0.64   0.64 × 500 individuals = 320 individuals

For the Kenyan population,

p = 0.5; q = 0.5

p^2 = 0.25 = 0.25 × 200 = 50 individuals

2pq=0.5 = 0.5 × 200 = 100 individuals

q^2=0.25 = 0.25 × 200 = 50 individuals

Calculating the number of alleles in the mixed population:

$AA = (20 \times 2) + (50 \times 2) = 140$

$Aa = 160\ A + 00\ A = 260$

$a = 160\ a + 100\ a = 260$

$aa = (320 \times 2) + (50 \times 2) = 740$

In the mixed population:

$140 + 260 = 400\ /\ 1400 = 0.286 = p$

$260 + 740 = 1000\ /\ 1400 = 0.714 = q$

The number of individuals with the Ugh phenotype will be one-half of the individuals with the

AA (p^2) and Aa (2pq) genotypes.

$p^2 = (0.286)^2 = 0.082 \times 1000 = 82$

$2pq = 2\ (0.286)(0.714) = = 0.408 \times 100 = 408$

$(1/2)(408 + 82) = 245$

The total number of Ugh individuals is 245.

20-11.

a.   50% of the population shows high levels (over a broad range), so they must have either the 1/1 or 1/2 genotype. 50% have low levels so must have the 2/2 genotype. That means $q^2 = 0.50$, $q = 0.71$ and p (**frequency of allele 1**) = 1-q or **0.29**.

b.   In the babies born in the following year, **the allele frequency will be the same, 0.29**.

c.   In the population of babies, use the p and q allele frequencies to determine genotype frequencies.

p = 0.29; q = 0.71

$p^2 = 0.084$

$2pq = 0.412$

$q^2 = 0.50$

The last genotype is selected against. In a population of 1000 individuals, there are 500 babies with this genotypes. 10% of the population is attacked by the virus, and 80% of those infected will die. Therefore, 50 will be infected and 40 of these will die. 460 with genotype 2/2 live. The numbers after childhood will be:

460 with genotypes 2/2; 84 with 1/1; 412 with genotype 1/2. For allele 1, there are $84 \times 2 + 412$ or 580 / 1920 = **0.30**.

20-12.

a.   There are 150 mice.

$60\ t^+/t^+ = 120\ t^+$ alleles

$90\ t^+/t = 90\ t^+$ alleles $+ 90\ t$ alleles

$210\ t^+$ alleles / 300 total alleles $= \mathbf{0.7\ t^+}$ **allele frequency** (p), q $= t$ **allele frequency $= 0.3$**.

b. First determine the frequencies of the three genotypes if all lived, then remove the inviable mice from your calculations and recalculate the allele and genotype frequencies.

$p^2 = 0.49 = 49$ individuals

$2pq = 0.42 = 42$ individuals

$q^2 = 0.09 = 9$ individuals

$49 + 42$ survive $= 91$ total

$49 / 91$ are $t^+/t^+ = 0.54$

$42 / 91$ are $t^+/t = 0.46$

Of 200 individuals, 0.54 (200) or **108 are wild-type; 92 are tailless**.

c. Determine the number of alleles contributed by each population to calculate the allele frequency for the mixed population then the genotype frequencies from the mating. Since the homozygotes die, you will have to subtract out these mice and recalculate the frequencies of the genotypes.

	Dom1	Dom2

$t^+/t^+ = 16 \times 2 = 32\ t^+$ alleles $48 \times 2 = 96\ t^+$ alleles

$t^+/t = 48\ t^+$ alleles $36\ t^+$ alleles

    $48\ t$ alleles $36\ t$ alleles

$32 + 96 + 48 + 36 = 212 / 296\ t^+$ alleles $= 0.716 = p$

$84\ t$ alleles $/ 296 = 0.284 = q$

After mating, the following genotype frequencies are generated:

$p^2 = 0.512$

$2pq = 0.407$

$q^2 = 0.08$

The t/t (q^2) mice die, so there are $512 + 407 = 919$ total living progeny (assuming 1000 zygotes); $512 / 919 = \mathbf{0.557} =$ **frequency of t^+/t^+ mice in the next generation** and $407 / 919 = $

$\mathbf{0.443} =$ **frequency of t^+/t mice in the next generation**.

20-13. The frequency of colorblind males ($X^{cb}Y$) is equivalent to p since there is only one allele (one X) in these individuals. The frequency of colorblind females is $\mathbf{p^2}$ ($\mathbf{X^{cb}X^{cb}}$) $= \mathbf{0.0064}$.

20-14. CF/CF is 1/17,000 or 0.000059 = q^2. q is 0.0077; therefore p = 0.9923. 2pq (frequency of heterozygote carriers) = **0.015 or 15 in 1000** (Note: In the Caucasian population the frequency of carriers is much higher = ~1/20).

20-15.

a. The F_1 flies are heterozygous and produced in the F_2 a population of flies consisting of 250 vg^+ / vg^+, 500 vg^+ / vg, 250 vg / vg. The allele frequencies are 500 vg^+ + 500 vg^+ = 1000 / 2000 = 0.5 for each as they were in the F_1. When the vg / vg are dumped in the morgue the allele frequency in the mating population of F_2 flies changes so that there are now 250 × 2 = 500 + 500 = 1000 / 1500 or 0.666 vg^+ and 0.333 vg. That means that the genotype frequencies produced in the F_3 are $(0.666)^2$ = 0.444 for the vg^+ / vg^+; 2(0.666)(0.333) = 0.444 for vg^+ / vg and **0.111 for vg / vg.** The first two genotypes are wild-type so the **frequency of wild-type flies is 0.888.**

b. When the vestigial F_3 are dumped, the allele frequency shifts because the flies that are allowed to mate are 444 vg^+ / vg^+ and 444 vg^+ / vg. The **frequency of the vg^+ allele is** (2(444) + 444) / 1776 or **0.75. The frequency of the vg allele is 0.25.**

c. The genotype frequencies in the F_4 are calculated using the allele frequency of the F_3 generation. p^2 = $(0.75)^2$ or 0.5625; 2pq = 2(0.75)(0.25) or 0.375; q^2 = $(0.25)^2$ or 0.0625. The first two classes are wild-type, so 0.5625 + 0.375 = 0.9375. Now the vestigial winged flies are not dumped and random mating occurs. **The frequency of wild-type and vestigial genotypes will be the same in F_5 as in F_4 - wild-type = 0.9375, vestigial = 0.0625.**

20-16. The farther from equilibrium, the greater the Δq, so the population with an allele frequency of 0.2 will have the larger Δq.

20-17. A fully recessive allele is not expressed in a heterozygous organism, so there is no selection against the heterozygotes. **Selection against the homozygous recessive genotype will decrease the frequency of the recessive allele in the population, but it will never totally remove it, as the recessive allele is hidden in the heterozygote (see Figure 20.6).** In addition, a recessive allele sometimes confers an advantage when present in the heterozygote, as seen for the sickle cell allele in areas where malaria is prevalent. Finally, mutation can produce new recessive alleles in the population.

20-18. The equilibrium frequency will be different for the two populations. Equilibrium frequency is a balance between selection and mutation and the selection is very different in the two populations.

20-19.

a. **There is a founder effect of the descendants coming from a small number of individuals**, so whatever recessive alleles were present in the population are more likely to be combined. Therefore the frequency of some alleles and genotypes can be higher in that population. Other alleles may not have been included in that original gene pool.

b. **An advantage to studying the Finnish population is that there is genetic homogeneity and probably fewer genes** (potential modifiers) that may affect the trait and therefore can be more easily dissected. **A disadvantage is that some mutations that are present in general population may not be found in this small, inbred population** and therefore will not be identified in studies of Finns.

20-20.

a. Using genetic clones, **only environmental effects contribute to variation**.

b. Monozygotic twins are genetically identical so they can be thought of as genetic clones whereas dizygotic twins are genetically different. **Comparison of MZ and DZ twins provides an assessment of the effect of genes versus environment.**

c. Cross-fostering is removing offspring from a mother and placing with several different mothers to randomize the effects of different mothering environments. This is done to **reduce environmental effects** when determining heritability of a trait.

20-21. High heritability indicates that the phenotypic differences observed are due in large part to genetic differences. **Choice b.** would be true.

20-22.

a. **2n +1**

b. A formula here would be $1/4^n$. There would be four pairs of genes controlling kernel color.

20-23.

a. When examining crosses between heterozygous individuals, you can think of the allele frequency as being effectively 0.5. 32 cm represents the extreme phenotype, therefore the proportion of plants with this extreme phenotype are $(1/4)^4 = 1/256 = 0.0039$.

b. When the frequency of the alleles in the population is not uniform as it is in part a. of this question, you have to use the allele frequencies to determine the probability of obtaining a particular genotype. So, in the pool of gametes there is a (0.9)(0.9) probability that an *AA* genotype will form; a (0.9)(0.9) probability of a *BB* genotype; (0.1)(0.1) probability of *CC* genotype; and a (0.5)(0.5) probability of *DD* genotype. To get 32 cm all these allele combinations have to be present in one plant, so **the overall probability is the product of all of these: 0.0016**.

Chapter 21   Evolution at the Molecular Level

Synopsis:

This chapter takes a look at the evolution of the information of life, combining some of our recent knowledge of the structure and function of nucleic acids and the organization of information in the genomes with fossil record and paleontological evidence. Much of the information is speculative and hypotheses are proposed that are thought-provoking. Genome analysis provides information that drives the development of new hypotheses in this field.

Significant Elements:

After reading the chapter and thinking about the concepts, you should be able to:

♦ discuss the merits of DNA and RNA as information carrying molecules

♦ distinguish between synonymous and non-synonymous changes in DNA sequence

♦ describe how duplications can occur

♦ derive a molecular clock value and understand how to use it

♦ think creatively

Problem Solving Tips:

♦ Synonymous base changes in DNA do not change the amino acid that is encoded. Non-synonymous base mutations will result in a different amino acid being present in a to the protein.

♦ Duplications can occur by unequal crossing-over between repeated sequences or transposition of DNA.

♦ Molecular clock rates can be established by dividing the number of base changes by the length of time when the two organisms diverged (according to fossil or paleontological evidence).

Solutions to Problems:

21-1. a. **4**; b .**6**; c. **5**; d. **2**; e. **1**; f. **7**; g. **3**.

21-2. The fact that **the same genetic code directs the cellular basis of life in varied organisms** supports the unity of life hypothesis.

21-3. Statements a. and c.

21-4.

a.   RNA is a **relatively unstable** molecule, so would not be as good a storage information molecule as DNA. RNA **can be degraded by chemical or enzymatic hydolysis** (RNases). RNA is also **not easily compacted**.

b.   Because **RNA has only 4 "letters" in its alphabet, there is less variety than can be achieved in proteins**.

21-5.

a.   The **enzyme consists of an RNA molecule**.

b.   The **enzyme has both an RNA and a protein component**.

21-6. The differences should affect many processes. The **majority of the alterations should be effects on brain structure and function** since the differences between humans and chimps are primarily in the cognitive functions. One way that many functions can be affected is by **alterations in master regulatory genes** (changing function or creating new regulators) that control the expression of many other genes.

21-7.

a.   The rates of non-synonymous substitutions differ for these three genes because there is **a different constraint on the function of each of the proteins**. The function of the histones is very fixed in different organisms, so there is little room for variation, whereas growth hormone may have evolved to interact with other proteins that also have evolved.

b.   The rates of **synonymous substitutions** are more constant between different genes because these base changes **do not affect function of the gene product**.

21-8. The immunoglobulin genes encode subunits of antibodies and **a great diversity of antibodies is needed in the immune response**. The changes that occur are therefore evolutionarily advantageous and would be maintained rather than being selected against.

21-9. The maintenance at a relatively high level suggests that there is **some benefit to the *CF* allele in the heterozygous state**. Recent evidence indicates that the allele may be somewhat protective against certain diseases. The *CF* allele may be similar to the sickle cell allele that protects heterozygous individuals against malaria.

21-10. A single copy of a gene can be duplicated if it is surrounded by **small repetitive sequences and these sequences misalign during meiosis and crossing-over occurs**. The gene that lies between the misaligned sequences will be duplicated (**see Figure 13.10**).

21-11. Because the three-color vision genes are cross-hybridizing, it is l**ikely that they arose by duplication events and then diverged evolutionarily to function as different color receptors**.

21-12. In transposition, **a copy is transposed to a new location in the genome and therefore can be found at a very different location than the original copy. Unequal crossing-over produces extra copies adjacent to the original copy**.

21-13.

a.   The *A* allele in humans and the *M* allele in mice are 600 base pairs different and mice and humans are 60 million years apart in evolutionary time. The clock rate we can derive from this information is that over 1 million years, 10 bp on average were changed. The *A* allele has 3000 base differences compared to *Xenopus* (*X* allele). Using the mouse-human numbers, we estimate that **frogs and humans are 300 million years apart. That means that frogs and mice are separated evolutionarily by 240 million years**.

b.   **Two duplications** must have occurred to lead to alleles *B* and *C* in humans. Because the *C* allele has 300 bases different from *A* the **A - C duplication occurred 30 million years ago** and has undergone base changes over that time period. The *B* allele has fewer differences compared with *C* so was duplicated and evolved more recently than *C*. Using our clock rate, **the B allele arose by duplication 1 million years ago**.

c.   **The duplication of B resulted in a copy on a different chromosome** (or far enough away on the same chromosome so that it is unlinked). The duplication **could have been a transposition** type of event. Because *C* **is a tandem copy, it could have arisen by a misalignment of repeated sequences** around the original copy followed by a crossing over.

21-14. The **plasmid could have been introduced** (via natural transformation or conjugation- see chapter 13 for more information on these processes) **from another species in the wild**.

21-15. This aberrant phylogeny of the glucose-6-phosphate isomerase could indicate that **this particular gene was introduced from a different species**.

21-16. Chromosome mutations have the potential for much more dramatic effects. For example, regulatory regions of one gene get juxtaposed to regions of other genes; several genes can be lost by deletion.

21-17. a. **exons**; b. **genes**.

21-18.

a. **In vertebrates there are four *HOX* gene family clusters** compared with one cluster in *Drosophila*, for example. This fits well with the tetraploidization hypothesis since the two doublings should lead to four copies of the genes.

b. The variation within the four gene family clusters **could be the result of duplications within the gene family in certain lineages**.

21-19. LINES or SINES can mediate genome rearrangements by unequal crossing-over between elements and they can contribute regulatory elements adjacent to a gene.

21-20. The size of dinucleotide repeat sequences can change **by unequal crossing-over between the repeated sequence within the element**. There is also some variation that can occur by slippage during DNA replication.

21-21.

a. **SINES or LINES.**

b. **centromeric satellite DNA**; although we don't know the precise function, it appears to be needed for proper centromere function in higher eukaryotes.

21-22. a. **2**; b. **4**; c. **1**; d. **3**.

Reference A   *Saccharomyces cerevisiae*: Genetic Portrait of Yeast

Synopsis:

Saccharomyces cerevisiae is a model eukaryotic cell for studying basic cell processes. The following features of yeast genetics and molecular biology make it a powerful model system.

- It is easy to screen for rare mutations.

- Tetrad analysis allows the direct analysis of all the products of a single meiosis in one ascus. $2^+:2^-$ segregation indicates that a mutation in a single gene causes the phenotype under study.

- Mutants can be isolated in either haploids or diploids. Recessive mutations are immediately recognized in haploids. If mutations are being sought in genes controlling an essential process, mutagenesis can be done in diploids. The recessive mutation can then be identified in the haploid cells after sporulation. Sporulation (meiosis) can be induced in a culture by specific growth conditions.

- Yeast cells are easy to transform with DNA.

- Several plasmid vectors have been developed that can easily be grown in either yeast or bacteria. Artificial linear chromosomes allow the study of chromosome structure in yeast.

- Yeast chromosomes can be separated by size using pulse-field gel electrophoresis.

- Gene replacement allows a researcher to change a gene specifically by transforming with a replacement copy (engineered variant) that is forced to recombine into the yeast genome and in doing so to replace the original copy.

Significant Elements:

After reading the chapter and thinking about the concepts, you should be able to:

- predict the effects of mutations in the components of the mating system (production of pheromone, recognition of pheromone, signal transduction) and be able to apply these analytic principles to other processes

- test mutations to determine if they are in the same or different genes (complementation analysis); determine if mutations in different genes act in the same or different pathways (epistatic analysis)

- think about how genetic analysis is done in this organism, including: isolation of mutants, characterization of mutations (complementation and epistasis analyses), mapping a gene, cloning a gene, and making a mutant cloned copy of a gene to reintroduce into the genome

Problem Solving Tips:

♦ Reason through the problems; get used to making hypotheses and interpreting how results fit a hypothesis

♦ Some of the longer problems in this chapter are real research problems that were carried out essentially as described to you. They require that you integrate your knowledge of classical genetics, molecular genetics, genome analysis. This is a good time to try to think like a geneticist. Think about how you would discover the genes involved and how their gene products may act- as regulators, messengers, structural components, etc.

Solutions to Problems:

A-1. a. **3**; b. **4**; c. **5**; d. **7**; e. **2**; f. **1**; g. **6**.

A-2. Assuming an average protein coding region size of 500 amino acids, there would be 1500 nucleotides per gene. 6000 open reading frames (protein coding regions) $\times$ 1500 = 9,000,000 bp or 9,000 kb in protein coding DNA. This value is **3/4 of the total genome size** (9000/12,000). This result means that yeast genes in general do not have very large introns, and that the genes are not separated by large expanses of DNA. By way of comparison, similar calculations done for the human genome would show that <10% of human DNA is protein-coding.

A-3. You could isolate DNA from cultures of each of the ten clones, use pulse-field gel electrophoresis to separate the chromosomes by size and compare the sizes of chromosome XII in each isolate.

A-4.

a. If the histidine biosynthesis genes are coregulated, the mRNA for each gene should be induced by the same conditions. For amino acid biosynthesis genes the genes are usually turned on by starvation for the appropriate amino acid. Cells would be grown with and without histidine in the medium, the RNA from the two cultures isolated and separated by electrophoresis, and the RNAs on the gel would then be transferred to nitrocellulose filters where they could be probed with DNA representing each of the biosynthetic genes. In other words, this would constitute a **Northern blot experiment. If the various genes were coregulated, the mRNA for each would probably be induced by starvation for the amino acid.** (Note: Coregulation of genes in yeast is not the same as cotranscription as seen in prokaryotic operons. In prokaryotic operons, all the genes in the operon are cotranscribed into a single mRNA. Operons are found almost exclusively in prokaryotes. In eukaryotic cells such as yeast, genes that are coregulated transcriptionally are transcribed into different mRNAs.)

b. **You might expect to find similar DNA sequences in the regulatory regions to which transcriptional regulators would bind so as to regulate expression.** In this way, one or more regulatory proteins would be able to regulate the expression of several genes simultaneously.

A-5.

a. You could **separate all the chromosomes by pulse-field gel electrophoresis,** transfer the separated chromosomes to nitrocellulose filters (so as to **make a Southern blot**), and then do **a hybridization** to the blot **using probe DNA containing a subtelomeric repeat sequence.**

b. You could **use the cloned piece of DNA as a probe in a hybridization with a filter containing DNA from yeast chromosomes that had been separated by pulse field gel electrophoresis.** Note that this problem is of greater historical than present-day interest, because not only has a physical map of the yeast genome been made, but the entire nucleotide sequence of the *S. cerevisiae* genome has been determined. Thus, if you have a cloned piece of DNA and want to know what chromosome the cloned DNA came from, the simplest method would now to determine the DNA sequence of a small part of the clone and match it on the computer (with a BLAST search) to the yeast genome.

A-6.

a. **The 2^+:2^- segregation pattern is characteristic of a monohybrid segregation of the alleles of a gene from a heterozygote during meiosis.** Because the analysis is done in *Saccharomyces cerevisiae*, a fungal species in which the four meiotic products are packaged together in an ascus, the ratio is easily seen. The 2:2 segregation also indicates that the strain is heterozygous.

b. The screen identified the chromosome segregation defect in a diploid cell, which we know from the sporulation results is heterozygous. The phenotype seen in a heterozygous diploid is the dominant phenotype, so in this case **the mutant allele is dominant.**

A-7. The chromosomes containing the MAT locus could align in any combination during meiosis, leading to the following combinations of *a* and α alleles in the diploid spores: *a/a* **which has an *a* mating phenotype,** *α/α* **which has an α mating phenotype, and** *a/α* **which is a nonmating sterile strain.** We assume here that in the first meiotic division in the tetraploid, the four chromosomes pair as two bivalents (**see <u>Chapter 13</u>**). The actual proportions of the three types of diploid spores you would expect is more difficult to calculate. If the MAT locus were located exactly at the centromere, you would expect the ratio to be 1 *a/a* : 4 *a/α* : 1 *α/α* (**see <u>Figure 13.33</u>**). However, as shown on **<u>Figure A.3</u>**, the MAT locus is actually located about 25 cM from the centromere of chromosome III. As the problem did not ask for these proportions, we will not attempt this difficult calculation here (discretion is the better part of valor).

A-8.

a.   **ii, iv, v**

b.   **i, ii, iv, vi**

c.   **i, iii, iv, vi**

A-9.

a.   In an cell, α1 + Mcm1 protein act as a positive regulator of α-specific gene transcription; α2 + Mcm1 protein act to repress *a*-specific genes. In an *a* cell, Mcm1 protein alone transcribes *a*-specific genes. In an *a/α* strain, Mcm1 has no effect.

b.   In the α *mcm1⁻* strain, there would be no transcription of α-specific genes and no repression of *a*-specific genes, so the strain behaves and mates as an *a* strain. In the *a mcm1⁻* strain, there would be no *a*-specific gene expression (for example, no *a* pheromone), so the strain would be sterile. The *mcm1⁻* mutation would have no effect in an *a/α* strain. It would be sterile as is the $Mcm1^+$ *a/α* strain.

A-10.

a.   **Complementation.** You are in essence putting two mutations into the same *a/α* diploid cell, and then asking what phenotype these diploid cells have in terms of their growth on the four different substances. If the diploid fails to grow on one of the intermediates, it means that both mutations affect the same gene that determines one of the steps in the pathway.

b.   **The 100 haploid spores that grew on minimal medium must contain wild-type alleles of two different genes.** The $1^+/1^-\ 2^+/2^-$ diploid, for example, would produce haploid spores with four different genotypes: $1^+\ 2^+$ (which grows on minimal), $1^+\ 2^-$, $1^-\ 2^+$, $1^-\ 2^-$ (none of which grow on minimal media) if the genes are unlinked. One-quarter of each type of spore would be produced, so there would be 100 $1^+\ 2^+$ spores out of 400 spores. Mutations that are in the same gene or are in closely linked genes will only produce $1^+\ 2^+$ and $1^-\ 2^-$ spores by recombination between the mutations; this will be a low frequency event dependent on the distance between the mutations. **The combinations of mutations that only produce a low number of spores (< 5) must be closely linked mutations.**

c.   **There are 4 complementation groups: [7,3]; [4,5]; [2,6]; [1].** These are identified by looking at the growth of the diploids on different intermediates.

d.   sleepan ⟶ happan ⟶ dopan ⟶ sneezan
       [7, 3] [4, 5]   [1]   [2, 6]

Because none of the mutants grew on sleepan, this is the earliest intermediate. On happan, [7,3] and [4,5] grow, but because these mutations are in two different complementation groups they represent two different genes. **The enzyme that converts sleepan to happan is therefore composed of at two different subunits (heterodimer).** Dopan is the next intermediate, with the gene identified by mutation 1 involved in the production of dopan. The gene identified by mutations 2 and 6 encodes the enzyme to convert dopan into sneezan.

A-11.

a.   **You could do a Southern hybridization using the *SUC2* DNA as a probe in a hybridization with a filter containing genomic DNA from the other *Saccharomyces* strains.** Alternatively, and more rapidly, you could design PCR primers that would amplify the *SUC2* gene, and then attempt to amplify homologous DNA from the other strains. If the *SUC2* gene were present in one of the other strains, you would expect to amplify a DNA fragment of the same size as that amplified in the first strain. This would only be possible if the DNA sequences corresponding to the PCR primers were identical in the two strains.

b.   **pseudogenes**

c.   **There are two transcripts made from the *SUC2* locus.** The smaller one is produced when glucose is present or absent; the larger one is present only when glucose is absent. The larger transcript requires the *SNF1* gene product, but the smaller one does not require *SNF1*. **The data imply that it is the larger transcript that encodes the functional SUC2 protein.**

A-12.

a.   A complementation test would be done by **crossing haploid strains to each other and examining the phenotype in the diploid.** The mating type of the haploids crossed to each other would have to be opposites so that *a* and α cells were mated. You would ensure that the original haploid parents would not be exposed to UV light before or during the mating (so that the cells would remain alive). To examine the phenotype, it would be best to examine replica plates of diploid colonies; one replica exposed to UV light and the other grown without exposure. Replica plating would establish that failure to grow was in fact due to the UV treatment.

b. **Three loci.** Mutations 1, 2, 4 are in the same gene; 3, 5 are in another gene; 6 is in another gene.

c. **The data imply that there are 3 different pathways.** Mutations 1 and 3 are in genes in two different pathways; 3 and 6 are in different pathways; and 1 and 6 are also in different pathways. Thus each mutation affects a different repair pathway. You should note that the theory underlying this problem is somewhat simplistic. The theory assumes that each of the mutations is a null mutation that completely blocks the function of an entire repair pathway. Thus, mutations in two different genes involved in the same repair pathway would have the same effect on UV sensitivity in a double mutant as in either single mutant. Only when two different independent pathways are shut down would the UV sensitivity of the double mutant exceed that of the single mutant. However, one could easily imagine that a particular mutation would only partially inhibit a pathway rather than completely shut it down.

d. **Conduct a Northern blot experiment that compares the mRNA from cells grown under normal conditions with the mRNA from cells of the same genotype exposed to UV.** Isolate the RNA independently from both populations of cells, separate the RNAs in each sample according to their sizes by gel electrophoresis, transfer the RNAs to filters, and hybridize the blots using specific gene probes to determine if expression of the genes is regulated by the UV exposure.

e. You would first transform the library into yeast cells, so that you would have a collection of colonies, each of which harbored a different piece of yeast DNA fused to *lacZ*. You would then compare the expression of *lacZ* in your library of clones before and after exposure to agents causing DNA damage. Any of the colonies that show increased expression of β -galactosidase (as seen by an increase in blue color on petri plates containing the substrate X-gal) after exposure to these agents would contain fusions in which a regulatory region of a UV-inducible yeast gene had been cloned adjacent to the *lacZ* gene.

Reference B   *Arabidopsis thaliana*: Genetic Portrait of a Model Plant

Synopsis:

Arabidopsis thaliana serves as a useful plant model for genetic and molecular analyses. While many of the basic cellular mechanisms are the same in plants and animal cells, the events during development in plants are quite different from animals. Even so, the unique developmental events in plants often have underlying mechanisms that are analogous to those found in animals, fungi, etc. The following list of characteristics of *Arabidopsis* indicates why it is a useful model for plant development, and points out some genetic characteristics and manipulations that are unique to this plant.

♦ *Arabidopsis* has a fast generation time compared to other plants.

♦ Each plant can produce a large number of seeds, which is advantageous for mutant screening.

♦ Researchers can manipulate whether a plant undergoes self-fertilization or an outcross.

♦ Mutant screens can be done using T-DNA or other transposable elements to create mutations. This facilitates the identification of the altered gene at the DNA level.

♦ Very little repetitive DNA is present in *Arabidopsis,* so the genome is very small for a plant. Most other plants have very large genome, mostly because they contain massive amounts of repetitive DNA.

Significant Elements:

After reading the chapter and thinking about the concepts, you should be able to:

♦ think about how genetic analysis is done in this organism including: mutagenesis, isolation of mutants, characterization of mutations (complementation and epistasis analyses), study of mutant phenotypes, and gene

♦ understand how to interpret data obtained with molecular methods such as gene cloning, PCR, nucleic acid hybridization, the use of DNA markers for mapping, the use of fusions to study gene expression, and making a mutant copy of a cloned gene for reintroduction into the genome

♦ describe the steps that would be involved in a search for the gene affected by a T-DNA induced mutation

Problem Solving Tips:

♦ Remember that a seed contains a plant embryo. When seeds are mutagenized, you are not mutagenizing gametes. Gametes are produced in the reproductive structures of an adult plant.

♦ Some of the longer problems in this chapter are real research problems that were carried out essentially as described to you. They require that you integrate your knowledge of classical genetics, molecular genetics, and genome analysis. This is a good time to try out thinking like a geneticist. Consider how you would discover the genes involved, and how you would determine how the products of these genes might act - whether as regulators, messengers in a signal transduction system, structural components, etc.

Solutions to Problems:

B-1. a. **3**; b. **1**; c. **4**; d. **5**; e. **2.**

B-2.

a. The shorter generation time of *Arabidopsis* **allows the geneticist to study many more generations in less time.** In corn or peas, a researcher would be able to generate at most a few generations in a year.

b. The very large number of seeds produced is **an advantage for a geneticist searching for rare mutations.**

c. The small genome size is **advantageous for molecular analysis,** in particular for cloning genes. There is very little repetitive DNA to have to deal with. Historically, the small genome size made it feasible to determine the complete DNA sequence of the genome; *Arabidposis* is the first plant for which this goal has been achieved.

B-3. Most of the extra DNA in tobacco and pea is repetitive DNA that presumably is not necessary for the physiology of the plant. **Most repetitive DNA does not encode genes**, so it does not contribute to the proteins made in the organism. In large part due to the paucity of repetitive DNA, genes in *Arabidposis* are packed more closely together (with less intervening DNA), and the number and sizes of introns are smaller, than in other plant species.

B-4.

a. **Plants of different ecotypes that show the extreme phenotypes (the largest compared to the smallest seed set) could be crossed to produce hybrid F_1 plants, that are then self-fertilized to produce the F_2.** If the plants in the F_2 generation show three discrete phenotypes, there is a single gene that determines the trait (that is, the three phenotypes would be associated with the

AA, Aa, and *aa* genotypes). If these plants displayed a few more clear-cut phenotypic classes, then two or three genes would be involved. If the F_2 plants have a range of phenotypes in a continuum, the trait is determined by several genes and might also have an environmental component. (**Review Chapter 3** for a description of how multiple genes and the environment can interact to produce a range of phenotypes.)

b. **Different ecotypes contain different alleles of molecular markers** that can be used to identify regions containing genes involved in determination of the phenotype. (**See Chapter 20** for a discussion of quantitative trait analysis).

B-5.

a. Endosperm is $3n$ and is formed by fertilization of one sperm nucleus and two $1n$ nuclei in the ovule. The genotype of the endosperm will be similar that of the embryo, in the sense that the endosperm produced from an outcross will yield a combination of genetic information from both parents.

b. The zygote is $2n$ and is formed by fertilization of one sperm nucleus with the $1n$ egg; the genotype will again be a combination of genetic information from both parents.

c. The embryo sac cells are $2n$ and they are part of the female parent plant. The embryo sac cells will have the genotype of the female parent plant on which they are found, and thus have a very different genotype than the embryo.

B-6. A geneticist usually wants to study mutations that are in the germ cells and can therefore be transmitted to progeny and maintained as stocks for further study. The cells that were mutagenized are the embryonic cells in the seed that will divide and develop into the plant. If the mutagen were applied very early, soon after fertilization, it might only affect one of the two DNA strands on a chromosome in the original zygotic cell, in which case one daughter cell would be mutant and the other wild type after the first mitotic division. Alternatively, if the mutagen were applied later in embryonic development, the mutation causing the phenotype might have occurred in only one of many embryonic cells present at that time. In either case, the plant growing from the mutagenized seed would be a mosaic, with some cells carrying the mutation, and others being wild type. The first generation mutations might only affect tissues that could not become germ cells, and the trait would therefore not be transmissible.

B-7. a. **2**; b. **3**; c. **1.**

B-8.

a. The steps you would apply are in order: **iv, iii, i, ii.** Step iv would suggest a candidate gene that might be involved in the process you wanted to study. Step iii provides a group of mutant lines, one of which might harbor a mutation in the candidate gene. Step I would identify the particular mutant line with a T-DNA insertion in the vicinity of the candidate gene; you would be looking for a PCR product that could only be made from genomic DNA in which T-DNA were adjacent to the candidate gene. Once you found such a mutant line, you would want to make the mutation homozygous as in step i, so as to determine whether there was a recessive phenotype associated with the mutation in the candidate gene.

b. **The technique you would use is summarized on <u>Figure B.12.</u>** The DNA you would transform into *Arabidopsis* would have an inverted duplication of the entire gene of interest or even of a part of the gene. This inverted duplication would be under the control of a promoter that can be expressed in all or most *Arabidopsis* tissues. Cells bearing the transgene will express an RNA molecule that will fold on itself to form a double-stranded RNA (dsRNA) with a hairpin loop between the inverted repeats. The *Dicer* complex within *Arabidopsis* cells will cleave this long dsRNA into 21 bp fragments of dsRNA, promoting the degradation of the homologous mRNA in the cell. Thus, the expression of the gene of interest will be silenced, even though a true mutation in that gene is not available.

B-9. You could do manipulated crosses where you collect pollen from the sterile plant and use it to fertilize a wild-type plant. The reciprocal cross would also be done: pollen from a wild-type plant crossed with the sterile mutant. From the results you would know if the defect was in the male or female reproductive structures (assuming that one of the crosses would produce progeny while the other cross could not).

B-10. To look for an homolog for a gene in other plants, you would **perform a whole genome Southern blot**. You would first make genomic DNA from the petunias, snapdragons, and potatoes; cut these genomic DNAs with restriction enzymes, separate the resultant restriction fragments by gel electrophoresis according to their size, and then transfer the DNA fragments to nitrocellulose filter paper. You would then do a Southern blot hybridization using the *APETALA* gene from *Arabidopsis* as a probe. Two other approaches are also conceivable. First, if other investigators have determined the sequences of parts of the genomes of these other plants, or EST sequences from individual cDNA clones from these organisms, you could perform computerized BLAST searches to see if any of these nucleic acid sequences could encode a protein related to that encoded by *APETALA*. Second, you could try to amplify *APETELA*-like sequences from the genomes or cDNA libraries of the other species using PCR. However, this would only work if the protein were highly conserved in evolution; even if all the amino acids encoded by the sequences in the PCR primers were identical to those in the *Arabidopsis* protein, you would still have to account for the degeneracy of the genetic code and devise sets of primers that could account for all the possible codons for those amino acids.

B-11.

a. **You could examine the gravitropic response in an auxin mutant** (defective in auxin production). If the gravitropic response is lacking in an auxin mutant, the response must be either directly or indirectly mediated by auxin.

b. There are a several different ways you could examine the distribution. You could **perform a Northern blot** analyzing the expression of each auxin-regulated mRNA. You would cut the plant up into different parts (such as the hypocotyls, flowering stalk, and root), and then grind up each tissue sample in order to isolate its RNAs. You would then fractionate these RNAs according to their sizes on a gel, transfer the RNAs to nitrocellulose filter paper, and hybridize using an auxin-regulated gene as a probe. For a more accurate picture, you could use DNA for an auxin-regulated gene as a probe against preparations of tissue from the plant (that is, **an *in situ* hybridization**). Or, you could make a fusion between an auxin-regulated gene and a reporter gene such as GUS or luciferase (which emits light when given substrate luciferin), transform the fusion back into the plant, and look at the distribution of where the reporter gene is expressed in the intact plant.

B-12.

a. The *ctr1⁻* constitutive mutation in a signaling pathway should not be affected by what is happening in the ethylene biosynthesis. This is because in mutant plants, the pathway is turned on independently of the production of ethylene. Thus, **there would still be expression of the ethylene response** in *ctr1⁻* mutants even if ethylene biosynthesis were inhibited.

b. Because inhibitors of biosynthesis affect these mutants, the **mutations are in genes in the biosynthetic pathway**.

c. **You could look for additional copies of the *ETR1* gene by Southern hybridization using the *ETR1* gene as a probe in a hybridization to a blot containing genomic *Arabidopsis* genomic DNA.**

d. The protein encoded by *APETALA2* is known to be a transcription factor, and the information provided indicates that the EREBP proteins are also transcription factors that bind to bases in the promoter region of several ethylene-responsive genes. **These proteins might thus share some similar features (types of motifs) that would allow DNA binding or transcriptional activation/repression.**

e.   **To control expression of the ethylene response genes, you would need to make a fusion of an inducible regulatory region with the ethylene response gene you want to induce. The inducible regulatory region would have to be one that you could turn on in the plant cells where you wanted expression.** Identifying a suitable inducible regulatory region might take some effort. For example, you might want to look at many plants transformed with a modified T-DNA that carries a promoterless *GUS* gene reporter. In each transformed plant line, *GUS* will be expressed in specific parts of the plant based upon regulatory regions in the vicinity of the T-DNA insertion. This "enhancer-trapping" methodology is shown in **Figure B.10**. You would look for a plant that displayed *GUS* expression in the tissues in which you wanted to induce the ethylene response gene. You would then take the regulatory region near the T-DNA insertion, fuse it with the ethylene response gene, and transform the construct back into *Arabidopsis*.

Reference C   *Caenorhabditis elegans*: Genetic Portrait of a Simple Metazoan

Synopsis:

♦ The nematode *C. elegans* has become a very important model for genetic analysis in multicellular organisms based on the following characteristics:

♦ Researchers can grow large numbers of worms quickly and thus screen for rare mutants.

♦ Self-fertilization occurs in hermaphrodites. This greatly simplifies the identification of recessive mutations, because a hermaphrodite heterozygous for the mutation can be allowed to self-fertilize so as to produce homozygous recessive progeny. This is much harder in other animal species where matings would have to be set up between two heterozygous individuals bearing the mutation. In other species, the two individuals with the mutation may have very different genetic backgrounds, which could potentially cause complications; but in self-fertilizing nematodes, this is not a problem.

♦ *C. elegans* has a small genome size, which means that gene cloning and genome analysis is easier. In fact, the first species of multicellular animal with a completely sequenced genome was *C. elegans*.

♦ Because the worms are essentially transparent, investigators can easily view internal organs and cells in a living worm via microscopy.

♦ The precise lineage of cells during development is defined and known, allowing the design of intricate experiments to investigate almost any developmental event.

♦ Researchers can kill specific cells via laser ablation at any point in development, thus defining the importance of those cells and their descendents to various biological processes.

♦ Free duplications (extrachromosomal pieces of DNA) can be introduced to create mosaic worms in which some cells contain the free duplication while others do not.

Significant Elements:

♦ After reading the chapter and thinking about the concepts, you should be able to:

♦ set up crosses to construct strains of worms with specific genotypes

♦ interpret the results of selfing in a hermaphrodite and outcrosses between hermaphrodite and male.

♦ think about how genetic analysis is done in this organism, including: isolation of mutants, characterization of mutations (complementation and epistasis analyses), mapping a gene, cloning a gene, and making a mutant cloned copy to reintroduce into the genome

Problem solving tips:

♦ When a hermaphrodite population is mated with a male population, the progeny will arise from self-fertilization of hermaphrodites as well as the outcrosss of males to hermaphrodites.

♦ Males (XO) spontaneously arise by a low frequency of non-disjunction during gametogenesis in XX hermaphrodites.

♦ Some of the longer problems in this chapter are real research problems that were carried out essentially as described to you. They require that you integrate your knowledge of classical genetics, molecular genetics, and genome analysis. This is a good time to try to think like a geneticist. Think about how you would discover the genes involved and how their gene products may act- as regulators, messengers, structural components, etc.

Solutions to Problems:

C-1. a. **2**; b. **5**; c. **4**; d. **7**; e. **6**; f. **1**; g. **3**.

C-2. Five times the genome size (100Mb) is 500 Mb or 500,000 kb. With an average cosmid size of 50 kb, you would need **10,000 clones** to make sure that you have a coverage of 5 times the genome.

C-3. Hermaphrodites have the advantage of self-fertilization. It is therefore easier to recover recessive alleles than if there was only uncontrolled outcrossing. (In other animals, you would require an additional generation of crosses to obtain zygotes homozygous for a recessive mutation because it is necessary to first get the mutation in individuals of both sexes.) In addition, mutants with some pretty severe defects (such as paralysis) can be isolated and maintained because the worms do not have to move to mate.

C-4.

a.   **ii, iv** [combining two mutations into the same strain might allow you to infer if the genes work in the same pathway by looking at the phenotype of the double mutant (ii) or to do complementation analysis (iv)]

b.   **ii, v** [RNAi could be used to phenocopy a mutation in a gene; if RNAi of two genes yielded the same phenotype, the genes could possibly be in the same pathway (ii); and you would be able to tell if any gene subjected to RNAi affected vulval development (v)]

c. **i, ii, v** [A suppressor mutation might bypass a normal mechanism disrupted by a mutation in another gene (i), or it might be a mutation in a gene that works in the same pathway (ii) or is otherwise needed for proper vulval development (v)]

d. **iii** [Cell ablation is specifically used to determine the importance of a cell for development (iii)]

C-5. Choice b would be expected if trans-splicing occurred. As shown on **<u>Figure C.3</u>**, trans-splicing in *C. elegans* involves either of two splice-leader sequences that are added on to the 5'-ends of many different mRNAs. In addition, the genome contains very few copies of the genes encoding the splice leader sequences.

C-6. You would want to ablate the anchor cell at a time after the anchor cell and the vulval precursor cells have been formed, but prior to the time at which the fate of the vulval precursor cells has been established. As described in the text on, this would be during the L3 larval stage.

C-7. Increasing the amount of *lin-4* mRNA would probably reduce the Lin14 protein level further because *lin-14* mRNA is believed to be a negative regulator of *lin-14* mRNA translation. This could lead to cells differentiating at an inappropriately early time since Lin14 protein concentration acts as a clock during development. The cells would differentiate precociously at an inappropriate time of development earlier than would occur in wild type. This should create a T cell lineage diagram similar to that seen in a *lin-14* loss-of-function mutation, as shown in **<u>Figure C.16a</u>**.

C-8. You could probe DNA from newly-induced mutants and from wild-type Bergerac worms with *unc-22* cloned DNA on a whole genome Southern blot. If there was an insertion of a Tc1 element, the fragments that hybridized in the mutant and wild-type would be different. You would then want to confirm that the new fragments seen in the mutant strain DNAs also contain Tc1-homologous sequences to show that the insert is within the *unc-22* gene. This could be accomplished by rehybridizing the blot with cloned Tc1 DNA. Alternatively, you could use PCR with one primer consisting of a Tc1 sequence and the other being an *unc-22* sequence. If a fragment is produced using these primers, the Tc1 element must have transposed in the vicinity of the *unc-22* gene. Because you don't know at the outset the location and orientation of the Tc-1 insert with respect to the *unc-22* gene, you might need to try several pairs of PCR primers in order to see a PCR product fragment.

C-9.

a. **Cross the wild-type males to the homozygous *dpy-10⁻ / dpy-10⁻* hermaphrodites** (or to homozygous *daf-1⁻ / daf-1⁻* hermaphrodites; it doesn't actually matter). Either of these is the only kind of cross you could perform in any event. In the answers to the remainder of this problem, we will assume that the first cross involved *dpy-10⁻ / dpy-10⁻* hermaphrodites; you should remember that these animals are also *daf-1⁺ / daf-1⁺*, and that the *daf-1⁻ / daf-1⁻* hermaphrodites are also *dpy-10⁺ / dpy-10⁺*.

b. **The wild-type male progeny of the first cross would be crossed to the *daf-1⁻ / daf-1⁻* hermaphrodites.** In other words, the second cross would be *dpy-10⁺ / dpy-10⁻* x *daf-1⁻ / daf-1⁻*.

c. **The final step is the self fertilization of hermaphrodite progeny from cross 2.** You will have to pick several hermaphrodites because half of the them will be doubly heterozygous as desired (that is, with genotype *dpy-10⁺ / dpy-10⁻ ; daf-1⁺ / daf-1⁻*), but the other half of the hermaphrodites will be *dpy-10⁺ / dpy-10⁺ ; daf-1⁺ / daf-1⁻*. These two types of worms cannot be distinguished because they are both wild-type hermaphrodites. Because only half of the wild-type hermaphrodites will have the genotype required to produce a doubly homozygous worm, only (1/2)(1/4)(1/4) or **1/32 of the progeny of this final self-fertilization will have the *dpy-10⁻ / dpy-10⁻ ; daf-1⁻ / daf-1⁻* genotype desired**. Almost all of the progeny will be hermaphrodites because the last step is a self cross rather than an outcross.

C-10.

a. The first cross is wild-type males (*unc54⁺ daf-8⁺ / unc54⁺ daf-8⁺*) x *unc54⁻ daf-8⁺ / unc54⁻ daf-8⁺* hermaphrodites. To set up the second cross, **you would use the wild-type male progeny of this cross (*unc54⁺ daf-8⁺ / unc54⁻ daf-8⁺*)** and cross them to the *unc54⁺ daf-8⁻ / unc54⁺ daf-8⁻* hermaphrodites.

b. The progeny from the hermaphrodite self-fertilization are *unc54⁺ daf-8⁻ / unc54⁺ daf-8⁻*. The progeny from the outcross are

Males: *unc54⁺ daf-8⁺ / unc54⁺ daf-8⁻* and *unc54⁻ daf-8⁺ / unc54⁺ daf-8⁻*
Hermaphrodites: *unc54⁺ daf-8⁺ / unc54⁺ daf-8⁻* and *unc54⁻ daf-8⁺ / unc54⁺ daf-8⁻*
All of these progeny are phenotypically wild-type.

c. **The wild-type hermaphrodites produced in part b will be selfed to get the double homozygotes.** The hermaphrodites that will be able to produce the desired animals are those heterozygous for both genes (that is, the *unc54⁻ daf-8⁺ / unc54⁺ daf-8⁻* animals). However, one-half of the wild-type hermaphrodites produced by the cross described in part b are not heterozygous for both genes; instead, they are $unc54^+$ $daf\text{-}8^+$ / $unc54^+$ $daf\text{-}8^-$. As these genotypes cannot be distinguished phenotypically, **only one-half of the self-crosses of wild-type hermaphrodites will be able to make doubly homozygous progeny.**

d. **Because the *daf-8* and *unc-54* genes are linked, there has to be recombination (crossing-over) between these genes during meiosis to generate gametes containing both *unc-54⁻* and *daf-8⁻* on the same chromosome.**

e. The genes *daf-8* and *unc-54* are 18 map units apart, so 18% of the <u>gametes</u> will be recombinant. Of these recombinants, 1/2 (= 9%) will contain the $unc54^-$ $daf\text{-}8^-$ combination of alleles. The probability of two gametes with this genotype combining is (.09)(.09). Because only one-half of the wild-type hermaphrodites are the double heterozygotes as described in part c, **1/2 (.09)(.09) or .004 or 0.4% of the progeny will have the desired double homozygous genotype.**

C-11.

a. The candidate gene approach would involve testing the suspected gene (here, the previously identified K^+ channel gene) in the mutant worm to determine if the defect is in the channel gene. **The candidate gene would be cloned from the mutant or amplified from the mutant by PCR, and the sequence of the gene from mutant animals would be determined. A change in the sequence of that particular K^+ channel gene in the mutant suggests that the suspected gene may be involved.** This evidence would be stronger if the mutation were a nonsense or frameshift mutation that would cause a truncation of the channel protein. This is because it is hard to predict if a particular missense mutation is or is not compatible with the function of the protein, while it is highly likely that a truncated protein would be nonfunctional. If you could examine a number of independent alleles, all of which showed a different change in the sequence of the gene, the case for saying that lesions in the K^+ channel gene were responsible for the phenotype would be even stronger. **More conclusive evidence would come from introducing a wild-type copy of the cloned candidate gene into worms by transformation and look for rescue of the phenotype. Another alternative is to use the RNA interference approach. You would inject worms with double stranded RNA (made in vitro) corresponding to the gene and see if this results in a phenotype similar to that of the mutant.**

b. **You could determine if the Tc1 insertion occurred in the ion channel gene by Southern hybridization analysis.** The probe would be a normal copy of the gene channel gene and you would look for an alteration in the restriction pattern (an addition of DNA equivalent to the size of the Tc1 element) of the channel gene compared to wild-type. **Alternatively, you could use**

PCR with one primer consisting of a Tc1 sequence and the other primer being a sequence from the ion channel gene. If amplification occurs and a fragment is produced, the Tc1 element must be in the vicinity of the ion channel gene.

c. **The mutation would probably have a dominant phenotype because it does not result in lack of a channel but produces a channel with an altered function.** In the heterozygote, it is likely that some of the channels in each cell would remain open inappropriately if those channels were constructed from the mutant protein.

d. You could compare the phenotypic response to isoamyl alcohol and diacetyl alcohol in the wild-type and *egl-2⁻* mutant strains. That is, would the worms be attracted toward or repulsed away from these two substances? For example, a finding that wild-type worms were attracted toward isoamyl alcohol, while *egl-2⁻* mutant worms did not react to this substance would support the conclusion that the K⁺ channel encoded by the *egl-2* gene was involved in chemosensory function in the AWC neurons. Such a result would be consistent with the fact that the *egl-2* gene is in fact expressed in these cells.

C-12.

a. **The wild-type mouse *netrin-1* gene could be cloned and transformed into an *unc-6⁻* / *unc-6⁻* mutant worm to determine if the mouse gene could substitute for the worm gene and correct the uncoordinated phenotype.** The coding region of the mouse gene should be cloned downstream of a copy of the promoter of the worm gene to ensure that the mouse gene would be transcribed in the right tissues and in the right amounts.

b. **You would compare the migration of these other cells in *unc-5⁻* and *unc-6⁻* homozygous mutants with their migration wild-type worms.** This analysis would of course require that the mutant animals survive; this is indeed the case because the *unc* mutations only cause a lack of coordination.

c. **You could identify interacting proteins by isolating suppressors of the partial loss-of-function *unc-5⁻* mutations. The new mutations, in interacting genes, compensate for the defect in the *unc-5⁻* mutants.** As described in <u>Chapter 19</u>, the genes identified in this type of suppressor screen could encode proteins that interact directly with the UNC-5 receptor, but they could also encode proteins that have different functions. For example, a gain-of-function mutation in a gene encoding a transcription factor that would be the ultimate target of a signal transduction pathway initiated by binding of UNC-6 netrin to the UNC-5 receptor would suppress the *unc-5* mutation. Such a transcription factor might not directly touch the UNC-5 protein; instead, its activity might be only indirectly regulated by UNC-5. Much additional research would be required to understand how the suppression takes place.

d. **You would clone the protein-coding part of the *unc-5* gene next to a promoter that was expressed at all times in all cells (a housekeeping type of gene, for example).**

Reference D *Drosophila melanogaster* : Genetic Portrait of a Fruit Fly

Synopsis:

Drosophila has been a very important model system for studying chromosome structure and development because of the ease of doing genetic and cytogenetic analyses in this organism. Our understanding of pattern development, an underlying theme of development, is based on the pioneering genetic studies in *Drosophila*. Many features of *Drosophila* biology have been used and further developed as research tools.

♦ As in other model organisms, the life cycle is rapid, allowing many generations to be studied in a short amount of time.

♦ Banding patterns in polytene chromosomes allow the gross-level mapping of chromosomes and chromosomal rearrangements.

♦ Small genome size and a small number of chromosomes in the genome (4) makes cytological and mapping experiments easier, and also facilitate the construction of genetic stocks.

♦ Balancer chromosomes, having multiple inversions, a dominant genetic marker, and a recessive lethal allele, allow the maintenance of a recessive lethal mutation on the other homolog.

♦ P element mutagenesis creates mutations that are tagged with a transposon, making cloning of the gene easier.

♦ Enhancer traps can be used to identify genes with specific expression patterns (e,g., in specific tissues or at certain times in development).

♦ Mosaic flies can be created during development by X-ray induced mitotic recombination. Within the last few years, researchers have improved the efficiency of mitotic recombination by exploiting the FLP/FRT site-specific recombination system in clever ways.

♦ Ectopic expression of a gene (expression in a different cell or at a different time than normally occurs) can be created by introducing a transgene linked to a promoter that is expressed in the desired way.

Significant Elements:

After reading the chapter and thinking about the concepts, you should be able to:

♦ set up crosses using balancer chromosomes

♦ create maps by crossing deletions and point mutations

♦ analyze results of in situ hybridization of probes to mutant and wild-type embyros

♦ distinguish between maternal effect mutations and zygotic mutations

♦ design experiments to look for homologous genes in other organisms and test the functionality of those genes in *Drosophila*

♦ think about how genetic analysis is done in this organism including: isolation of mutants, characterization of mutations (complementation and epistasis analyses), mapping a gene, cloning a gene, making a mutant cloned copy and reintroducing it into the genome

Problem solving tips:

♦ Draw out chromosomes in parents and in their gametes to determine the genotypes and phenotypes of progeny expected.

♦ In balancer chromosomes, the dominant mutation is a marker that indicates which flies received the balancer; the recessive lethal mutation means that a fly receiving two copies of the balancer will not live; the multiple inversions prevent crossing-over between the balancer and a normal chromosome.

♦ P flies contain P elements; M flies lack P elements. Fertilization with a sperm from a P male and an egg from an M mother results in hybrid dysgenesis, the transposition of the P element in the germ line of the resultant embryo.

♦ Enhancer trapping is used to search for regulatory regions whose patterns of expression are recognized by expression of nearby transpositions of the bacterial *lacZ* gene.

♦ A gene product is cell autonomous if it affects only the cell in which it is made.

♦ If a female fly is homozygous for a maternal effect mutation, all of her progeny will affected, regardless of their genotype. This is because the mutation interferes with the deposition of materials into the eggs that are needed subsequently for proper development of the progeny. If the defects in the progeny are sufficiently severe so that they cannot develop into adults, their mothers are classified as sterile.

♦ Some of the longer problems in this chapter are real research problems that were carried out essentially as described to you. They require that you integrate your knowledge of classical genetics, molecular genetics, genome analysis. These problems give you more practice in thinking like a geneticist. Think about how you would discover the genes involved and how their gene products may act- as regulators, messengers, structural components, etc.

Solutions to Problems:

D-1. a. **4**; b. **5**; c. **1**; d. **6**; e. **3**; f. **2**.

D-2.

a. This experiment measures **complementation,** because the two mutations are present in the progeny being examined but there has been no chance for meiotic recombination yet. As you will see from the answers to parts b and c, the data table allows you to determine the phenotype of animals with the two mutations: do these flies live or die?

b. The cross between mutant 1 and mutant 2 generates four genotypic classes of progeny: *m1 / m2*; *m1 / TM3*; *m2 / TM3*; TM3 / TM3. The first class will be wild type if the mutations are in different genes and therefore complement one another. However, if *m1* and *m2* are in the same gene, the phenotype of animals in this class will be mutant. The second and third genotypic classes will have the Stubble phenotype because of the presence of one copy of the TM3 balancer. The fourth class, with two copies of the TM3 balancer, will be dead because of the recessive lethal phenotype of Stubble. The 103 progeny that are not wild-type in this cross are of the second and third genotypic classes, and therefore have the **Stubble phenotype.**

c. **The 50 zygotes that do not live are homozygous for the balancer, TM3 / TM3.**

d. **There are 6 genes represented by these 10 mutations.** Mutations in the same gene do not complement each other and therefore there will not be any wild-type progeny produced. Crosses that produce roughly 50 wild-type progeny are those in which the two mutations complement each other. **The complementation groups are thus [10,1]; [9,5]; [7,8,4]; [6]; [3]; [2].**

e. This second experiment measures **complementation** between the recessive lethal mutations and the deletions. The data table is asking whether animals whose genomes contain both a particular recessive lethal mutation and a particular deletion survive.

f. If no wild-type progeny were produced, there was no complementation between the point mutation and the deletion mutation. Therefore the mutation maps within the deleted area; a fly with both the recessive lethal mutation and the deletion has no wild-type information for an essential gene. To determine a map, it is easiest to convert the information in the table into a different form showing which mutations map within which deletions. It is also useful to parse the mutations into their complementation groups as in part d, because if a deletion fails to complement one point mutation in the gene, it should fail to complement all of the point mutations in that gene. (That is, the deletion chromosome is devoid of that gene's function.)

Deletion	Genes mapping within the deletion
A	[2]; [4, 7, 8]; [6]
B	[1, 10]; [2]; [3]; [4, 7, 8]; [6]
C	[1, 10]; [3]; [5, 9]
D	[3]; [4, 7, 8]; [6]
E	[1,10]; [3]; [4, 7, 8]

To determine the order, you can start by comparing different deletions, looking for a common set of deleted genes. For example, deletion A and deletion D both remove [4, 7, 8] and [6], but deletion A also removes [2] while deletion D also removes [3]. This implies that [4, 7, 8] and [6] are adjacent to each other, with [2] on one side and [3] on the other. To figure out which side is which, you could look at deletion E, which removes [1,10], [3], and [4, 7, 8]. This implies that the order must be [2] [6] [4, 7 8] [3] [1, 10]. The only missing piece is [5, 9]; because deletion C removes [1, 10], [3], and [5, 9], the implication is that [5, 9] is closer to [1, 10] than to [2]. **The order of the genes is thus shown on the figure below, which also indicates the approximate extent of the five deletions.**

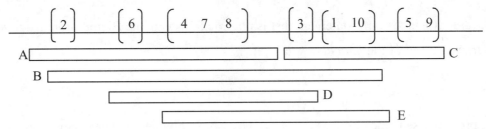

There are a few aspects of this figure that are worth noting. First, the map would be equally correct if it were flipped 180 degrees; we cannot ascribe a left-right order to the map because it is not related to anything else. Second, we don't know the order of the mutations within genes like [4, 7, 8] defined by several mutations. The only way this could be determined would be by recombination experiments. Third, the extent of the deletions shown is only approximate. For example, deletion E does not have to remove all the DNA corresponding to the locations of all three mutations 4, 7, and 8. If the deletion removed any functional part of that gene, it would be unable to complement any of the mutations within the gene.

D-3. There are several possible ways to do this - we outline only one such method here. You could start with male flies that had one or more insertions of the w^+ marked P elements on the X chromosome. These male flies must have a source of transposase elsewhere in the genome. There will be P element transpositions in the germ line of these males. You now mate these males to females who are homozygous for w^- mutations on their X chromosomes. In the next generation, you look for male flies that have pigmented eyes indicating the presence of the w^+ marked P element. These flies must have gotten their X chromosome from their mother, so the pigmentation must have arisen from transposition in the father's germ line that created sperm with the w^+ P element on an autosome or on the Y chromosome. To distinguish the autosomal inserts from the Y chromosome inserts, you take these pigmented males and again mate them to w^-/w^- females. If the P insert is autosomal, some males and some females will have pigmented eyes. If the P insert is on the Y, all the males but none of the females will have pigmented eyes.

D-4. Only the animals described by **statement c** would show somatic phenotypes caused by hybrid dysgenesis (**review** __Figure D.7__). P element transposition will occur in the germ line of the F_1 progeny of a male containing a P element (P male) crossed to a female lacking P elements (M female). Phenotypes in somatic cells will thus be manifested only in the progeny produced from gametes made in such a germ line.

D-5. Statement e is the only statement that is not true.   Instead, the Bicoid protein activates the zygotic transcription of the *hunchback* gap gene.

D-6.

a. The promoter sequence itself is recognized by RNA polymerase. The region also contains binding sites for transcription factors such as Bicoid that ensure that the *hb* gene is transcribed only in zygotic nuclei in the anterior part of the egg. The promoter region also must have binding sites for other transcription factors that ensure the *hb* gene is transcribed in the proper cells in the mother so that it is deposited uniformly in the egg before fertilization.

b. The coding region contains codons that specify the amino acids in Hunchback that comprise DNA binding domains and domains involved in the transcriptional regulation of gap and pair rule genes.

c. The sequence could be needed for translational repression (carried out by Nanos protein), thereby preventing translation of the maternally supplied *hb* mRNA in the posterior portion of the embryo.

D-7. The gap and pair rule genes act before cellularization so the gene products can freely diffuse in the syncytium in the embryo. The protein products of most gap and pair rule genes are transcription factors. In contrast, **the segment polarity genes act after cellularization of the embryo,** when intrasegmental patterning is determined by the interaction between ligands made in one type of cell and receptors found on nearby cells. Most of the segment polarity genes therefore encode proteins involved in signal transduction, only a few of which are transcription factors.

D-8. The cytoplasm from the anterior of a wild-type embryo could be injected into the anterior end of a *bicoid* mutant embryo to see if there was rescue of the mutant phenotype. You would want a control that would indicate that the physical act of injection does not cause rescue. The control would be injection from a *bicoid* mutant embryo into a mutant embryo. **Purified *bicoid* mRNA injected into the anterior end of a *bicoid* mutant embryo would be a more definitive experiment** indicating that *bicoid* alone is sufficient to rescue. Finally, **purified *bicoid* mRNA could be injected into the posterior end of a wild-type embryo.** If *bicoid* is an anterior determinant, there should be two anterior ends developing.

D-9. These results indicate that 1) maternally-supplied *hunchback* mRNA is completely dispensable and 2) that the function of the Nanos protein is only needed to restrict the translation of the maternally-supplied *hunchback* mRNA in the posterior of the egg. Development is fine if there is no maternally-supplied *hunchback* mRNA, so if there is no *hunchback* mRNA in an embryo, Nanos is not needed and can be defective (mutant) without showing any effect. If there is too much maternally-supplied *hunchback* mRNA, this swamps out the Nanos protein supplied even by wild-type mothers. The result is too much Hunchback protein in the posterior of the egg, and this prevents proper abdominal development. This is a peculiar situation : the fly doesn't really need *hunchback* maternal mRNA at all, but it makes it anyway. Because it makes this unnecessary *hunchback* maternal mRNA, it now needs *nanos* and the other posterior group genes to prevent its translation in the posterior of the egg. This illustrates that evolution does not always come up with the simplest solution to a problem, just one that happens to work. You should note that Hunchback is still needed for proper development even if the maternal hunchback mRNA is not; this is because *hunchback* must still be transcribed from zygotic nuclei so it can perform its role as a gap gene.

D-10.

a.   Absence of *knirps* function has no effect on Hunchback protein distribution but does affect Kruppel protein localization. More specifically, these results show that the wild-type Knirps protein is needed to restrict the posterior limit of the zone of Kruppel expression. This suggests that the wild-type Knirps protein negatively regulates the expression of Kruppel, at least in this part of the embryo.

b.   Hunchback protein would be seen throughout the embryo because there is no Nanos protein to inhibit its translation (see <u>Problem D-9</u> above).

D-11.

a.   There is more than one progenitor cell that divides to form the cells within any one facet. If the cells in a facet were restricted to the descendents of a single progenitor cell, you would not expect to see any mosaic facets. And because all possible combinations of mosaics were observed, there are no restrictions that demand particular sets of cells come from the same progenitor.

b.   These results show that the expression of the *white* gene is cell autonomous. The phenotype of any cell (red versus white) is dictated by the genotype of that cell for the *white* gene, and the phenotype of that cell is not influenced by the genotype of any neighboring cell.

D-12.

a.   To solve the mapping, look first at the deletions. If no non-Cy progeny are obtained (as shown by an entry of "N" in the table), there was no complementation and the deletions must have been overlapping. Such a result would indicate that both deletions must have removed in common one or more genes essential for survival. Deletion A overlaps B and deletion A overlaps C, but B and C do not overlap. The map position of the point mutations with respect to the deletions can be determined by looking for non-complementing lethal mutations ("N") or the uncovering of a sterile phenotype ("S") when crossed to the deletion. Mutation 4 is uncovered as a sterile by combining with deletion A or deletion B, so the mutation must be located in the region of overlap between those two deletions. Similarly, mutation 1 fails to complement deletion A or deletion C, so the gene is located in the region of overlap between these two deletions. Mutations 5 and 6 (which do not complement each other for fertility) are uncovered by deletion C only, and mutation 3 is uncovered by deletion A only. Mutations 2 and 7, which do not complement each other for viability and therefore must be in the same gene, are also not complemented by deletion B. The map summarizing these findings is:

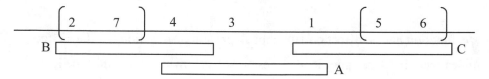

b.   **There are five genes: [1]; [2, 7]; [3]; [4]; [5, 6].** You can sort the mutations into complementation groups by looking at the pairwise complementation data. Any entry in the table of "N" or "S" indicates a lack of complementation between the mutations. Any entry of "F" means the mutations can complement each other.

c.   **Three genes ([3]; [4]; [5, 6]) are needed for female fertility** as indicated by the "S" (sterile) phenotype that occurs when any of these mutations are combined with a deletion that removes that gene. **If the mutant females** (females homozygous for a female sterile mutation, or heterozygous for two different female sterile mutations in the same gene, or heterozygous for a female sterile mutation and an overlapping deletion) **make normal-looking unfertilized eggs that become fertilized but arrest their development soon afterward, the mutations are maternal effect lethal mutations. If the eggs laid by these females are abnormal or are unable to be fertilized, the genes disrupted by the female sterile mutations are involved in oogenesis.**

d.   **There are two zygotic lethal genes ([1]; [2,7])** as indicated by the lack of viable non-Cy progeny (the "N" phenotype) in crosses of the point mutations with the deletion strains. You should recognize that "zygotic lethal genes" is shorthand for genes whose function in the nuclei of the embryo is essential for the embryo's survival and development into an adult.

e.   **You could not determine map distances between mutations in the same gene because double mutants of genotype [2 / 7] do not survive, while the female double mutants of genotype [5 / 6] are sterile.** In order to do recombination analysis, female double mutants would have to survive and be fertile, since recombination would be assessed in their progeny. (You couldn't determine the map distance in the progeny of [5 / 6] males; even though they survive and might be fertile, because recombination does not occur in *Drosophila* males.)

D-13.

a.   By obtaining a *lacZ* insertion into a developmental gene, **you could study the time of expression during development and the cells in which the gene is expressed.** This is the basis of "enhancer trapping" (see **<u>Figure D.10</u>**), which assumes that the *lacZ* gene will be controlled by the enhancers that regulate the developmental gene into which it inserts. The expression of *lacZ* should thus be turned on at the same time and in the same place as the developmental gene. This appears to be true in most, but not all cases.

b.   **The female flies would have to be heterozygous and could be maintained in the presence of a balancer chromosome** corresponding to the chromosome on which the mutation was located. That is, the fertile females allowing continual regeneration of the stock would be *mutant / Balancer* heterozygotes. **You could study the oogenesis phenotype by looking at the egg chambers in *mutant / mutant* females in the same population; these females would be**

sterile. You would discriminate between these females by virtue of the dominant marker on the balancer chromosome. The males in the stock would be *mutant / Balancer* and *mutant / mutant*; both genotypes would be viable and fertile. Animals of either sex of genotype *Balancer / Balancer* would die early in development, so you would not see any such adults.

c.   **You could compare the pattern of *bicoid* mRNA localization in *homeless* vs wild-type embryos by *in situ* hybridization using a *bicoid* DNA probe; the pattern seen in wild-type embryos would be the control.** An additional control would be to examine the localization of other mRNAs in the embryo such as the *nanos* message, to make sure that the effect of the *homeless* mutation is specific to *bicoid* and that the mutation does not simply result in the non-specific degradation of many RNAs in the embryo.

d.   **You would make females of the genotype centromere - ovo^D - ho^+ / centromere - ovo^+ - ho^-. You would then induce mitotic crossing-over in these females as shown in <u>Figure D.14</u>.** The females would normally be unable to produce ovaries because of the ovo^D mutation, but they could produce an ovary if mitotic crossing-over <u>within the germ line</u> yields a cell that is homozygous for both ovo^+ and the *homeless* mutation. If all females with such a germ line clone are sterile, this implies that the *homeless* gene function is needed in the germ line. If these females are fertile, it implies that *homeless* gene function can be supplied by non-germ-line cells (that is, by the somatic follicle cells).

e.   **Ras is needed for several different pathways, but the level of Ras protein required is different in the various signaling pathways.**

D-14. Note that the *bruno* gene has been cloned, but that there are no known mutants of this gene, while mutants of *arret* have been isolated, but the gene has not yet been cloned. The simplest test would be to clone the wild type bruno gene, and to insert it into a vector for P element-mediated transformation. This would be transformed into flies, and then crosses would be made to **establish whether the $bruno^+$ transgene can rescue the fertility of *arret* homozygotes.** If this were the case, you would then ask whether the *oskar* mRNA was properly localized in eggs produced by *arret* mutant mothers. As a supplementary experiment, you could clone the *bruno* gene from *arret* mutants, and sequence these genes to determine if there are potentially significant mutations of the *bruno* gene in *arret* mutants.

Reference E   *Mus musculus*: Genetic Portrait of a House Mouse

Synopsis:

The most significant methods developed for analysis of genes and their function in the mouse are the transgenic technologies. These techniques and the other characteristics of the organism listed below have increased out abilities to use the mouse as a model for human disease as well as for the study developmental questions.

- The organization of the mouse genome is similar to that of humans. This is most obvious in the conserved synteny between the species in genetic maps of the chromosomes.
- Because humans and mice have similar physiologies, models for many human diseases can often be created in the mouse.
- Methods exist for the creation of transgenic mice in which a piece of foreign DNA is injected into the nucleus of an egg, where it inserts into the genome.
- It is possible to perform targeted mutagenesis in ES cells so as to knockout specific genes and then observe the effect on the organism.

Significant Elements:

After reading the chapter and thinking about the concepts, you should be able to:

- understand the consequences of transgenic technology and targeted mutagenesis + what happens within the genome
- think through crosses from generation to generation
- analyze patterns of mRNA and protein expression and describe that regulation or type of mutation based on these patterns
- think about how genetic analysis is done in this organism including: isolation of mutants, characterization of mutations (complementation and epistasis analyses), mapping a gene, cloning a gene, making a mutant cloned copy and reintroducing it into the genome

Problem Solving Tips:

- Transgenic technology can result in the addition of an extra copy of the wild-type gene to the genome, or the introduction of an extra mutant copy (add-on type of analysis). Another type of transgenic technology is targeted mutagenesis, in which a mutated copy of a gene replaces a normal copy.

♦ After replacement of one copy of a gene with a mutated copy, mice containing the replacement on one copy of the homolog must be mated to obtain a homozygous knockout mutant mouse and study the gene's function.

♦ Analysis of patterns of expression using a reporter gene such as *lacZ* can give clues as to the function of a gene.

♦ FISH (fluorescent *in situ* hybridization) allows one to quickly locate the chromosome on which a gene is located. The mouse genome has recently been substantially sequenced, so another way to locate the region from which a cloned sequence of DNA originates is to conduct computerized homology (BLAST) searches against a mouse genome database.

♦ Some of the longer problems in this chapter are real research problems that were carried out essentially as described to you. They require that you integrate your knowledge of classical genetics, molecular genetics, and genome analysis. These problems give you practice in thinking like a geneticist. Think about how you would discover the genes involved and how their gene products may act- as regulators, messengers, structural components, etc.

Solutions to Problems:

E-1. a. **3**; b. **5**; c. **1**; d. **6**; e. **7**; f. **2**; g. **4**.

E-2.

a. **Because the mouse genome is so large, many more clones have to be screened to find a single copy gene of interest.** The large genome size of mouse is due to a large amount of repetitive DNA. This can create problems for finding a piece of unique DNA to use for finding a specific clone and can create some problems for mapping clones within the genome.

b. **Mouse is particularly useful as a mammalian model system for human diseases, human development, and other physiologies shared between the two species.** In addition, progress in the completion of the mouse genome project means that some of the experimental difficulties mentioned in part a will present less of a barrier in the future.

E-3.

a. **Yeast.** Because this is a general question about Cl⁻ channel function, it could be addressed in yeast where the creation of specific mutants is easier and quicker.

b.   **Mouse.** Because the question is specifically addressing a mammalian disease (cystic fibrosis) and mammalian physiology, the question should be studied in mice. In fact, it would be impossible to perform such studies in yeast because they are unicellular.

c.   **Yeast.** In yeast the genetic experiments and gene replacements are much easier to do and quicker to analyze.

d.   **Yeast and mouse.** You might want to start the drug discovery process in yeast where it is easy to test several drugs quickly, but after finding good candidates, you could use the mouse to determine if the new drugs were efficacious in the treatment of disease, or if they caused any other ill effects in mammals.

E-4.

a.   **Perform FISH (fluorescent *in situ* hybridization) using a piece of the mouse gene as a probe versus a human chromosome spread.** Alternatively, if you know any of the DNA sequence of the mouse gene, you could search computer databases of the human genome for homologs. (The alternative of finding linked molecular markers is much more time and resource intensive.)

b.   **Use the mouse gene as a probe against a human genomic library or a library specifically containing chromosome 8 clones.** The clone may contain other genes as well. The fragment(s) containing the gene could be identified by further hybridizations to blots of the restriction digested clone. As an alternative, if you found a human homolog in database searches as in part a, you could then design PCR primers to amplify the human gene from human genomic DNA.

E-5. The fertilized egg in mammals can split into two separate cells or two separate parts during the first divisions, and because the cells have not irrevocably assumed specific fates during these early stages, the two cells or embryo halves can develop into all parts of the body. **In other words, the early embryonic cells in mammals are totipotent.** In *Drosophila,* however, the early embryo is a syncytium in which the nuclei (subsequently cells) are in a common cytoplasm (**review Reference D**). **The *Drosophila* egg already has a polarity even prior to fertilization, so division of the embryo will result in two half-embryos, each of which is missing information important for development of the whole organism.**

E-6. The ES cells contain a different coat color allele than the cells of the blastocyst with which they will be combined. The chimeric embryo that develops will have some cells of one coat color and other cells of the other coat color. This makes it easy to recognize those mice that developed from chimeric embryos.

E-7.

a. **Because there are no -/- mice, the *CTF* gene seems to be an essential gene.**

b. Apparently, **the product of the wild-type *CTF* gene affects growth in both males and females and is necessary for fertility in males.**

c. (We assume here that you do not already know the phenotypes associated with the three possible *CTF* genotypes.) **A genetic way to determine the *CTF* genotype of the mice would be possible if there were a closely linked marker causing an incompletely dominant or codominant visible phenotype.** The CTF^+ chromosome in both parents would be adjacent to one marker allele, while the chromosomes bearing CTF^- in both parents would have the other marker allele. **If no such marker were available, you would need to genotype the animals at the DNA level.** Because the problem states that you know where the gene is expressed, you would almost certainly know the molecular identity of the gene and the nature of the CTF^- mutation. **You could then use any of the genotyping methods described in <u>Chapter 11</u> on DNA extracted from blood or the end of the tail.** For example, you could PCR amplify the *CTF* gene from this DNA and hybridize it with allele-specific oligonucleotides.

E-8.

a. **If mutant A is in the *RAR* gene, it must be a point mutation that does not alter the size of the RAR protein but does alter its function, such as a nonconservative missense mutation.** Alternatively, it is possible that the limb development abnormalities are cause by mutations in another gene.

   Mutant B is a point mutation that blocks in a step of gene expression subsequent to the transcription and splicing of the *RAR* mRNA. This could be a nonsense mutation early in the gene's open reading frame mutant, a missense mutation encoding an aberrant protein that is degraded as soon as it is made, or a mutation affecting translation of the mRNA (perhaps a missense mutation altering the initiation codon).

   In mutant C there is a truncated RNA, but a normal DNA, so there appears to be altered RNA transcription or processing. **Mutation C could introduce a premature transcription termination signal into the gene, or perhaps more likely, the mutation might prevent the inclusion of an exon into the mature mRNA during splicing.**

   In mutant D, the DNA and RNA are both altered, suggesting that **there is a deletion in the DNA that would have to remove all or part of one or more protein-coding exons.**

In mutant E, no RNA or protein are made and the DNA is altered, so **the regulatory region (transcription start) may be deleted.**

Mutant F appears to be a deletion of the entire gene, because the gene does not appear on a Southern blot.

b.   **A cross could be done using a marker closely linked to the *RAR* locus** to see if the defect cosegregates with the marker. If the defect was in a different gene, it would not cosegregate with the marker. Alternatively, if you had a mouse strain like B-F that you knew displayed limb deformities because of a mutation in the *RAR* gene, you could perform complementation analysis to see if the mutation in A failed to complement the *RAR* mutation in the other strain. Complementation analysis is made possible here because you know that all of the limb deformity phenotypes are recessive.

c.   **You could clone the *RAR* gene next to a neuron cell specific promoter and create a transgenic mouse that now expressed *RAR* in neurons.**

E-9.

a.   **If the mutation was due to an insertion of the transgene, the MMTV *c-myc* gene should segregate with the phenotype.** That is, all subsequent animals that had the limb deformity should have the *c-myc* fusion and vice versa. The presence of the *c-myc* fusion could be recognized by Southern or PCR analysis.

b.   **Clones containing the *c-myc* fusion could be identified by hybridization of MMTV sequences versus a library of genomic clones produced from the cells of the mutant mouse.** The DNA surrounding the MMTV *c-myc* fusion in this clone would be the gene of interest.

c.   **The sequence of the gene into which the MMTV *c-myc* fusion inserted could be analyzed in the *ld* mutant to determine if there were mutations in the gene.** Alternatively a clone containing only the wild-type copy of the gene into which the MMTV *c-myc* fusion inserted could be be injected into a mouse heterozygous for the *ld* mutation. If the homozygous *ld / ld* progeny that had also received the wild-type transgene are phenotypically wild-type, the transgene rescues and therefore must correspond to the *ld* gene.

d.   **The presence of transcripts in both embryo and adult is still consistent with role in embryonic limb development.** The expectation would be transcription in the embryo. The presence of transcripts in adults may indicate that the gene also plays a role in adults as well.

E-10.

a. **The data are consistent with the hypothesis** because the level of Ob protein is low in the animals that are starved and highest in those that were force-fed the high calorie diet and therefore should be satiated.

b. **To determine if the Ob gene is transcriptionally regulated, the Ob DNA would be used as a probe vs RNA from starved, normal, and force-fed mice** (a Northern blot experiment). If regulation occurs at the transcriptional level, you would expect the lowest levels of *Ob* mRNA in the starved animals and the highest levels in the force-fed mice.

c. If the mRNA levels are the same while protein levels are different, **the regulation must either be at the translational level** (so that the rate of *Ob* mRNA translation is highest in the force-fed mice) **or at the level of Ob protein stability** (the protein would be more likely to be degraded in the starved mice).

d. **Mutant A** has the normal sized RNA, so the mutation cannot be a large deletion or an alteration in the RNA processing. The **mutant phenotype may result from an alteration in the amount or amino acid sequence of the Ob protein. Mutant B** has no RNA, so it could be **a deletion of the entire gene or a defect in the regulatory region that blocks *Ob* transcription. Mutant C** has a larger than wild-type transcript, so it **could have a defect in RNA processing** (for example, it does not splice out an intron). Alternatively, it could involve a deletion of DNA that fused the *Ob* gene to another gene, producing one large hybrid transcript. **Mutant D** has a smaller than normal transcript which could mean there is **a deletion in the gene or an altered processing (splicing) signal**.

e. **1. B** (Fat cells are the cells where you expect the *Ob* gene would be expressed.) **2. A** (You want to express the gene in mouse cells, so should use a mouse promoter you can control. The yeast *LEU2* promoter would probably not work in mammalian cells in any event.) **3. A** (Addition of metal (Zn) will induce expression of the fusion gene because the promoter is that for the metallothionein gene.)

f. **You would expect animals without the Ob receptor to be obese.** Lacking the Ob receptor, mice would not be able to receive the signal from the Ob protein that they were full. The animals would thus not realize that they were satiated and would probably would continue to eat and become obese. This phenotype should in theory reflect that of mice homozygous for null mutants in *Ob*, assuming that the only ligand for the receptor is the Ob protein, and the only receptor for the Ob protein is the one you knocked out.

g. Injection of the Ob protein into mutant mice where Ob is not normally made could in theory rescue the phenotype, if the injected protein were able to be transported to the brain. That is, **the**

mice might not continue to eat large quantities of food, and they would therefore not become obese as would the uninjected mutants. It is possible that the regulation of Ob protein levels might be critical- in the absence of a feedback mechanism, mice injected with too much Ob might stop eating even when they need food, and therefore starve to death.